LIST OF FIGURES

LIST OF TABLES

LIST OF ACRONYMS

ABP	Average boiling points
AIChe	American Institute of Chemical Engineers
AIT	Autoignition temperature
API	American Petroleum Institute
AQUIRP	Auto/Oil Air Quality Improvement Research Program
ASTM	American Society for Testing Materials
ATF	Alternative fuels
BFG	Butane free gasoline
BL	Battery limits
BOP	Biological oxygen demand
BOV	Blending octane value
BP	Boiling point
BPD	Barrels per day
BSFC	Brake specific fuel consumption
BTX	Benzene, toluene, and xylenes
C	Carbon atoms
CAA	Clean Air Act
CAFE	Corporate Average Fuel Economy
CCR	Continuous catalyst regeneration
CDU	Crude distillation unit
CFRR	Cash flow rate of return

CI	Cetane index
CNG	Compressed natural gas
COD	Chemical oxygen demand
CR	Catalytic reforming
DAO	Deasphalted oil
DC	Delayed coking
DCF	Discounted cash flow
DCFRR	Discounted cash flow rate of return
DIB	Deisobutanizer
DIH	Deisohexanizer
DIP	Deisopentanizer
DMF	Dimethylformamide
DMSO	Dimethylsulfoxide
DOE	Department of Energy
DON	Distribution octane number
E & C	Engineering and construction
EPA	Environmental Protection Agency
ETAE	Ethyl tertiary amyl ether
ETBE	Ethyl tertiary butyl ether
FBP	Final boiling point
FC	Fluid coking
FCC	Fluid catalytic cracking
FCCU	Fluid catalytic cracking unit
FEON	Front end–octane number
FM	N–Formylmorpholine
G & H	Gary and Handwerk

HC	Hydrocracking
HDS	Hydrodesulfurization
HF	Hydrofluoric acid
HOC	Heavy oil cracking
IBP	Initial boiling point
IGCS	Integrated gasification combined cycle
LHSV	Liquid hourly space velocity
LNG	Liquified natural gas
LP	Linear program
LPG	Liquified petroleum gas
LV	Liquid volume
MABP	Molal average boiling point
MeABP	Mean average boiling point
MidBP	Mid boiling point
MON	Motor octane number
MMT	Methylcyclopentadienyl manganese tricarbonyl
MW	Molecular weight
NAAQS	National Ambient Air Quality Standards
NFRCI	Nelson–Farrar Refinery Construction Index
NPRA	National Petroleum Refiners Association
NRS	Non–renewable source
OGJ	Oil & Gas Journal
PC	Personal computer
PFD	Process flow diagram
PONA	Paraffins, olefins, napthenes, aromatics
PVF	Present value factor

Resid	Residual oil
RFG	Reformulated gasoline
ROI	Return on investment
RON	Research octane number
RONC	Research octane number clear
RS	Renewable sources
RVP	Reid vapor pressure
SCFB	Standard cubic feet per barrel
SDA	Solvent Deasphalting
SEE	Standard error of the estimate
SSF	Saybolt seconds furol
TAME	Tertiary amyl methyl ether
TAN	Total acid number
TBP	True boiling point
TDS	Total dissolved solids
TEL	Tetra ethyl lead
TML	Teta methyl lead
TOC	Total organic carbon
UOP	Universal Oil Products
VAPB	Volumetric average boiling point
VB	Visbreaking
V/L	Vapor/liquid
VCM	Volatile combustible matter
VOC	Volatile organic compounds
WABP	Weight average boiling point

ACKNOWLEDGMENTS

I am indebted to my good friend Bob Jones for the helpful encouragement, suggestions, and constructive criticism he has provided. I am even more indebted to my wife, who for more than fifty years has indulged my fascination with refinery product yields and property data.

My Houston colleague Ed Swain has been a frequent source of market information and my new friend and computer mentor John Christie was instrumental in "polishing up" many of the figures/line drawings. I would also like to express my pleasure in working with the very competent editor Linda Robinson.

PREFACE

This book is an outgrowth of data collected and compiled over many years and correlated as the need arose. It presents, in an organized way, yield data (plus other useful information) on the important petroleum refinery processes practiced commercially today.

The use of the data in evaluating technology from an economic standpoint is illustrated. The reader is referred to other works for detailed descriptions of the technologies considered. The emphasis in this book is on what products a process yields and their properties, and not on how this is accomplished.

The techniques used to correlate the data are described, and the use of the correlations is illustrated with examples. This will enable the readers to make and to use their own correlations if they wish.

SECTION A:

INTRODUCTION

CHAPTER 1

INTRODUCTION

Just as it is said, "Necessity is the mother of invention," it may be said that necessity was the mother of this book. During more than 55 years in the engineering profession, the author has often realized a need for the "tools" contained in this "toolbox."

In the Preface to the first edition of *Petroleum Refinery Engineering*, W. L. Nelson stated, "The current literature contains much that is useful, but the literature is so voluminous that it may be useless to a busy engineer unless it is presented in an organized manner."[1] The primary purpose of this book is to present yield data for refinery processes in an organized manner so that they are useful to the reader in the performance of process comparisons or process economic studies of various types, and to show how to use this information in performing such studies.

Very early in the author's career as a process engineer he found himself faced with a large amount of plant data from the operation of a particular process. To determine the optimum mode of operation, it was necessary to correlate the yields of the products from the process to some property of the feedstock or of one of the products in order to estimate the relative economic values of the various operations of that process.

Later it became necessary to compare competing processes. This necessitated yield correlations for all processes involved in the comparison. Still later, there was a need to develop processing schemes of varying complexity and to make economic comparisons.

3

The result is a collection of yield correlations for all of the important, commercially established petroleum refinery processes. Some readers may balk at the empirical methods employed in this work, preferring a purely analytical approach. Again quoting Nelson, "The history of industrial development shows that commercial plants are usually built before the theory of the process is fully understood."[1] Many of the processes are far too complex to lend themselves to the simple type of description that is needed for many of the engineer's purposes, such as a preliminary comparison of proposed processes or process schemes.

Each correlation in the book is accompanied by operating requirements (utilities, catalyst, chemicals, etc.) plus the capital cost of a unit of typical size. Some engineers think in terms of dollar-per-barrel costs for process plants. At best such numbers can be correct for only one size of each type of plant. What was a good number last year, five years ago, or longer, could be greatly in error today. The cost-curve type of estimate employed here is far more suitable for preliminary studies and offers the advantage of consistency over a wide range of sizes for a given process.

This technique is best described by the following equation[2] for plants A and B:

$$\frac{C_A}{C_B} = \left[\frac{Q_A}{Q_B}\right]^X$$

where C represents cost, Q represents capacity, and X is the "Lang" exponent.[3] This equation would describe a straight line on a log-log plot with the slope of the line equal to X. In plotting actual cost data on log-log paper, a curved "best" line is sometimes indicated. This may mean that the cost exponent for the particular process varies with capacity. It could also signify that the plants differ in more ways than just capacity.

The yield correlation technique employed here involves bringing together data from various sources and handling them in a consistent manner. If data from one or more of the sources differ significantly from the average for the group, this will be evident from a plot of the data or from the calculated deviations from the regression (correlation) line.

Using the Nelson-Farrar Refinery Construction Cost Indexes[4] and appropriate "Lang" cost-capacity exponents, the capital cost of a unit of given size can be translated to a different time and size. The Lang exponent may be assumed to be 0.6 or "six-tenths" unless specified otherwise for a given process.

With yields, product properties, operating requirements, and capital costs in a single source, the user has all of the information required to perform preliminary economic evaluations of single processes or complete refinery process schemes.

Accompanying each process yield correlation is a simplified process flow diagram and a brief description of the process. These process descriptions may contain occasional references to typical operating conditions. This however is not a primary purpose of the book. The principal focus is on what a process produces (product yields and properties) and not on how it produces them. More complete process descriptions can be found in a number of books.[5,6,7,8]

Fully worked out examples are provided to illustrate the use of the correlations and the economic data presented. The quality of these correlations is adequate for preliminary economics and to be incorporated in a computer model of a refinery for simulation or for linear program (LP) optimization. For definitive process comparisons or for actual plant design, however, a basic process design should be obtained from a qualified engineer or through an appropriate licensor if licensed technology is used. This may require pilot plant tests on the proposed feedstock(s).

The methodology employed in deriving the correlations is explained in sufficient detail in order that the readers may be able to make their own correlations if they choose. Finally, this book will prove very helpful to refinery engineers, refinery planners, refinery management, engineers with engineering and construction (E & C) firms, cost engineers, consultants to the refining and chemical process industries, market researchers, and college students in process engineering and engineering economics courses.

Two chapters, "Transportation Fuels" and "The Environment and the Refiner," were not included in the original plan of this book. However, with the growing public concern about what is happening to our environment, it was decided they were needed. Emissions from mobile sources (using transportation fuels) represent about half of the undesirable pollutants going into our atmosphere.[9] Refineries not only produce these fuels, but also produce atmospheric emissions and solid and liquid wastes that need to be disposed of in an ecologically acceptable manner.

The book has been designed to be useful to a wide range of readers with varying backgrounds:

- Chapter 2 describes the regression techniques employed to derive the yield correlations presented in chapters 8 through 16. These techniques will be new to some readers.

- Chapters 3, 4, and 5 provide background material for those less familiar with petroleum refining, its problems, its products, and their properties.

- Chapter 6 discusses the impact of transportation fuels on the environment and what changes in refinery installations and processing may be required to produce more environmentally-friendly operations and products.

- Chapter 7 describes how to obtain crude oil product yields from assay data supplied by others.

- Chapters 8 through 17 present correlations developed by the author for the principle refinery conversion processes.

- Chapters 18 through 29 cover processes that involve relatively simpler reacting systems with little variation in their outcome. Therefore, typical yields and properties are cited.

- Chapter 30 has been inserted in this revision. It deals briefly with waste water treatment and waste disposal.

- Chapter 31 discusses the anomalous behavior encountered by the refiner in the blending of some products from the streams produced by the refinery. Methods of surmounting these problems are presented.

- Chapter 32 demonstrates how to utilize the material contained in the previous chapters to perform comparisons, technology evaluations, conceptual process designs, and feasibility studies.

Notes

1. Nelson, W.L., *Petroleum Refinery Engineering*, 1st ed., McGraw-Hill Book Company, New York, 1936

2. Nelson, W.L., *Oil & Gas Journal*, January 4, 1965, p. 112

3. Lang, H.J., *Chemical Engineering*, June 1948, p. 112

4. Published in the first issue each month in the *Oil & Gas Journal*

5. Meyers, R.A., *Handbook of Petroleum Refining Processes*, McGraw-Hill Book Company, New York, 1986

6. Gary, J.H., and Handwerk, G.E., *Petroleum Refining Technology and Economics*, 2nd ed., Marcel Dekker, Inc., New York, 1984

7. Leffler, W.L., *Petroleum Refining for the Nontechnical Person*, 2nd ed., PennWell Books, Tulsa, 1985

8. Shaheen, E.I., *Catalytic Processing in Petroleum Refining*, PennWell Books, Tulsa, 1983

9. Kuhre, C.J., and Sykes, J.A., Jr., *Clean Fuels from Low Priced Crudes and Residues*, AIChE Meeting, New Orleans, March, 1973

CHAPTER 2

CORRELATION METHODOLOGY

The literature contains much data on pilot plant and commercial plant operation of most of the petroleum refinery processes of interest. However, when it comes to correlating empirical data, one rarely is satisfied with the quantity or coverage (range of feeds and/or conditions) of the data. Yet, these data are of little value until gathered together and correlated in a meaningful way.

Rarely will one find a set of data fitting exactly the particular set of conditions of interest (feedstock, product octane, product smoke point, etc.). In general it is better to take information from a good correlation than to use isolated sets of data. This is particularly true when evaluating process results over a range of values (e.g., product octane). The absolute values from the correlation may be somewhat in error, but the differences between points should be very meaningful.

For our purposes, a set of data consists of the simultaneous, steady-state yields of products (and their properties) when a unit is processing a particular feedstock at a fixed set of conditions—temperature, pressure, type of catalyst (if any), space velocity, etc. Each product yield or property constitutes a point of data in the set.

A first step in correlating sets of gathered data is to tabulate the data with a row for each set and a column for each variable. Each product is a dependent variable, but may also be an independent variable at times. Other independent variables can be type of feed, and/or one or more properties of the feed, or of a product (gravity, boiling range, characterization [K] factor, etc.).

9

As we would expect (and shall see later), the actual operating results from the literature will be scattered to some degree or other. This is due to the complexity of the systems involved (a multitude of species of hydrocarbons) and, of course, uncertainties in observations (errors in measuring, reading, recording), failure to attain true steady state, etc.

In correlating these data, we are attempting to find a relatively simple expression (equation) to characterize the relationship between two or more variables in a very complex system. From a consideration of the chemistry involved, possible reaction mechanisms and kinetics, we may infer a possible relationship between a set of variables. A plot of the data on this basis will indicate by the pattern of the points (trend and scatter) how well the assumed relation fits the data.

Ordering the data (arranging by increasing or decreasing order) in terms of one of the variables often helps to indicate a possible correlating parameter. Plotting an independent variable against one of the dependent variables may indicate the type of relationship (linear, quadratic, exponential, etc.) between them—if any.

Once a possible relation is detected, the equation describing this apparent relationship is usually determined by linear regression analysis—or multiple linear regression analysis if more than one independent variable is involved.

Before electronic computers (mainframe or personal computer [PC]), this was a very tedious process—even with a calculator. Now however, with ready access to PCs with very high speed and capacity, regression analysis is quick and easy.

Spreadsheet programs such as Lotus, QuatroPro, Excel, etc., provide great flexibility in the arrangement and manipulation of data (moving columns, transforming data, etc.) and provide for automatic plotting of data in addition to regression analysis capabilities.

In developing each correlation in this book, an attempt was made to discover a single independent variable as a basis for correlation. The degree to which this was successful will be apparent from the graphs on which both the raw data and the regression lines have been plotted. Frequently, it was necessary to employ two or more independent variables to obtain a satisfactory correlation. The results are summarized in the following tabulation:

Process	Independent Variables	
	Yields	*Properties*
Solvent deasphalting (SDA)	1	2
Visbreaking (VB)	1	1–2
Delayed coking (DC)	1–2	1–2
Fluid coking (FC)	1	1–2
Fluid catalytic cracking (FCC)	1–2	1–2
Heavy oil cracking (HOC)	1–2	1
Hydrocracking (HC)	1–2	1–2
Hydrodesulfurization (HDS)	2	2
Catalytic reforming (CR)	1–5	1–2

Parameters Used in Correlating Process Yields

A review of the literature reveals some consistency in the correlation parameters used by the author and others:

Process	*Author*	*HPI*	*G & H*
SDA	Wt% DAO	na	na
VB	Wt % Conv.	nC_5insol	na
FC	Wt % CCR	Sed Cont na	na
DC	WT% CCR	WT% CCR	WT% CCR
FCC	LV% Conv	LV% Conv	LV% Conv
HOC	LV% Conv	na	na
HC	LV% Gaso	LV% Lt HC	LV% Lt HC
HDT	Feed API WT% S	% Desulf	na
CR	Reformate RON Feed N+2A	Reformate RON Feed N+2A	Reformate RON Feed K

11

where:

HPI represents HPI Consultants, Inc.

G & H represents Gary and Handwerk

DAO represents deasphalted oil

CCR represents continuous catalyst regeneration

LV represents liquid volume

A General Data Correlation Procedure

A step-by-step procedure for performing data correlations follows:

- Enter data in a spreadsheet format with a column for each variable (yield or property) and a row for each set of data.

- Select a column for the dependent variable (product yield or property) to be correlated with some feed property.

- Select a column(s) for the independent variable(s), feed property(s) usually. Note that the independent variables must be in adjacent columns, since the range selected cannot be interrupted. One of the big advantages of spreadsheet programs is that columns can be moved easily.

- Delete any row where there is no entry for one or more of the variables selected, since there can be no empty cells in the selected columns.

- At this point, any column of data may be manipulated:

    1. A variable may be ordered (put in ascending or descending order).

    2. Any variable may be transformed—into a logarithmic value, a trigonometric function, a higher or lower power, etc.

- After indicating the location on the spreadsheet for the regression results to be displayed and whether an intercept is to be calculated or the line forced through the origin, the regression may be performed. There are times when you will know that the regression should pass through a certain point, such as the origin, but the regression results may indicate a better fit (over the range of the data) when a finite interception is computed.

The regression output will give the value of the constant, the coefficient(s) of the variable(s), the coefficient of correlation, and the standard error of the estimate of the dependent variable and of each of the coefficients, also the number of points and the degrees of freedom.

(NOTE: The coefficient of determination, R^2, is a measure of the variation in the dependent variable explained by the derived regression equation. The closer R^2 approaches 1.0, the less will be the scatter of the data points about the calculated regression line. The standard error of the estimate of the dependent variable is approximately equal to its standard deviation. In like manner, the standard error of a coefficient is a measure of the confidence in the value of that coefficient.)

Having the equation of the regression line, one may calculate values of the dependent variable for each of the sets. The difference between the calculated value and the corresponding "observed" value may then be calculated. The magnitude of the differences may point to certain data that do not fit with the rest of the population. Reference to the source of these data may suggest reasons for discarding these data. A plot of the data points together with the regression line will give a visual indication of the appropriateness of the relation selected to represent the data.

It is not necessary to have a regression program to obtain the same results. They can be calculated from the sums of the individual variables, of their squares, and of their cross products. In the case of a first order or linear equation, this requires the sums of the following:

X, Y, XY, X^2, Y^2 and N (the number of points)

For a second order equation, 10 such sums are required—for third order, 15 values are needed. With so many regression programs available, some in the public domain, it is hardly practical to go through such a long and tedious procedure. Use of a regression program can expedite the user's work and permit the user to focus on the relationship represented by the data and its significance.

Significance of Results

The yield of full-boiling range gasoline in FCC has been chosen to illustrate the significance of a regression analysis and the use to which it may be put. Figure 2–1 is a plot of the data (382 points) for gasoline yield (Y) from FCC vs. conversion (X) together with the regression line for the equation:

$$Y = a + bX + cX^2$$

Assuming for practical purposes the standard error of the estimate (SEE) of Y is equal to the standard deviation (D) and that for a normal distribution, 95% of the data should lie within plus or minus 2D of the regression line, Figure 2–2 is a plot of Figure 2–1 with lines of plus 2SEE and minus 2SEE added defining the 95% probability band for the data.[1] This band is sometimes referred to as the error band. Points lying outside this band are known as "outliers" and may be disregarded in further regression of the data. These points result from errors in measurement of variables, errors in recording of data, or because the data do not fit in the remaining population.

The simplicity of this correlation is all the more impressive when one considers the very large number of variables at play in the FCC process:

- Boiling range of gasoline
- Composition of feed
- Type and activity of catalyst
- Catalyst to feed ratio

14

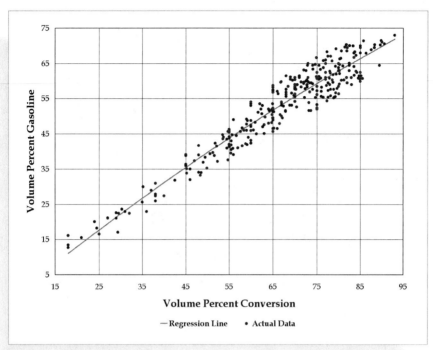

Fig. 2–1 FCC Gasoline Yield Data

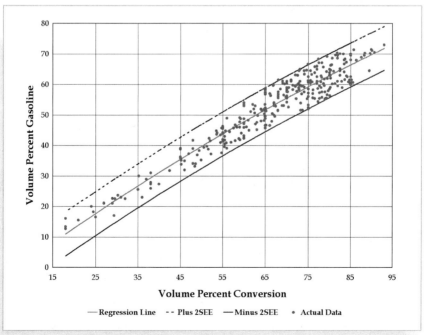

Fig. 2–2 FCC Gasoline Error Band

15

- Hetero elements such as S, N_2, Ni, V, Na, Fe, and As in the feed
- Reactor geometry
- Carbon on regenerated catalyst

Application to an existing process unit

This relationship for an existing unit may be obtained by passing a "best" line parallel to (if not coincident with) the regression line and passing through actual operating data for that unit. Figure 2–3 shows such a plot where actual plant data have been plotted. Thus by measuring the deviation (d) of this line from the regression line, a constant is obtained that may be used to "tune" the correlation to the actual unit:

$$Y = (a+d) + bX + cX^2$$

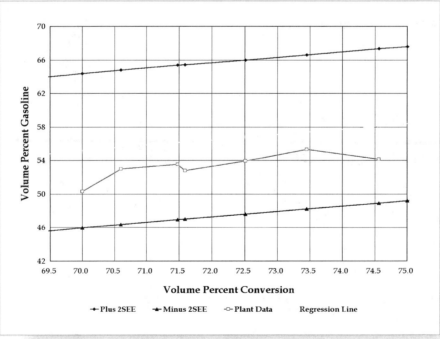

Fig. 2–3   FCC Gasoline Yield Tuning

Notes

1. McElroy, E.E., *Applied Business Statistics*, Holden-Day, Inc., San Francisco, Second Edition, 1979, p. 293

Barish, N.N., *Economic Analysis for Engineering and Managerial Decision Making*, McGraw-Hill Book Co., Second Edition, 1978, p. 597

CHAPTER 3

CRUDE OILS, HYDROCARBONS, AND REFINERY PRODUCTS

Petroleum Supply/Demand Picture

The U.S. has become increasingly dependent on imported crude oil. Crude oil imports exceeded 50% of crude charged to U.S. refineries in 1993 and are still increasing (Fig. 3–1).[1] As a result largely of stringent environmental regulations, the number of operating refineries in the U.S. has decreased from 303 in 1981 with a combined crude charge capacity of 18.5 million barrels per day, to 161 in 1999 with a charge capacity of 15.3 million barrels per day.[2]

During this period, the number of refineries operating in the non-communist world increased from 440 to 535 and the capacity of U.S. refineries decreased from about 44% of the non-communist capacity to about 26%.[3]

Also, during this same period, the demand for refined petroleum products in the U.S. increased from less than 16 million barrels per day to more than 18 million barrels per day. With negligible increase in the importation of products, this increase in demand has been met essentially by increasing refinery utilization from 70% of capacity to more than 95% (Fig. 3–2) and by increasing severity of operations (increasing conversion of residual oils to lighter oils).[4] See the following tabulation:

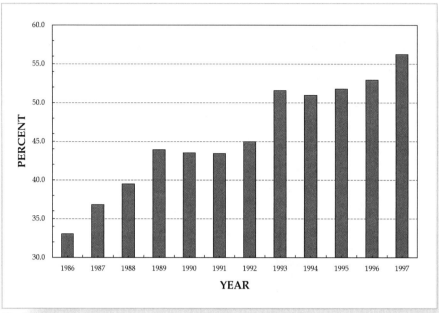

Fig. 3–1 Percent Imported Crude Oil Charged to U.S. Refineries

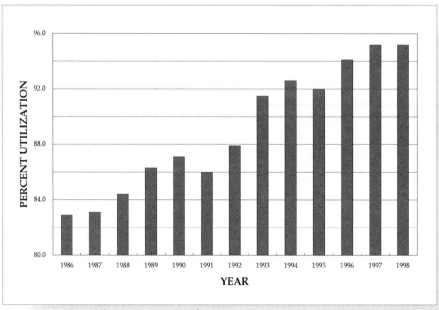

Fig. 3–2 Percent Utilization Rate (U.S. Refineries)

Product Yields as Percent of Crude Charged[5]

Year	Gasoline	Jet Fuel	Distillate	Residual Oil
1988	55.2	10.9	23.5	10.4
1998	56.6	11.2	24.5	5.9

It is evident from the above that the slight increases in lighter products have resulted from destruction of residual oil. Demand for residual fuel oil decreased as a result of environmental constraints and the increased availability of competitively-priced natural gas.

Nature of Crude Oils

Crude oils are complex mixtures of hundreds of different species of chemical compounds. Most of these compounds are hydrocarbons, principally paraffins (alkanes), naphthenes (cycloparaffins), aromatics, or combinations of these, such as alkyl naphthenes, alkyl aromatics, and polycyclic compounds. Another type of hydrocarbon found in refinery products—as a result of chemical reaction during processing—is the olefin (alkene). It is generally assumed olefins are not present in crude oil, that olefins reported in a crude assay arise due to decomposition during the distillation of the crude.

In addition, a number of elements (other than carbon and hydrogen) appear in crude oils. Principal among these is sulfur in the form of free (elemental) sulfur, hydrogen sulfide (H_2S), mercaptan (RSH), thiophene, etc. Sulfur is very troublesome in petroleum products, causing corrosion, producing sulfur dioxide (which is toxic and is the main component of acid rain) when burned, and in the form of H_2S is an insidious poison. In sufficiently high concentration, H_2S paralyzes the olfactory nerves so that its victim is unaware of its presence. Sulfur is also a poison to some catalysts. The petroleum industry is a large producer of sulfur as a product from treating operations that convert H_2S to elemental sulfur.

Nitrogen is another undesirable element occurring in crude oil since it, too, is a poison to some catalysts. Other catalyst poisons include arsenic, vanadium, and nickel. Vanadium can also cause corrosion and in the form of V_2O_5 acts as a flux that can cause furnace refractories to flow.

21

Compounds containing oxygen, notably naphthenic acids, are corrosive, particularly when the process temperature is 430°F to 750°F and the stream velocity is high, as in nozzles, transfer lines, and return bends.

Finally, one other problem material is salt (primarily chlorides, sulfates and carbonates of sodium, calcium, and magnesium) that occurs in crude in the form of fine droplets of brine or minute particles of salt. The salt content of a crude is expressed as pounds of salt (NaCl equivalent) per thousand barrels of crude. Salt by itself is bad enough, causing corrosion and depositing in equipment—particularly on heat transfer surfaces. However, in combination with H_2S and H_2O a vicious corrosion cycle occurs on carbon steel and even stainless steel. Later in the book, we shall see how the refiner deals with these and some other problems.

There has been an understandable tendency among refiners (with a free choice) to select from the crudes available, those with the lower amounts of these difficult materials. As a result, the reserves of more desirable crudes decline while the reserves of less desirable crudes grow. On average, crude oils available for import are heavier and contain more of these contaminants than do domestic crudes. As the U.S. has become more dependent on imported crude, these effects have been reflected in the quality of the average crude refined in the U.S., progressively heavier and containing more sulfur and the other undesirables (Figs. 3–3 and 3–4).[6]

Properties of Hydrocarbons

The many hydrocarbons in crude oil vary widely in physical and chemical properties. It is these differences that determine the ways in which they are processed in the refinery and finally the ways in which finished products from the refinery are employed.

Figure 3–5 demonstrates the close correlation of the boiling point (BP) of a hydrocarbon with its number of carbon atoms per molecule for different series of hydrocarbons. This permits the use of the number of carbon atoms as a parameter in lieu of BP.[7]

Figure 3–6, containing data calculated by author, shows how the weight ratio of carbon to hydrogen (C/H) varies with number of carbon atoms for different series of hydrocarbons. The primary purpose of this graph is to

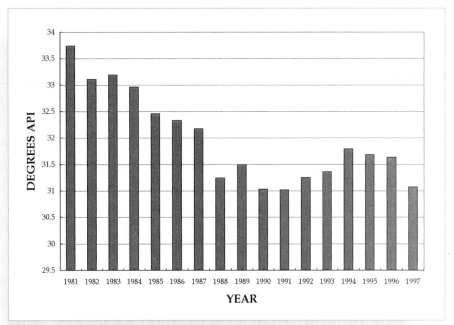

Fig. 3–3   API Gravity of Crude to U.S. Refineries

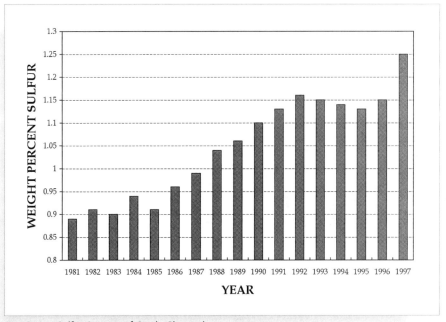

Fig. 3–4   Sulfur Content of Crude Charged

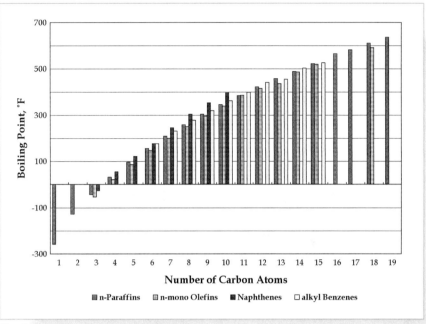

Fig. 3–5   Boiling Point vs. No. of Carbons

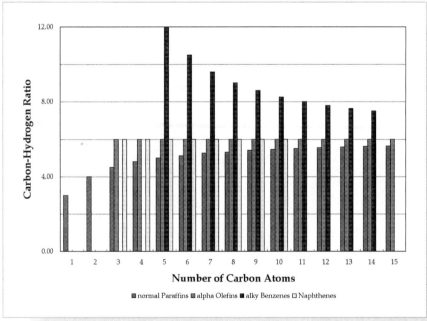

Fig. 3–6   C/H Ratio vs. No. of Hydrocarbons

show the marked difference in C/H between the alkylbenzenes and the other hydrocarbons. This will be referred to later in our discussion of processing.

One familiar characteristic of petroleum products (and thus of hydrocarbons) is their combustibility. Figure 3–7 shows the close correlation between the flash point of a hydrocarbon and its number of carbons atoms.[8] This should not come as a surprise in view of the close correlation between BP and Cs, since the flash point is the temperature at which sufficient vapor is generated to support combustion when ignited.

The upper and lower explosive limits of normal paraffins (as representative of the various series) are plotted against Cs in Figure 3–8.[7,8,9] The area between the lines is the explosive range. Below the lower limit, the mixture is too lean to flash—above, too rich.

The autoignition temperature (AIT) is the temperature at which a vapor will ignite spontaneously (in the absence of a flame). The anomalous behavior of aromatics with respect to AIT is demonstrated in Figure 3–9

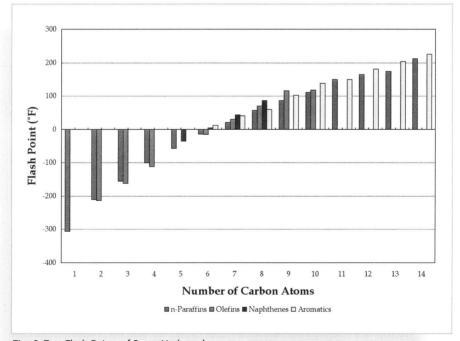

Fig. 3–7   Flash Points of Some Hydrocarbons

25

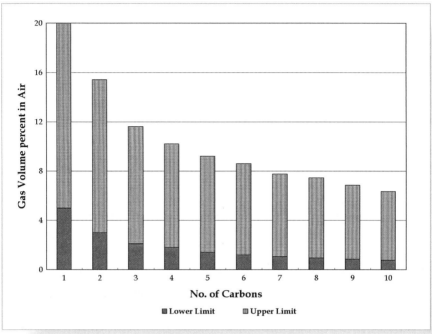

Fig. 3–8   Flammability of Normal Paraffins

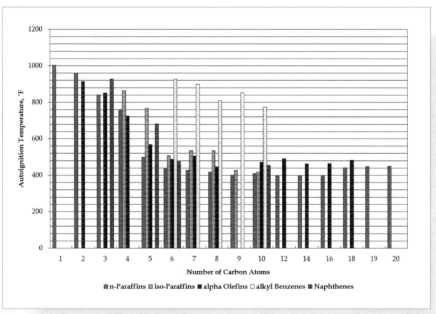

Fig. 3–9   Autoignition Temperature, °F

showing that the other hydrocarbons have significantly lower AITs for the same number of Cs.[8,9] It should be noted that a gasoline or naphtha could ignite spontaneously on coming into contact with a 600°F surface.

The specific gravity of crude oil and petroleum products is generally expressed as degrees API (American Petroleum Institute), which is defined by the following equation:

$$°API = \frac{141.5}{sp.\ gr.} - 131.5$$

where the specific gravity is the ratio of the density of the material at 60°F to the density of water at that same temperature. The gravity of water at 60°F is 10° API.

A calculated value that has been used widely as a parameter for correlating properties of petroleum products is the Watson characterization factor or K factor. The defining equation is:

$$K = \frac{MeABP^{1/3}}{sp.\ gr.}$$

where MeABP is the mean average BP of the fraction in degrees Rankine. MeABP is the arithmetic average of the molal average BP (MABP) and the weight average BP (WABP). A correlation of WABP, MABP, MeABP, and cubic average BP in terms of the volumetric average boiling point (VABP) and the slope of the ASTM distillation was presented by Watson and Nelson.[10]

Figure 3–10 is a plot of K factors for petroleum fractions calculated using the above equation. Figure 3–11 shows the variation of K with the number of carbon atoms for the different series of hydrocarbons.[7] Figure 3–12 shows cetane number vs. number of carbon atoms for various series of hydrocarbons. Cetane number is defined in chapter 5 under the discussion of diesel fuel characteristics.

Oftentimes the engineer has the distillation and gravity of a material but needs additional property information. Fortunately, the properties of petroleum fractions and the interrelationships between these properties have been investigated extensively. Some of these relationships have been

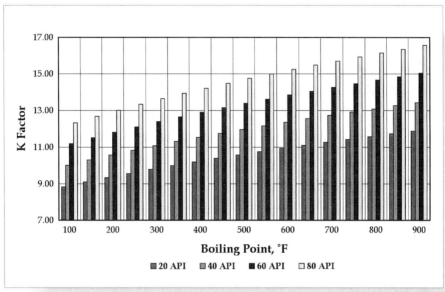

Fig. 3–10 K Factors for Petroleum Fractions in terms of Boiling Point and API Gravity

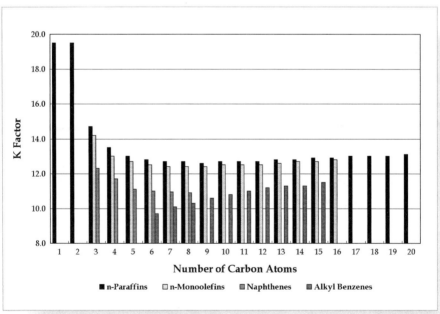

Fig. 3–11 K Factor vs. No. of Carbon Atoms

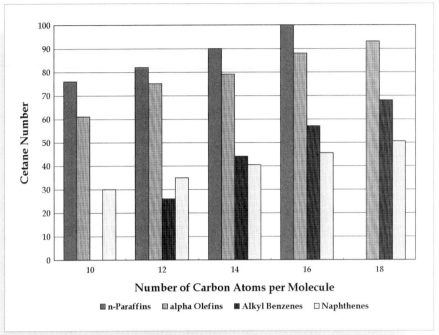

Fig. 3–12 Cetane Numbers of Some Hydrocarbons

presented in the form of graphs or nomographs, to be found in various references, such as the graphs provided by American Society for Testing Materials (ASTM) in chapter 5 showing sensitivities of various series of hydrocarbons and the graphs in chapter 30 showing some relationships for jet fuels. The API Data Book contains numerous property correlations.

The early refiners produced their finished products directly from crude oil by distillation alone—and not very good distillation at that. Except for some residual fuel oil, the products were quite different from those produced in today's very complex refinery. With strict environmental regulations in effect and crude's higher sulfur content, products from crude distillation today require further (sometimes multi-step) processing to produce acceptable finished products.

Refinery Products

Of the multitude of products produced by the refiner today, only those listed below are considered in this work. They are listed in order of increasing boiling temperatures, since distillation is the primary means employed for separating refinery products. The values are general and are included only for comparative purposes.

	Boiling Range, °F	Av. MW	No. Cs
LPG	-44–31	44–58	3–4
Gasoline	31–400	100–110	4–11
Kerosene, Jet fuel	380–520	160–190	10–15
Diesel fuel	520–650	245	15–20
Atmos. gas oil	650–800	320	20–25
Atmos. Resid	800+	—	25+
Vacuum gas oil	800–1,000	430	25–50
Vacuum resid	1,000+	800+	50+
Petroleum coke	2,000+	2,500+	200+

Such specialty products (of relatively low amounts) as lubes, greases, carbon black, petrochemical feedstock, etc., are not considered.

Initially, the refiner distilled crude oil to produce kerosene in competition with coal oil. After the explosion of several oil lamps, it was realized that a quantity of lower boiling material would need to be removed before recovering kerosene. This material began to be used in the horseless carriage. As the number of automobiles grew, the demand for gasoline grew. In a short while, gasoline became the major product produced by the refiner.

At first, product quality demands were made primarily by the auto-
mobil ufacturers. More recently, competition between refiners and came a factor as well. Still more recently, the government in he EPA and state regulatory agencies has become the primary changes on the refiner.

Notes

1. "Forecast and Review," *Oil & Gas Journal*, in January each year plus EIA, Table A5, "Annual U.S. Petroleum Supply and Demand," October, 1997

2. "Midyear Forecast," *Oil & Gas Journal*, in July each year

3. "Annual Refinery Survey," *Oil & Gas Journal*, in December each year

4. "Midyear Forecast," *Oil & Gas Journal*, in July each year

5. "Worldwide Refining," *Oil & Gas Journal*, in December each year plus EIA, Table 19, "Percent Refinery Yield of Petroleum Products by PAD and Refining Districts," each year

6. Swain, E.J., *Oil & Gas Journal*, September 9, 1991, pp. 59-61 plus October 5, 1998, ff 43

7. American Petroleum Institute, *API Technical Data Book— Petroleum Refining*, 1987

8. Dean, J.A., *Lange's Handbook of Chemistry*, 13th ed., McGraw-Hill Book Company, New York, 1985

9. Hercules Incorporated, Data Guides, Tech-Notes and Service-Notes, *Hazard Evaluation & Risk Control Services*, Rocket Center, W. Va, 1987

10. Watson, K.M., and Nelson, E.F., *Industrial and Engineering Chemistry*, vol. 25, 1933, p. 880

CHAPTER 4

REFINERY PROCESSING: AN OVERVIEW

From its very beginning, the petroleum refining industry has been changing. The following presents the principal milestones in the development of the industry.

At first a refinery consisted of a simple batch (differential) distillation in which successively higher boiling hydrocarbons were vaporized, condensed, and segregated according to the boiling ranges of kerosene, gas oil, and fuel oil. Continuous distillation was soon adopted.

The discovery of thermal cracking (a time/temperature dependent decomposition of large molecules into smaller, more desirable molecules) made it possible for the refiner to meet the growing demand for gasoline with a better (higher octane) product.

The addition of lead alkyls proved a relatively inexpensive way to upgrade gasoline.

Then the catalytic era began. Catalytic polymerization provided a way to utilize the light olefins produced in thermal cracking (principally propylene and butylenes) to produce a high-octane gasoline material. Catalytic cracking was a much-improved way (over thermal cracking) of producing olefins, gasoline, and distillates from gas oils. Catalytic alkylation was developed as a way to combine isobutane with light olefins to produce very high octane gasoline. Initially, this alkylate was slated for aviation uses. Catalytic reforming followed as a means of upgrading the octane of gasoline range materials principally by converting naphthenes to aromatics. Combined with solvent extraction of reformate, this provided a source of benzene, toluene, and the xylenes. Catalytic isomerization permitted conversion of normal C_4, C_5, and C_6 paraffins to their more desirable "iso" forms.

As shown in the previous chapter, the higher value products (gasoline, jet fuel, and diesel) have average molecular weights (MW) below 300 while gas oil and residues have much higher MWs and represent from 25 to 50 volume percent of a crude oil. Therefore, it is of interest to the refiner to upgrade these lower-value, high MW materials into higher value materials. (In this discussion, MW is used for convenience.) The real objective is to bring the boiling ranges into those of the desired products.

Another need in the processing of certain oils is the lowering of the C/H ratio. As shown in Figure 3–4, the C/H of polycyclics (increasing in occurrence in the higher boiling fractions of crude) are 12+ resulting in C/H in gas oils and residues of 8+. The C/H desired in the higher value products is in the 5 to 7 range. There are two ways that have been developed to lower the average MW of an oil and to lower its C/H as well. These are:

- Removal of carbon by forming coke (a high MW, high C/H solid)
- Direct addition of hydrogen

The first is easy, relatively inexpensive, and is exemplified by DC, SDA, FC/flexicoking and the coke deposited on FCC and other catalysts. The second is more difficult and is much more expensive. It is HC.

Figure 4–1 is a plot of the time–variation in crude running capacity of U.S. refineries.[1] Figure 4–2 is a similar plot of the capacities of catalytic cracking, CR, alkylation, and HC.[1] These figures show very little change in capacities in the past 10 years. Catalytic cracking has endured more than 55 years and has been and remains the conversion workhorse of the refinery. It has attained and maintained this position by virtue of continual improvement in mechanical design and catalyst development. CR has similar distinction in its very important niche. Though less important (capacity-wise), alkylation and isomerization promise to become more significant in the reformulated gasoline era.

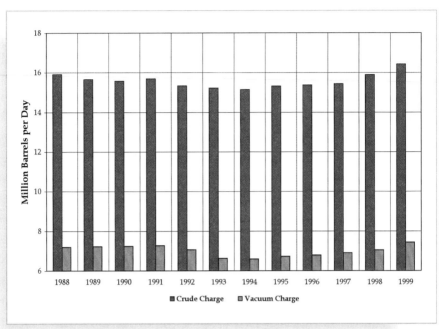

Fig. 4–1   Refinery Process Capacities (Millions of Barrels per Day)

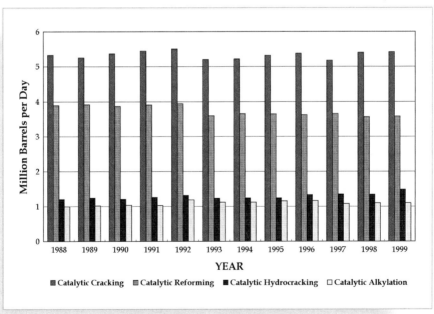

Fig. 4–2   Catalytic Process Capacities (Millions of Barrels per Day)

Catalytic HC offers flexibility (in product yields) not available in other processes, but at a price that inhibits its growth.

The increases in capacity of the various processes have been made to meet the demands of the marketplace.

Thus, the refinery has grown from a single simple distillation unit to a complex that might contain as many as 10 to 15 types of process units:

General Type	Examples
Separation	Crude distillation
	Absorption
	Adsorption
	Extraction
Reducing ave. MW	Visbreaking
	Delayed coking
	Fluid/flexicoking
	Fluid catalytic cracking
	Heavy oil cracking
	Hydrocracking
Quality improvement	Hydrotreating
	Catalytic reforming
	Isomerization
Increase ave. MW and quality	Alkylation
	Catalytic polymerization
Product blending treating auxiliary processes	Hydrogen generation
	Sulfur recovery
	Water treatment
	Waste disposal

Organizational Changes

Historically, the petroleum refining industry has been one of the highest volume, lowest margin industries in the world economy.

"Current economic conditions in the industry (excess crude supply and reduced product demand in some areas) coupled with continually increasing regulatory requirements in the U.S. have resulted in very difficult business conditions for the refining and marketing companies. There is a clear recognition that fundamental changes must be made in the way these downstream companies operate. The industry has shown financial results which are judged to be unsatisfactory in terms of common economic yardsticks. The basic problem is low return on investment (ROI) in refining and marketing." [2]

Margins for most refiners have become negative (at times) and many will not be able to survive unless rescued by some form of reorganization. As a result, the world petroleum industry is undergoing significant reorganization in the form of acquisitions, mergers, joint–ventures, etc.

Refinery Process Schemes and Capacities

Bruce Burke of Chem Systems says that in the past five years 45% of the U.S. refining capacity has changed hands; refineries that can form alliances through equity partnerships or mergers will continue while others will be forced out of the industry.[3]

Table 4–1 lists the process configurations for typical refineries in the U.S., Europe, and the Asia/Pacific area.[4] The numbers are for percent of crude charge capacity. It will be noted that the U.S. refineries average a much higher conversion capacity than do the other areas.

A better appreciation of this difference can be obtained by looking at the Nelson Refinery Complexity Indices for a number of regions:

Region	Index
Middle East	4.2
Latin America	4.7
Africa	3.3
Europe	6.5
Asia	9.0

YEAR	CRUDE CHARGE	THERMAL CRACK	CAT REF	CAT CRKG	HC	HYDRO TREAT	ALKY
U.S. REFINERIES							
1999	100	13	22	33	9	65	7
1998	100	12	22	34	7	66	
1997	100	12.3	26.1	33.6	8.8	106.2	7.6
1996	100	12.3	23.6	52.9	8.7	105.6	7.0
1995	100	12.7	23.8	34.8	8.1	67.0	6.9
1994	100	8.0	15.8	20.5	4.9	39.9	4.0
1993	100	11.8	23.6	34.2	8.1	79.9	6.7
1992	100	13.1	26.4	52.4	13.1	63.5	7.1
1991	100	13.1	25.2	35.2	13.1	62.8	13.2
1990	100	13.1	26.0	34.6	8.0	62.0	6.6
1989	100	13.2	25.2	52.2	13.1	78.6	6.6
1988	100	12.3	26.6	53.1	7.6	79.1	6.7
EUROPEAN REFINERIES							
1999	100	13	15	16	4	56	2
1998	100	13	16	16	5	56	2
1997	100	13.8	15.3	15.9	4.9	66.8	1.8
1996	100	14.3	16.6	16.0	4.8	53.0	1.8
1995	100	14.3	16.6	16.0	4.8	53.1	1.8
1994	100	13.1	16.0	14.9	3.7	51.2	1.6
1993	100	16.3	15.6	16.2	4.1	50.1	2.0
1992	100	13.7	16.8	16.5	4.2	47.8	1.5
1991	100	13.2	16.7	14.8	4.2	46.2	1.4
ASIA/PACIFIC REFINERIES							
1999	100	4.7	10.3	13.3	3.9	40.1	0.7
1998	100	5.2	10.9	14.4	3.2	42.4	0.7
1997	100	5.5	12.6	14.1	3.3	43.2	0.8
1996	100	3.5	8.1	8.7	2.1	32.4	0.5
1995	100	3.5	8.1	8.7	2.1	22.5	0.5
1994	100	3.8	14.8	8.6	3.7	45.1	0.5
1993	100	2.0	5.1	8.1	1.3	20.4	0.3
1992	100	3.9	15.5	8.3	3.9	38.1	0.7
1991	100	8.0	9.3	16.0	2.8	39.7	0.6

Table 4–1 Average Process Configurations as Percent of Crude Charge

38

Region	Index
C.I.S.	3.8
Other	5.3
Canada	7.1
U.S.	9.5

These calculated values are due to Johnston.[5]

Note: The index is obtained by multiplying the capacity of each process unit divided by the crude capacity and multiplied by a complexity factor for that process. The resulting factor is the cost of the unit relative to that of the crude unit with the same capacity.

Two simplified block flow diagrams illustrate possible process combinations:

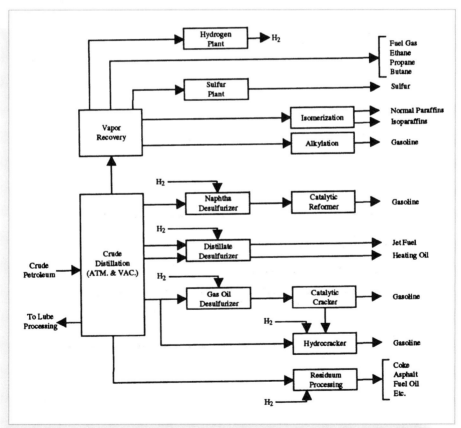

Fig. 4–3   Petroleum Refinery (Gasoline and Fuel)

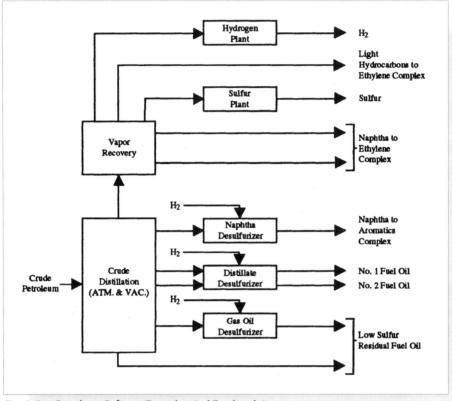

Fig. 4–4 Petroleum Refinery (Petrochemical Feedstocks)

Product slate

The change in product demand in recent years is illustrated graphically in Figure 4–5 and numerically in Table 4–2.[6] It is evident that the most significant change has been in the slight increase in gasoline and distillate demand and the decrease in demand for residual fuel oil. Figure 4–6 presents these same data as millions of barrels per day. The volume of gasoline sales in millions of gallons per day is depicted in Figure 4–7.

Trends in processing

In addition to continued improvement in traditional processes, several new technologies have been adopted by the refiner. These include cogeneration of electric power and manufacture of oxygenates.

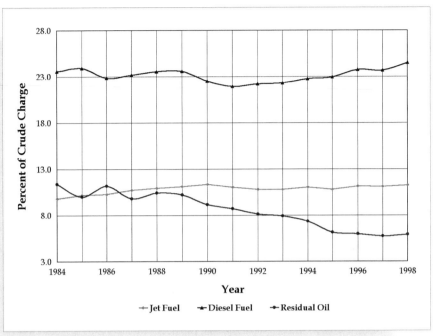

Fig. 4–5 U.S. Refinery Product Slate (Percent of Crude Charge)

YEAR	GASOLINE	JET FUEL	DISTILLATE	RESID
1984	55.6	9.8	23.6	11.4
1985	56.5	10.2	23.9	10.0
1986	54.6	10.3	22.9	11.2
1987	55.9	10.7	23.2	9.8
1988	55.5	10.9	23.5	10.4
1989	55.2	11.1	23.6	10.2
1990	54.5	11.3	22.5	9.2
1991	54.4	11.1	22.0	8.7
1992	55.0	10.8	22.2	8.1
1993	55.0	10.8	22.3	7.9
1994	54.8	11.0	22.8	7.4
1995	55.8	10.8	23.0	6.2
1996	55.6	11.1	23.7	6.0
1997	55.5	11.1	23.7	5.7
1998	56.6	11.2	24.5	5.9

Table 4–2 U.S. Product Demand as Percent of Crude Charge

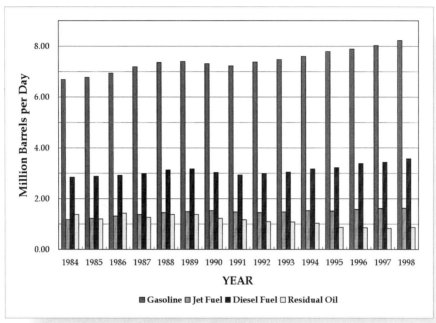

Fig. 4–6 U.S. Refinery Product Slate (Millions of Barrels per Day)

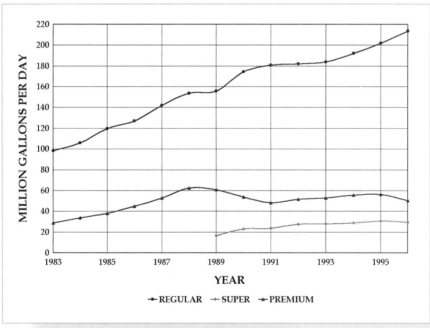

Fig. 4–7 Gasoline Sales Volume (Millions of Gallons per Day)

Cogeneration of power by integrated gasification combined cycle (IGCC) uses low value problem streams such as asphalt, vacuum residual, visbreaker tar, and coke to produce steam, electricity, and synthesis gas.

Some researchers are studying the biocatalytic desulfurization of gas oils and the bio–destruction of methyl tertiary butyl ether (MTBE).

Natural gas is receiving great attention as a source of organic liquids in addition to its direct use as a fuel.

Lubricants

Catalytic hydroprocessing has become competitive with solvent refining as a way to produce lube oils. In spite of their higher costs (3 to 10 times that of mineral oils), synthetic lubes are making inroads in the lube market (about 5% in 1998) due to their superior properties.

Pipelines and terminals

Pipelines are carrying more different products resulting in more interface material to be processed or blended off within product specifications. Close monitoring and fast response coupled with careful pipeline batch scheduling are required to minimize the quantity of interface. Electronic analyzers and blending controls are becoming commonplace at terminals, since most proprietary additives are blended at the terminal (this technique is called "splash blending").

Notes

1. Figures 4–1 and 4–2, "Annual Refinery Survey," in December issue of *Oil & Gas Journal* each year

2. Jones, J.R., talk before "Post Merger Integration Symposium," World Trade Institute, New York City, September, 1998

3. Anon., *Oil & Gas Journal*, April 20, 1998, p. 33

4. *Oil & Gas Journal*, "Worldwide Refining," in December each year.

5. Johnston, David, *Oil & Gas Journal*, March, 18, 1996, ff. 74

6. "Forecast and Review," *Oil & Gas Journal*, in January each year plus EIA, Table A5, "Annual U.S. Petroleum Supply and Demand," October, 1997

CHAPTER 5

ENERGY RESOURCES AND TRANSPORTATION FUELS

Energy Sources

Transportation fuels are but one of the energy demands in the U.S. that are competing for the various resources available. In 1996, these energy demands were met as follows:[1]

Non-renewable sources (NRS)	% of NRS	% of Total
Natural gas	26	
Petroleum	42	
Nuclear	8	
Coal	24	
Total NRS	100	92

Renewable sources (RS)	% of RS	
Solar	1	
Geothermal	5	
Biomass	41	
Wind	<0.5	
Hydroelectric	53	
Total RS	100	8

Though biomass (principally from wood) and hydroelectric account for practically all the renewable sources, solar and wind are experiencing modest growth. It is said utility scale wind plants can generate electricity for 3.5 to 5 cents per kWh.[2] Due to reduc-

45

tion in the average price of crystalline silicon modules, solar photovoltaic generators are down to about four dollars per kW.[1]

Alternative Fuels

The government relations committee of the American Institute of Chemical Engineers (AIChE) formed the Alternative Fuels Task Force to study this subject. In the report by this group,[3] they state that the U.S. will continue to depend heavily on petroleum products for the foreseeable future but that an analysis is warranted of fuels that could supplement or replace conventional gasoline and diesel fuels.

"The study suggests that, while no alternative fuel is a panacea for all problems, CNG, LPG, and RFG present the best overall alternatives to conventional gasoline based on current technology. These three alternative fuels (ATFs) provide environmental benefits at a relatively low fuel cost."

Eight key characteristics were considered for each fuel:

- Fuel cost—1995 average U.S. Gulf Coast wholesale market price per gallon adjusted to an equivalent heating value of a gallon of conventional gasoline

- Vehicle cost—The lower conversion or replacement cost of an existing gasoline vehicle to one that uses the alternative fuel

- Energy dependence—Qualitative effect for each fuel of its reliance on imported energy

- Net energy efficiency—Comparison of energy consumed in the production and distribution of each fuel with the energy available from its use

- Greenhouse emissions—Emissions for the life–cycle of each fuel

- Non–greenhouse emissions—Hydrocarbon emissions from production, distribution, fueling, and incomplete combustion of each fuel

- Infrastructure—Existing infrastructure currently available for production, distribution, and retail sale of each fuel

- Driveability—Factors such as vehicle range and refill or recharge time

The resulting averages fuel performance indices for the fuels studied were as follows:

Fuel	Index
Conventional gasoline	3.7
RFG	3.8
Ethanol	2.6
Methanol	3.1
Electric	3.2
Compressed natural gas (CNG)	4.1
Liquified petroleum gas (LPG)	4.0

Though CNG and LPG have the highest average ranking, they fall short on availability and customer convenience.

The Energy Information Administration of the Department of Energy reports estimates of alternative–fueled vehicles in use in the U.S., and their consumptions in thousands of gallons of equivalent gasoline (Tables 5–1 and 5–2).[4]

Table 5–3 compares some of the properties of two vegetable oils that have been studied as substitutes for or extenders of diesel oil.[4]

Gasoline

Gasoline is the largest volume single material produced by the U.S. refining industry, equal to about 56% of the crude oil refined. Most of the research and development efforts of the industry have been devoted, directly or indirectly, to increasing the yield and quality of gasoline. Therefore, it is fitting that we dwell awhile on its properties and characteristics.

Octane number. The most familiar property of gasoline, aside from its flammability, is its antiknock index or octane number. Octane number is a

Fuel	1992	1993	1994	1995	1996	1997
LPG	221000	269000	264000	259000	266000	273000
LNG	23191	32714	41227	50218	62805	81747
CNG	90	299	484	603	715	956
85% Methanol	4850	10263	15484	18319	19636	19787
100% Methanol	404	414	415	386	155	130
85% Ethanol	172	441	605	1527	3575	5859
95% Ethanol	38	27	33	136	341	341
Electric	1607	1690	2224	2860	3306	3925

Table 5–1 Number of Alternative-Fueled Vehicles in Use in the U.S.

Fuel	1992	1993	1994	1995	1996	1997
LPG	208142	264655	248467	232701	238681	244659
LNG	10825	21603	24160	35162	50884	81736
CNG	585	1901	2345	2759	3233	4702
85% Methanol	1069	1593	2340	3575	3832	3850
100% Methanol	2547	3168	3190	2150	2150	338
85% Ethanol	21	48	80	190	190	728
95% Ethanol	85	80	140	709	709	1803
Electric	359	288	430	663	663	1001

Table 5–2 Estimated Fuel Consumption (Equivalents of Thousands of Gallons of Gasoline)

measure of a fuel's tendency to knock in a test engine compared to other fuels. The knock sound in an engine is due to the fuel burning too rapidly (exploding) rather than burning slowly over much of the power stroke. Thus, low–octane fuels burn more rapidly then do high–octane fuels.

The octane number of fuel is equal to the percent "isooctane" (2,2,4–trimethylpentane) in a mix with normal heptane that has the same knocking tendency as the fuel being tested. A value of 100 was arbitrarily assigned to isooctane; zero octane, to n–heptane. These two materials are known as primary reference fuels. In reality, most octane tests are performed using secondary reference fuels that have been rated against primary reference fuels.

Actually, there are several octane numbers associated with a given gasoline. The one posted on the pump from which a gasoline is dispensed is really the average of two of these numbers: Its research octane number (RON) and its motor octane number (MON). This average is normally referred to as "R + M over 2" or:

$$\frac{R + M}{2}$$

This value is sometimes incorrectly referred to as the gasoline's road octane number. Whereas RON and MON are determined on a laboratory engine, road octane is properly determined by actual road test under specified conditions. Sometimes laboratory tests using a dynamometer (to vary the load on the engine) are designated as road octanes.

Both RON and MON can be determined on the same engine but under different operating conditions, primarily speed. The RON is run at low speed (600 rpm) to simulate city driving at low speed with frequent acceleration. The MON is run at 900 rpm to approximate highway driving. The relationship between the road octane of a fuel and its RON and MON varies with the car in which the road number is determined.

In one series of tests,[5] cars requiring premium gasoline appeared to have a nearly equal appreciation of both RON and MON. In other words the road octane required by the engine for satisfactory performance correlated well with R + M over 2. In this same series of tests, cars designed for regular gasoline displayed a greater, though variable, appreciation of MON than RON.

No single equation has been found that correlates the RON and MON of a gasoline with its road octane in all cars. As a compromise and in the interest of simplicity, the Federal Trade Commission stipulated that the average octane (R + M over 2) be posted on the pumps dispensing gasoline.

Scientists at Socony Mobil (now Mobil) developed an octane number method called distribution octane number (DON).[6,7] This method employed the same engine as used for research and motor numbers but with a specially designed manifold in which the lower boiling vapor could disengage from the higher boiling liquid portion of the gasoline. This was an attempt to simulate the maldistribution that can occur in small engines with a single carburetor and manual transmission. It was found that the resulting DONs correlated very well with road ratings during low–speed acceleration.

Another attempt at characterizing road performance by laboratory test is the front–end octane number or FEON that involves the distillation from the whole gasoline of the more volatile portion up to a certain volume percent or to a certain vapor temperature. The RON of this lower boiling material is then determined. The difference between the RON of the whole gasoline and this portion is referred to as R-100 when the gasoline is distilled to a vapor temperature of 100°C.[8] It is referred to as ΔR75 when the RON is determined on the first 75% of the gasoline distilled off.[9]

Like the DON, FEON appears to correlate well with road octane performance in representatives of particular segments of the car population.

Figure 5–1 traces average octane values for regular and premium gasolines marketed in the U.S.[10]

A driveability index has been accepted by the automobile and petroleum industries as providing a good indication of the driveability performance of a gasoline:[11]

$$DI = 1.5 \bullet T10 + 3.0 \bullet T50 + 1.5 \bullet T90$$

where:

DI = Driveability Index

T_{10}, T_{50}, and T_{90} represent the temperatures in deg. F at which 10%, 50%, and 90% of the fuel is distilled

Other indices have been derived to account for the presence of oxygenates:

$$DI = 1.5 \bullet T10 + 3.0 \bullet T50 + T90 + 11 \bullet wt\% \ O_2 \ ^{[12]}$$

$$DI = 1.5 \bullet T10 + 3.0 \bullet T50 + 1.0 \bullet T90 + 7.0 \bullet vol\% \ Ethanol^{[13]}$$

It has been shown that the DI increases on average as the RVP of gasoline decreases. Auto manufacturers recommend a maximum DI of 1,200 to provide satisfactory driveability.

Another characteristic of gasoline that is of interest is its sensitivity, which is defined as RON – MON. A maximum value for sensitivity is sometimes specified for a gasoline. Figure 5–2 illustrates the sensitivities of the

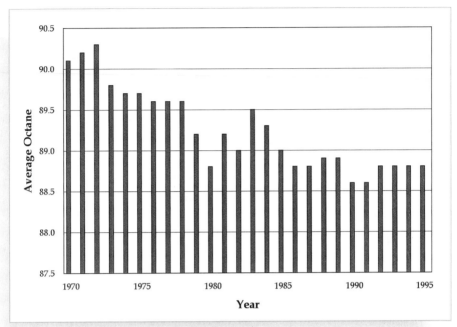

Fig. 5–1   Average Gasoline Octanes (Regular Gasoline)

principal hydrocarbon series of interest.[14] It will be noted that, except for paraffins, these hydrocarbons generally have positive sensitivities (i.e., RON is greater than MON).

The lead susceptibility of a gasoline is the increase in its octane (either research or motor) resulting from the addition of a given quantity of lead alkyl (TEL, TML, or a mixture of the two). TML has a lower BP (230°F) than TEL (392°F) and has been used primarily in Europe to help meet the high front–end octane demand of a predominantly high–compression, stick–shift car population. Since the use of lead in gasoline has been nearly phased out completely in the U.S., it will not be considered further in this book.

Volatility. The volatility of a gasoline is perhaps its most important single property. Without the formation of sufficient vapor at the existing ambient temperature, the engine won't start. At the other extreme, too much vapor can result in "vapor lock" where presence of vapor in fuel lines and fuel pump can curtail or completely prevent flow of liquid fuel.

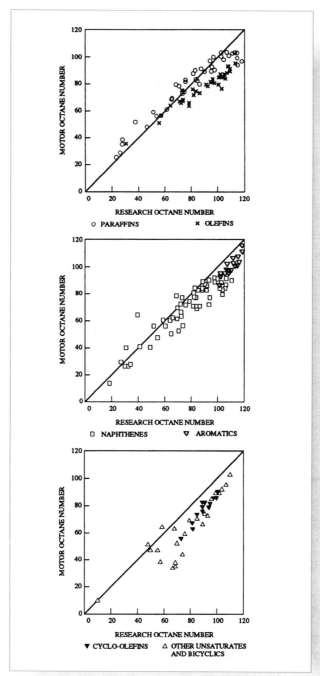

Fig. 5–2   Hydrocarbon Sensitivities (© ASTM; reprinted
with permission)

The common measure of the vapor forming tendency of a gasoline is its Reid vapor pressure (RVP). Figure 5–3 (data from reference 10) is a plot of annual average summer and winter RVPs for U.S. gasolines for a period of years. It is readily apparent that the winter values are 1.5 to 3.3 psig higher than the summer values.

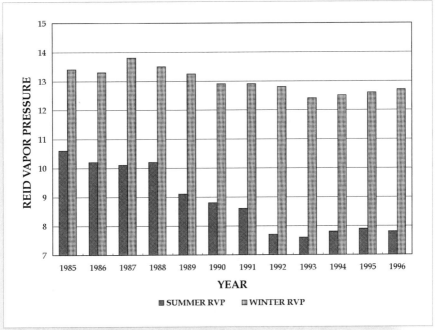

Fig. 5–3   Gasoline Vapor Pressures

Another means of measuring the volatility of a gasoline is by distilling it and noting the vapor temperature when successive percentages have been distilled, from the initial BP (IBP) to its final BP (FBP), by means of ASTM Method D 86.

Along with the RVP, this distillation describes the vaporizing characteristics of the gasoline. These data indicate not only the gasoline's engine starting capability and vapor lock tendencies, but also its ability to respond to fuel requirements during warm–up and acceleration.

Table 1 of ASTM Method D 4814 defines six volatility classes in which the RVP and certain distillation point temperatures are specified. The specifications include the following:

Class	AA	A	B	C	D	E
RVP Temp ∞F	7.8	9	10	11.5	13.5	15
Max 10%	158	158	149	140	131	122
Min 50%	170	170	170	170	150	150
Max 50%	250	250	245	240	235	230
Max 90%	374	374	374	365	365	365
Max EP	437	437	437	437	437	437

The schedule for the application of these volatility classes seasonally and geographically is given in Table 4 of ASTM D 4814.

It has been observed that vapor lock troubles can be expected when the vapor-liquid (V/L) ratio of a gasoline is 20 or more. The temperature at which this occurs can be determined by ASTM Methods D 2533 or D 5188 (see ASTM D 4814).

Components in gasoline. The gasoline produced by today's refiner is very different from that made by the first refiners. Instead of a single stream produced by distillation of crude oil, the refiner today may have available for blending into gasoline 10 or more different streams from the various processes in his plant or imported into his plant. There will probably be seasonal variations in the supply of these materials as well.

The gasoline blending problem is further complicated by the fact that the contribution of a component to the various properties (octane, RVP, distillation) of the resulting gasoline blend will vary with the types and relative proportions of the other constituents.

Fortunately, it has been possible to develop pseudo values for these properties of each constituent, differing from the values obtained in the laboratory on the constituent by itself, representing reasonably well the contribution of that constituent in the usual refinery blends. These pseudo properties are referred to as "blending numbers" and are used on a volume weighted basis to predict properties of gasoline blends. Table 5–3 presents a set of blending values for many of the components used by the refiner to produce the various gasoline blends.[15]

Figures 5–4 and 5–5[16] illustrate the change in the composition of gasoline over the years due to introduction of and improvements in refinery processes.

54

COMPONENT	RON	MON	Avg Oct
HYDROCARBONS			
Normal Butane	94.4	98.4	96.4
Isobutane	100.8	97.4	99.1
Butylenes	104.1	82.9	93.5
Natural Gasoline	73.5	72.5	73.0
Light Straight Run	59.2	59.9	59.6
Intermed. Straight Run	70.0	70.7	70.4
Heavy Straight Run	80.8	81.5	81.2
Light Naphtha	41.3	42.0	41.7
Reformate (80 RON)	80.7	74.3	77.5
Reformate (90 RON)	90.5	80.4	85.5
Reformate (95 RON)	95.2	83.4	89.3
Reformate (100 RON)	99.9	87.6	93.8
Light FCC Gasoline	94.7	79.3	87.0
Heavy FCC Gasoline	87.0	78.1	82.6
Light Hydrocrackate	86.6	83.0	84.8
Polymer Gasoline	96.0	83.0	89.5
Medium Hydrocrackate	76.5	71.0	73.8
Propylene Alkylate	91.7	92.2	92.0
Butylene Alkylate	94.3	94.5	94.4
Heavy Alkylate	85.4	82.2	83.8
OXYGENATES			
Methanol	133.0	99.0	116.0
Ethanol	129.0	96.0	112.5
MTBE	118.0	100.0	109.0
ETBE	118.0	102.0	110.0
TAME	111.0	98.0	104.5

SOURCE: EPA,"Costs and Benefits of Reducing Lead in Gasoline",
March 1984, EPA-230-03-84-005

Table 5–3 Average Octane Blending Values

Straight run (virgin) and thermal (from thermal cracking) stocks are very minor constituents in today's gasoline having been displaced by components produced by catalytic processes.

Engine characteristics. The theoretical thermal efficiency of the Otto cycle engine can be computed by means of the following expression:

55

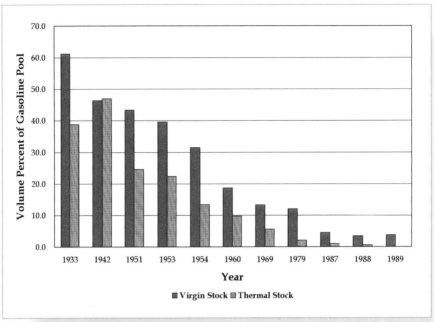

Fig. 5–4   Virgin and Thermal Gasoline Stocks

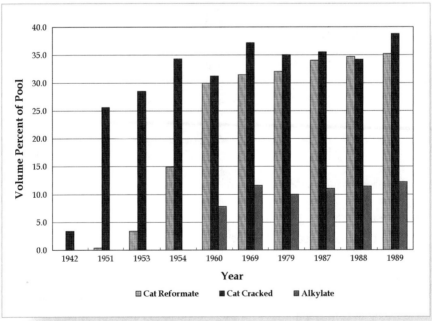

Fig. 5–5   Catalytic Gasoline Stocks

$$\% \; eff'y = 100 \cdot [1-(1/CR)^{(C_p/C_v-1)}]$$

where:

C_p/C_v is the ratio of the specific heat capacities at constant pressure and constant volume, respectively

CR is compression ratio

Assuming the working fluid to be essentially air, the specific heat ratio is 1.396. Calculating the efficiencies for a series of compression ratios and plotting the results yields the curve in Figure 5–6. The significant increase in theoretical efficiency possible by increasing compression ratio is readily apparent.

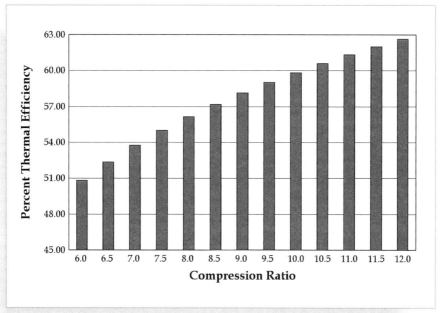

Fig. 5–6 Otto Cycle Efficiency

The amount of this thermal efficiency resulting in useful work is reduced by the energy rejected to cooling water, or air (25 to 30%), the energy rejected in the exhaust (25 to 30%), plus the energy lost due to radiation and friction, including wind resistance (10 to 20%). This leaves approximately 30 to 40% for propulsion of the vehicle.

In an SAE paper in 1965, Blomquist[17] presented data on brake specific fuel consumption (BSFC) for a range of compression ratios. Assuming an average lower heating value of 115,000 Btu/gal for gasoline and an average density of 6 pounds per gallon, percent useful work was calculated and is plotted in Figure 5–7.

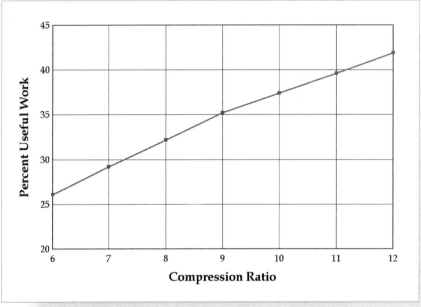

Fig. 5–7   Percent Useful Work

Using data from Blomquist's paper, rate of fuel consumption as gallons per hour was calculated and the results plotted in Figure 5–8. Morris, *et al.,*[5] summarized results of other similar studies. Their data recalculated as fuel consumption relative to a compression ratio of 12 at constant performance are plotted in Figure 5–9 along with Blomquist's data representing variable performance. In Volume 11 of the Third Edition of Kirk–Othmer, Lane calculated the rate of gasoline consumption (gallons per hour) versus highway speed on a level road at constant speed.[18] These calculations neglect the power required by power brakes, power steering, air conditioning, etc. The results are shown in Figure 5–10. If power accessories plus hill climbing, acceleration, and deceleration are taken into account the fuel consumption would be altered significantly.

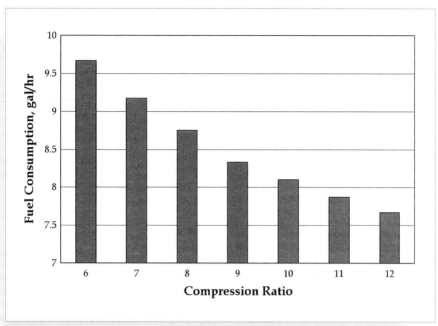

Fig. 5–8   Fuel Consumption

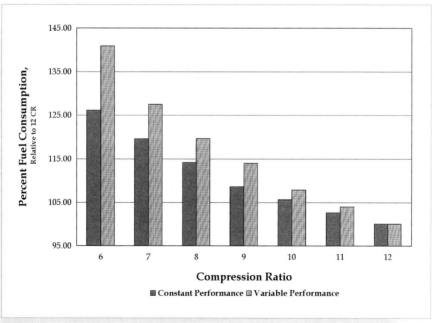

Fig. 5–9   Relative Fuel Consumption

59

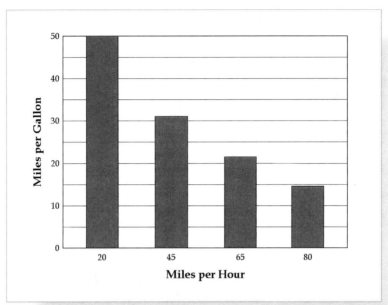

Fig. 5–10   Fuel Consumption vs. Speed

Figure 5–11 depicts the average car fuel economy in miles per gallon of gasoline versus model year.[19]

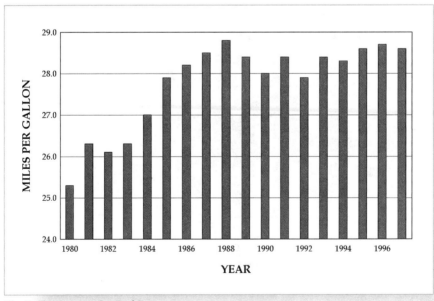

Fig. 5–11   Average Car Fuel Economy

Additives. In addition to the hydrocarbons comprising the bulk of blended gasoline, a number of other materials are added for various purposes:

- Antioxidants protect against gum formation and peroxides (knock inducers) formation.

- Metal deactivators tie up trace amounts of copper to prevent their catalyzing oxidation and gum formation.

- Antirust agents inhibit rusting, help prevent gummy deposits in the carburetor and combat carburetor icing.

- Anti–icing agents prevent ice formation in fuel lines and carburetor.

- Detergents (a.k.a. deposit control additives) prevent deposits in the carburetor or remove them from dirty carburetors.

- Upper cylinder lubricants lubricate cylinders and piston rings to limit deposits in the intake system.

- Dyes are used to distinguish between various grades and brands of gasoline. Originally they were added primarily to warn of the presence of lead alkyl compounds.

- Antidetonants (a.k.a. antiknock additives) are materials added to a gasoline to enhance its octane number to reduce knocking tendency. The mainstay here for many years has been the lead alkyls TEL and TML, but these are being phased out of use. Methylcyclopentadienyl manganese tricarbonyl (MMT) has been used as an antidetonant to a limited extent (particularly in Canada).

- More recently, oxygen–containing organic compounds (oxygenates) have been used for this purpose and to reduce emissions of volatile organic compounds (VOC)—essentially unburned hydrocarbons. Under present requirements of the amended Clean Air Act, in order to provide the necessary amount of oxygen, these materials will become a major constituent of future gasoline blends. Therefore, they should be considered as another type of gasoline blending component, rather than as an additive.

It is not surprising that the adjustment and condition of an engine can significantly affect the octane required by the engine. Increase in spark advance, temperature of coolant, or deposits in the combustion zone can increase the octane required. On the other hand, the reader may be surprised to know that increase in altitude or relative humidity decreases the engine's octane demand (see ASTM D 439).

Figure 5–12 illustrates how the composition of gasoline has changed over much of the life of petroleum refining in the U.S.[20]

Table 5–4 shows average properties for gasoline for recent years.[10]

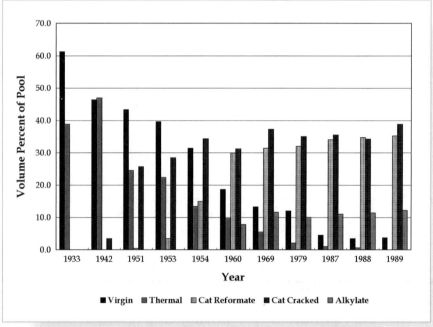

Fig. 5–12 Composition of Gasoline Pool

Diesel fuel

Diesel fuel is second in volume only to gasoline as fuel for internal–combustion engines. The diesel engine differs from the conventional gasoline engine in the way in which the fuel is ignited. In the gasoline engine, ignition is initiated by an electric spark—in the diesel, it results from the heat produced by compression of the fuel–air mixture.

YEAR	BENZENE VOL %	SULFUR WT %	OLEFINS VOL %	AROMATICS VOL %	AV. OCT.	RVP	ETHER VOL %
			REGULAR GRADE				
1991	1.61	0.038	11.8	28.1	87.1	8.6	1.0
1992	1.70	0.034	12.4	30.0	87.4	7.7	0.4
1993	1.40	0.036	11.8	30.4	87.4	7.7	1.2
1994	1.58	0.035	11.9	30.5	87.3	7.8	1.0
1995	1.33	0.029	11.8	31.0	87.3	8.0	0.8
1996	1.13	0.029	10.2	30.5	87.5	7.9	1.6
			PREMIUM GRADE				
1991	1.62	0.013	5.5	33.2	92.3	8.6	4.1
1992	1.55	0.012	6.1	34.3	92.3	7.7	2.0
1993	1.33	0.014	6.5	35.5	92.5	7.6	6.7
1994	1.52	0.013	6.5	33.8	82.5	7.8	3.9
1995	1.13	0.007	5.5	33.0	92.4	7.9	2.8
1996	1.04	0.010	4.7	33.1	92.5	7.8	3.0

Table 5–4 Some Gasoline Properties

In the diesel engine, air is drawn into a cylinder and compressed before fuel is injected into the cylinder. During compression, the air is heated to a temperature at which the fuel will spontaneously ignite after a short delay. The length of this delay varies with the types of hydrocarbons making up the fuel, increasing from paraffin to olefin to naphthene to aromatic.

Cetane number (ASTM Method D 613). The primary measure of diesel fuel quality is its cetane number. This is a function of the delay before ignition: the shorter the delay, the higher the quality or cetane number. Long delays cause rough engine operation, misfiring, difficult starting in cold weather, and smoky exhaust.

The laboratory rating of diesels is similar to the procedure used in determining the gasoline octane number. The paraffin hydrocarbon cetane (n–hexadecane) has been assigned the value of 100 cetane. A value of zero has been assigned to the bicyclic compound alpha methylnaphthalene. These two compounds are the primary reference fuels. Mixtures of the two are used to bracket the quality of the fuel being tested. Recently, the methyl-naphthalene has been replaced by the highly branched paraffin compound 2,2,4,4,6,8,8–heptamethylnonane with a cetane number of 15. This compound can be produced in very high purity.

The variation of cetane number with carbon number and type of hydrocarbon is shown in Figure 5–13.[21] The alkyl benzenes are obviously undesirable. The low carbon numbers have low cetane numbers and the high carbon numbers have high freeze points. Cetane numbers of diesels sold in the U.S. are in the 35 to 65 range, with most engine manufacturers specifying at least 45. The recent trend in cetane index (CI) and sulfur content of highway diesel is tabulated in Table 5–5.[22]

The following factors affect the problem of maintaining diesel fuel quality:

- The average API gravity of crude being processed is decreasing, corresponding to more dense or heavier crudes

- Cetane number of virgin diesel, in general, decreases as API gravity of crude decreases

- The heavier the crude, the more cracking is required to obtain a satisfactory yield of the lighter more desirable products

- Cetane numbers of cracked distillates are lower than those of virgin diesel from a given crude

As a result of this, we find progressively more cracked diesel of decreasing quality available for blending with a decreasing supply of virgin diesel that is also decreasing in quality.

The determination of diesel cetane number by engine (ASTM D 613) is much more expensive and difficult than the determination of gasoline octane by engine. Because of this, numerous attempts have been made to determine a satisfactory correlation between cetane number and one or more of diesel's other physical properties:

- Diesel index (DI) is equal to the product of aniline point times the API gravity divided by 100. This index was in use prior to adoption of the CI to which it is inferior.

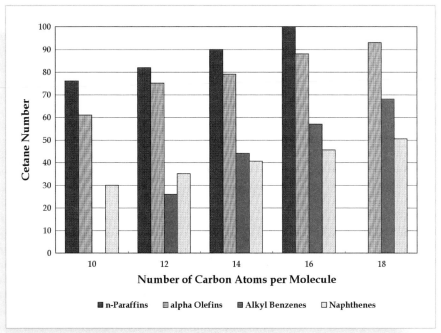

Fig. 5–13 Cetane Numbers of Some Hydrocarbons

YEAR	CETANE INDEX		WT % SULFUR	
	1-D	2-D	1-D	2-D
1986	47.5	46.8	0.105	0.260
1987	47.0	47.0	0.070	0.280
1988	47.5	45.5	0.060	0.260
1989	47.2	45.6	0.050	0.250
1990	47.5	45.0	0.040	0.255
1991	47.5	45.8	0.049	0.299
1992	47.6	45.0	0.056	0.286
1993	47.6	45.1	0.039	0.226
1994	48.7	47.6	0.023	0.032
1995	47.6	46.8	0.014	0.033
1996	48.3	47.9	0.019	0.033

Table 5–5 Highway Diesel Properties

65

- CI, (ASTM Method D 976) is a function of the API gravity and mid–BP (Mid BP) of the diesel:

$$CI = 0.49083 + 1.06577x - 0.0010552x^2$$

where:

$X = 97.833(\log \text{Mid BP})^2 + 2.2088(\text{API})(\log \text{Mid BP}) + 0.01247(\text{API})^2 - 423.51(\log \text{Mid BP}) - 4.7808(\text{API}) + 419.59$

This function correlates very well with the cetane number (CN) according to the following equation:

$$CN = 5.28 + 0.371\ CI + 0.0112\ CI^2$$

A plot of this equation can be found in chapter 30.

- Aniline point (AP), (ASTM Method D 611) has been correlated with CN using a fourth degree polynomial in AP:

$$CN = 16,419 - 1.1322\ (AP/100) + 12.9676\ (AP/100)^2 - 0.2050\ (AP/100)^3 + 1.1723\ (AP/100)^4$$

The relationship between the cetane number and aromatics content of some diesels is shown in Figure 5–14.[23]

Volatility. Some volatile compounds are needed in diesel for ease in starting. In general, the heavier compounds have higher heating values than the lighter compounds and for the same cetane number give better fuel economy. However, in addition to greater starting difficulty, too high a percent of heavies can cause high formation of deposits in the engine. Therefore, the refiner blends his available distillates to obtain the desirable volatility. Diesel fuels usually have an initial distillation point of about 320°F with a 90% point of about 550°F to 680°F depending on the grade.

Pour point (ASTM Method D 97) is the temperature at which flow of the material ceases. As the temperature of the diesel fuel decreases toward its pour point, pumping the fuel through supply lines, filters, and injectors

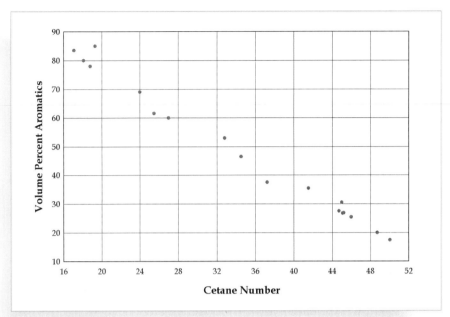

Fig. 5–14 Cetane No. vs. Aromatics

becomes more difficult. Naphthenes have low pour points, but also low cetane numbers. Paraffins have high pour points, but high cetane numbers. Therefore, the pour point specification for a given locale should be set as high as is compatible with expected winter temperatures, usually about 10°F below the expected low. There are additives available to improve pour point.

Cloud point (ASTM Method D 2500) is the temperature at which wax crystals begin forming, causing the fuel to become cloudy and clog lines and filters. Cloud point specification is usually less than 10°F above the pour point specification.

Viscosity (ASTM Methods D 88 and D 445) is another important flow property effecting pressure drop through the injection system and the degree of atomization of the fuel on injection. Different ranges of viscosity are specified for use in engines in different services.

Gravity (ASTM Method D 287) is an important property since the heating value of a fuel increases with its density. When specified, it is usually in the 30° to 40° API range.

Flash point (ASTM Method D 93) is the temperature to which the fuel must be heated to generate sufficient vapors for ignition to occur in the presence of a flame. Flash point specification varies with the grade of the diesel.

Carbon residue (ASTM Method D 189, Conradson carbon residue and ASTM Method D 524, Ramsbottom carbon residue) is the amount of carbon remaining after subjecting a sample to high temperature under specified conditions. It is an indication of the propensity of the fuel to deposit carbon in the engine.

Sulfur content (Various ASTM Methods) can cause corrosion and engine wear. Of greater concern today is the emission of SO_2 into the atmosphere as result of combustion of the sulfur content.

Ash content (ASTM Method D 482) is an indication of abrasive material remaining after combustion of the fuel and the erosion of fuel injectors that may result.

Various vegetable oils have been studied as possible substitutes (or extenders) for regular diesel fuel. Some properties are shown in Table 5–6.[24]

Jet fuel

Jet fuel is another very important transportation fuel. Though a minor amount is used in stationary engines, the bulk of jet fuel is used to power commercial and military aircraft.

Gas turbine fuel is a more general designation, since the stationary engines develop shaft power to turn electric generators, compressors or other equipment as contrasted with aircraft engines that develop thrust from exhaust gases passing through a nozzle at high velocity. In either case, useful work is obtained by expanding hot combustion gases. In this respect, the gas turbine is much like a steam turbine, but with a different working fluid.

Each type of gas turbine consists of three basic steps, with two of the steps common to both types:

PROPERTY	SOYBEAN	RAPESEED	No. 2 DIESEL
Ht of Comb'n (kJ/kg)	38460	35376	38537
Flash Point (deg C.)	171	83.9	80
Pour Point (deg C.)	-1.1	-9.4	-28.9
Cloud Point (deg C.)	-1.1	-2.2	-12.2
Viscosity (cs at 40 C.)	4.1	6	3.2
Sulfur (wt %)	0.04	0.005	0.29
Cetane	52	54.5	47.8

Table 5–6 Comparison of Conventional Diesel and Biodiesel

- Air compression
- Fuel combustion

The third step is the jet nozzle for thrust in the case of the aircraft turbine—the expansion turbine for shaft power in the stationary turbine.

The principal specifications for gas turbine fuels and the corresponding ASTM test methods include the following:

	D 1355	D 2880
Property	Aviation	Stationary
Aromatics	D 1319	—
Flash	D 56	D 93
Density	D 1298, 4052	—
Freeze point	D 2386	—
Smoke point	D 1322	—
Viscosity	D 445	D 445
Carbon residue	—	D 524
Pour point	—	D 97

The specifications for aviation turbine fuels are understandably more extensive than those for stationary turbines. Average properties for aviation turbine fuels are tabulated in Table 5–7.[25]

69

YEAR	FREEZE POINT, °F	WT % SULFUR	VOL % AROMATICS	SMOKE POINT, MM
JP-4 Military Aviation Turbine Fuels				
1989	-77	0.019	11.8	26.1
1990	-83	0.015	11.6	26.1
1991	-81	0.022	11.5	26.0
1992	-80	0.026	11.8	25.6
1993	-84	0.016	14.1	26.6
JP-5 Military Aviation Turbine Fuels				
1989	-56	0.019	17.3	20.6
1990	-57	0.030	16.6	20.6
1991	-56	0.039	18.5	21.1
1992	-55	0.041	17.1	21.0
1993	-56	0.034	19.5	21.0
1994	-58	0.025	19.0	20.6
1995	-53		21.0	20.0
JP-8 Military Aviation Turbine Fuels				
1996	-55	0.017	19.3	21.9
1997	-53	0.080	19.0	22.3
1998	-62	0.031	17.8	23.0
Jet A Commercial Jet Fuels				
1989	-48	0.048	18.0	21.9
1990	-48	0.047	19.0	21.8
1991	-47	0.051	19.4	22.1
1992	-48	0.062	18.8	22.1
1993	-47	0.049	19.4	22.2
1994	-48	0.070	18.9	21.4
1995	-50	0.061	19.0	22.0
1996	-53	0.062	18.1	21.5

Table 5–7 Average Aviation Turbine Fuels Properties

Figure 5–15 is a chart that could prove useful in estimating smoke point, luminometer number, aromatics content or hydrogen content of a distillate fraction when the aniline point and the gravity are known.[26]

70

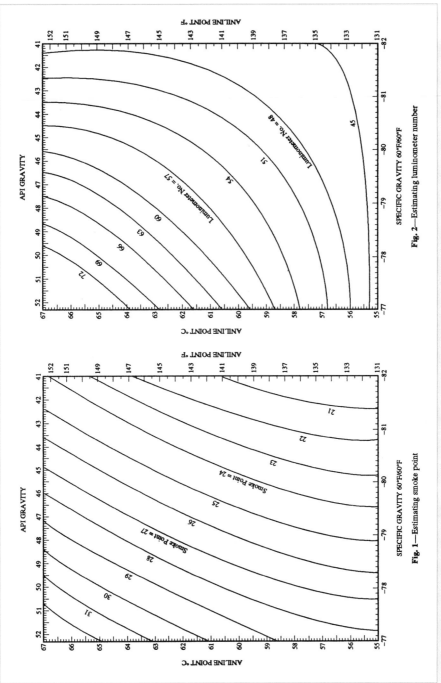

Fig. 5–15a   Quick Measure of Jet Fuel Properties (©ASTM; reprinted with permission)[14]

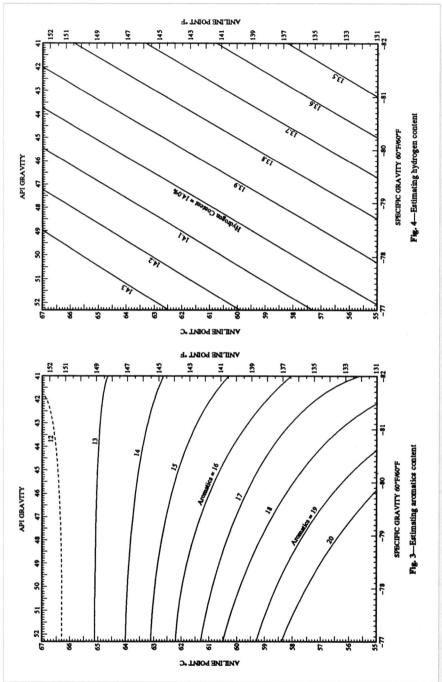

Fig. 5–15b   Quick Measure of Jet Fuel Properties cont'd (©ASTM; reprinted with permission)[14]

Tables of ASTM specifications for the transportation fuels plus some other refinery products are included in the Appendix.

The ASTM gives detailed requirements for six distilled heating oils in ASTM D 396. ASTM D 2069 Table 1 gives detailed requirements for 4 marine distillate fuels. ASTM D 2069 Table 2 defines 15 marine residual fuel oils. Tables showing these requirements can be found in the Appendix.

Notes

1. EIA, *Renewable Energy Annual,* Vol. 1, 1997

2. Dye, R. F., in *Flashpoint,* AIChE Fuels and Petrochemicals Division Newsletter

3. AIChE, "Alternative Transportation Fuels: A Comparative Analysis," September, 1997

4. EIA, "Alternatives To Traditional Transportation Fuels," DOE/ EIA–0585(95)

5. Morris, W.E., Rogers, J.D., Jr., and Pockitt, R.W., "1971 Cars and the New Gasolines," Society of Automotive Engineers, Mid–Year Meeting, Montreal, June, 1971

6. Foster, J.M., Goodacre, C.L., Griffith, S.T., and Lamarque, P.V., *Gasoline Antiknock Quality—Are Road Ratings Necessary?,* The Associated Octel Company Limited, London, July, 1964

7. Gerard, P.L., and Di Perna, C.J., "Multifaceted Octane Numbers for Diverse Engine Requirements," SAE International Automotive Engineering Congress, Detroit, January, 1965

8. Roselius, R.R., Gibson, K.R., Ormiston, R.M., Maziuk, J., and Smith, F.A., "Rheniforming and SSC—New Concepts and Capabilities," NPRA Annual Meeting, San Antonio, April, 1973

9. DuPont Technical Memorandum No. 357, Wilmington, February, 1962

10. Dickson, C.L., and Sturm, G.P., Jr., NIPER 178, 180, 188, 190, 198, 200

    Killen, P.J., *Oil & Gas Journal,* May 39, 1983, p. 86

    Shelton, E.M., Whisman, M.L., and Woodward, P.W., *Oil & Gas Journal,* August 2, 1982, pp. 95–99

    Unzelman, G.H., Oil & Gas Journal, April 7, 1986, pp. 88–95

11. Colucci, J., *Harts Fuel Technology & Management,* March, 1997, p.12

12. Anon.,"The International Fuel Quality Information Center," *World Refining,* January/February, 1999, p. 102

13. Colucci, J.M., Darlington, Tom, and Kahlbaum, Dennis, *World Refining,* January/February, 1999, p. 74

14. ASTM, *Knocking Characteristics of Pure Hydrocarbons,* API Research Project #45, ASTM Special Publication No. 225, Cincinnati, 1958

15. Nelson, W.L., *Oil & Gas Journal,* March 29, 1971, p. 79

    Smith, L.D., *Oil & Gas Journal,* June 24, 1985, pp. 95–97

    Unzelman, G.H., *Oil & Gas Journal,* April 4 and 17, 1988, pp. 35–41, 48–49

16. Anon., *Oil & Gas Journal,* June 18, 1990, pp. 49, 50

17. Blomquist, O.J., Good Gasoline Mileage is a Team Project, SAE Mid–Year Meeting, Chicago, May, 1965

18. Lane, J.C., "Gasoline and Other Motor Fuels," *Encyclopedia of Chemical Technology,* 3rd ed., Vol. 11, p. 675

19. American Automobile Manufacturers Association, *Motor Vehicle Facts & Figures 1997,* p. 80

20. Author's Files

21. Tilton, J.A., Smith, W.M., and Hockberger, W.G., *Industrial and Engineering Chemistry,* July, 1948, p. 1,271

22. Dickson, C.L., and Sturm, G.P., Jr., NIPER 187 and 195

    Eastwood, D., and Van de Veune, H., *Strategies for Revamping Distillate Desulfurizers to Meet Lower Sulfur Specifications,* NPRA Annual Meeting, San Antonio, March, 1990

23. Unzelman, G.H., *Oil & Gas Journal,* June 29, 1987, ff 55

24. Anon., "Vegetable Oils: from Table to Gas Tank," *Chemical Engineering,* February, 1993, p. 35

25. Dickson, C.L., and Sturm, G.P., Jr., NIPER 184 and 199

26. Jenkins, G.I., and Walsh, R.P., *Hydrocarbon Processing,* May, 1968, pp. 161–164

CHAPTER 6

THE ENVIRONMENT AND THE REFINER

The Problem

Apprehension about the degradation of the environment has grown rapidly in recent years to where it is now a matter of grave national and international concern. As a result of this concern, the refiner finds himself confronting two major areas of challenge:

- To produce transportation fuels meeting stringent requirements
- To operate a refinery within strict emission regulations

Since details of refinery operations are outside the scope of this book, these requirements will not be discussed in depth.

It is generally recognized that a major portion of our atmospheric pollution comes from transportation vehicles (mobile sources). As shown in Figure 6–1, vehicles are the primary source of CO while stationary sources are the main generators of particulates and of sulfur compounds.[1]

There has been a rapid growth in our reliance on the automobile as our principal form of transportation. Along with this growth was a demand by the driving public for increased performance by the automobile. This was accomplished by the automobile manufac-

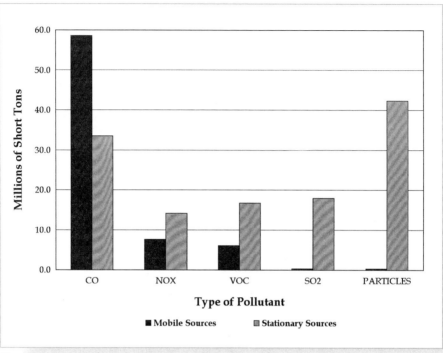

Fig. 6–1   Sources of U.S. Air Pollution

turers providing more powerful engines with greater displacement and higher compression ratio. These high compression engines required higher-octane fuel, supplied by a combination of the addition of lead alkyl antiknock agents, more butanes (thus increasing gasoline vapor pressure), as well as innovations and improvements in refinery processing.

A series of graphs (Figures 6–2 to 6–6) follows to illustrate these trends.

Regulations

The use of lead additives in gasoline soared to more than 400 million pounds in 1974. The Clean Air Act (CAA) of 1970 provided for the phasing out of lead in gasoline beginning in 1977. Since 1974, it plummeted to less than 100 million pounds in 1985. The 1990 Amendment to the CAA prohibits its use, so lead in gasoline has dropped essentially to zero, and ceases to be of further concern.

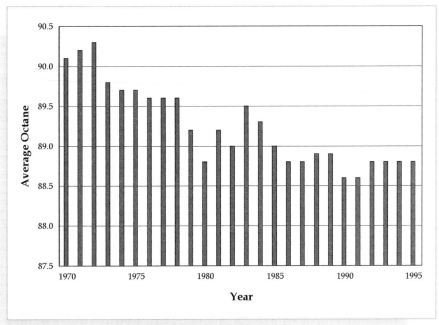

Fig. 6–2 Average Gasoline Octane (Regular Gasoline)[5]

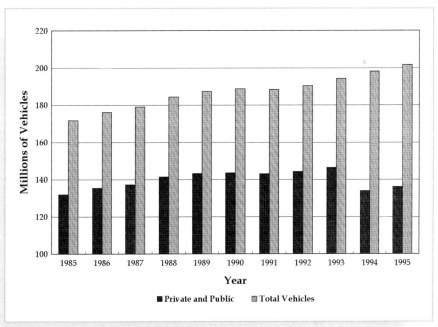

Fig. 6–3 U.S. Motor Vehicle Registrations[2]

79

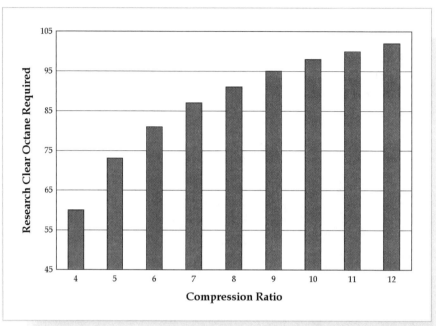

Fig. 6–4 RON Required by Compression Ratio[7,8]

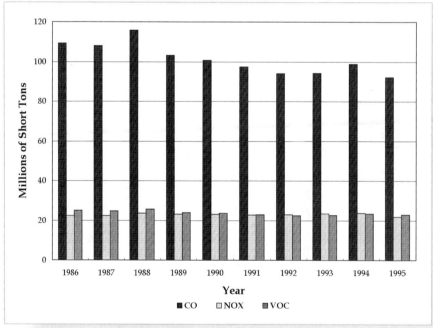

Fig. 6–5 Mobile Emission Reduction

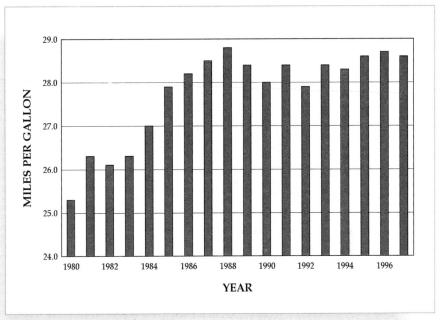

Fig. 6–6   Average Car Fuel Economy

The 1990 Amendment to the CAA further mandates reduction in the quantity and type of engine emissions by:

- Reduction in the vapor pressure (RVP) of gasoline
- Reduction in aromatics in general and benzene in particular, contained in gasoline

Both of these requirements result in lower octane in the lower boiling portion of gasoline—where octane demand is greatest.

Lowering vapor pressure means less butanes in gasoline. Butanes are premium blending components:

	Research Octane	*Motor Octane*	*Average Octane*
Normal butane	94.5	89.6	92
Isobutane	100.3	97.6	98.9

81

Aromatics are the principal source of octane in gasoline. Both gasoline from catalytic cracking and catalytic reforming are rich in aromatics and the principal components in the U.S. gasoline pool. The following data are for 1988:

Component	% Component in Pool	% Aromatics in Pool	% Olefin in Pool	% of Pool Octane
Reformate	34.0	21.3	0.2	35
Cat. Gaso.	35.5	10.4	10.2	34.5
Pool	100.0	32.0	10.8	100.0

The olefins are a secondary source of octane. They have higher RONs than the corresponding paraffins. Their motor octanes are essentially the same as those of the paraffins. Therefore, reduction of the olefins to the required level by saturation to paraffins would worsen the refiner's octane problem.

The 1990 Amendment to the CAA set a maximum value of 0.05 weight percent sulfur content and a minimum CI of 40 for diesel fuel by October, 1993.[9] The technology for accomplishing desulfurization of distillates to this level is well developed. California has added the further restriction of a maximum of 10 volume percent aromatics. Meeting the CI requirement involves adjusting cut temperatures, selecting from the available blending stocks, utilizing additives, and resorting to more severe hydrotreating.

The oxygenated and RFG programs require that gasoline must now have a minimum oxygen content. In the wintertime in CO non-attainment areas, this means a minimum of 2.7 weight percent oxygen (equivalent to about 15% MTBE or 7.6% fuel ethanol by volume), with a year-round minimum of 2 weight percent. Thirty percent of the oxygenate in RFG must be from RS (fermentation ethanol, ethyl tertiary butyl ether [ETBE], or ethyl tertiary amyl ether [ETAE]).

The regulations on volatility (RVP) reductions and VOC vary geographically and seasonally and are too detailed for this book.

Reduction in gasoline sulfur levels proposed by the National Ambient Air Quality Standards (NAAQS) are being challenged as unrealistic. It is "because 90% of the U.S. population, excluding California, lives in a county that meets all the NAAQS."[10]

Remedies

Omission of lead from gasoline permits the use of a platinum-bearing catalyst (which is poisoned by lead) in the muffler of the engine to reduce undesirable emissions.

Some solutions to the octane dilemma include:

- Add an oxygenate to the gasoline
- Isomerize the C_5 and C_6 paraffins
- Alkylate amylenes

Possible actions to reduce aromatics:

- Reduce reformer severity and FCC reaction temperature
- Tailor boiling range of reformate and FCC gasoline

The oxygenates are high-octane blending components that also reduce the amount of unburned hydrocarbons emitted by an engine.

Methyl tertiary butyl ether (MTBE) appears to be the favored oxygenate at present except in California where it is likely to be banned. However, further squeeze on gasoline vapor pressure will favor tertiary amyl methyl ether (TAME) due to its RVP blending value being lower than that of MTBE. The demand for MTBE in excess of domestic supply is being satisfied at the present time by importation.

The isomerization of pentanes and hexanes (C_5/C_6) would require a capital expenditure by most refiners. The reason for doing this is that the isoparaffins have significantly higher octane numbers than do the corresponding normal paraffins.

Whether amylenes can be alkylated by a given refiner depends on the availability of isobutane (usually limiting) and of capacity on the alkylation unit. Either situation would require capital expenditure and time—designing and building an isomerization unit to convert normal butane to isobutane and/or designing and building a new or expanded alkylation unit.

Reducing the severity of the reformer and the reaction temperature on the catalytic cracker could be accomplished immediately by the refiner, at the cost of lower octane.

In recent years there have been proposals suggesting a maximum aromatics content be specified for diesel fuel. Values as low as 10% have been discussed. To attain such low levels, particularly on cracked stocks, will require more severe hydrotreating than that required to reach the sulfur specification. As a result, much research effort has been devoted to developing the catalysts and operating conditions required. When needed, the technology will be available, and given reasonable lead time, so will the required facilities.

For the purpose of developing basic data on the effects of various components in gasoline on tailpipe emissions from modern vehicles, an auto/oil air quality improvement research program (AQUIRP) was begun by a consortium of companies. Members of this consortium include the 3 largest U.S. automobile manufacturers and 14 large oil companies. This comprehensive, multi-phase program was intended to provide a sound basis for making regulatory decisions aimed at improving "air quality by enhancing vehicle and fuel technology at the minimum cost to the public" (from report by Colucci and Wise).[11] Thus, refinery processing and gasoline blending may be effected directly by the results of this study.

Results

Figure 6–6[4] shows the estimated miles per gallon for the then existing car population for each of the years since 1980. The fuel economy (expressed as sales weighted miles per gallon) is shown for each model year (1980 through 1997). The dramatic improvement has been accomplished primarily by decreasing the size and weight of vehicles and retirement of older vehicles.

Some preliminary results of the first phase of this AQUIRP obtained on certain 1988 vehicles and a group of 1983 to 1985 models are displayed in Figures 6–7 and 6–8. The main lesson to be derived from this small sample of the program is that the problem is complex and a given change in gasoline composition may not have the same effect on emissions from all cars.

In spite of an increase in the number of vehicles in service and in the average number of miles traveled per year per vehicle, there has been some reduction in emissions from this source as shown in Figure 6–7.[12,13,14] This

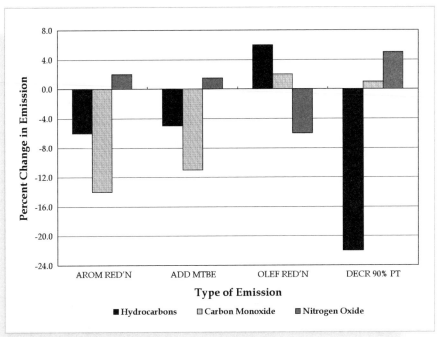

Fig. 6–7   Auto/Oil Air Quality Program (Current Vehicles)

reduction has resulted primarily from the imposition of federal standards on new cars and trucks (Table 6–1),[3] retirement of older, more polluting cars, and increased fuel economy due to implementation of corporate average fuel economy (CAFE) standards set by the Energy Policy and Conservation Act of 1975.

The EPA has reported that per NAAQS the following reductions were noted for the period from 1984 to 1993:

Pollutant	*Percent Decrease*
Ground level ozone	12
Lead	89
Sulfur dioxide	26
Carbon monoxide	37
Nitrogen dioxide	12
Particulates—less than 10 microns	20

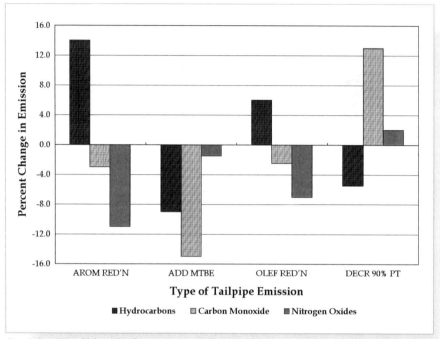

Fig. 6–8   Auto/Oil Air Quality Program (Older Vehicles)

MODEL	AUTOMOBILES			LIGHT-DUTY TRUCKS		
YEAR	HC	CO	NO$_x$	HC	CO	NO$_x$
PRECONTROL	10.60	84.0	4.1			
1968	6.30	51.0	6.0*			
1970	4.10	34.0				
1972	3.00	28.0				
1974	3.00	28.0	3.0			
1975	1.50	15.0	3.1**	2.00	20.0	3.1
1980	0.41	7.0	2.0	1.70	18.0	2.3
1981	0.41	3.4	1.0	1.70	18.0	2.3
1989	0.41	3.4	1.0	0.80	10.0	1.7

NOTES: Only years in which change occured are included.
  * NO$_x$ emissions increased with control of HCs & CO
  ** Change in test procedure

Table 6–1   Auto/Oil Air Quality Program (Older Vehicles)

Further, it reports that in 1984, 140 million Americans lived in ground level ozone non-attainment areas, but in 1993 this number had dropped to 43 million.[11]

Though of secondary interest in this book, mention should be made of alternative (to gasoline) fuels (discussed in chapter 5) that have been considered, tried, and in some cases required on a test basis. These include liquefied natural gas (LNG), CNG, electric power, and alcohols. Despite the favorable findings for some of these alternatives, gasoline promises to remain the dominant fuel for automobiles for years to come.

Federal regulations covering ambient air quality standards and allowable emissions from petroleum refineries date back to 1973. The principal air pollutants of concern are SO_x, NO_x, CO, hydrocarbons, and particulates. The oxides are present in flue gases from furnaces, boilers, and catalytic cracking unit regenerators. Hydrocarbon vapors arise from leaks, evaporation, storage, and handling. Particulates come primarily from cat cracker regenerators. Other concerns include odor and plume or fog from cooling water systems.

Water pollutants include oil, phenol, sulfur compounds, ammonia, chlorine compounds, chromium, and other heavy metals. These contaminants are expressed in terms of biological oxygen demand (BOD), chemical oxygen demand (COD), total organic carbon (TOC), total dissolved solids (TDS), and suspended solids.

Solid wastes are mainly spent cracking catalyst. Sludges are produced primarily by wastewater treatment facilities. Other concerns have to do with noise levels, radiation from flares, etc.

Notes

1. American Automobile Manufacturers Association, *Motor Vehicle Facts and Figures 1997*, p. 78

2. Ibid, p. 32

3. Ibid, p. 83

4. Ibid, p. 80

   Energy Information Administration, *The Motor Gasoline Industry: Past, Present and Future*, EIA Report DOE/EIA-0539, Washington, D.C., January, 1991

5. Shelton, E.M., Whisman, M.L., and Woodward, P.W., *Oil & Gas Journal*, August 2, 1982, pp. 95-99

   Grupa, L.M., Fields, P.W., and Yepsen, G.L., *Oil & Gas Journal*, February 2, 1986, pp. 98-102

6. Energy Information Administration, *Petroleum Marketing Monthly*, March, 1990, p. 6

7. Service, W.J., Payne, R.E., and Askey, W.E., *Oil & Gas Journal*, April 14, 1958, p. 91

8. *Encyclopedia of Chemical Technology*, 3rd ed., Vol. 11, p. 658

9. Ragsdale, R., *Oil & Gas Journal*, March 21, 1994, ff 57

10. Rhodes, Anne, and Chang, Thi, *Oil & Gas Journal*, March 23, 1998, p. 37

11. Colucci, J.M., and Wise, J.J., *Initial Results from the Auto/Oil Air Quality Improvement Research Program*, NPRA Annual Meeting, San Antonio, March, 1991

12. Schuller, R.P., Benson, D.E., and De Veirman, R.M., *Impact of Automotive Emissions on the Petroleum Industry*, UOP 1971 Technical Seminar, Arlington Heights, Illinois

13. Anderson, R., *Reducing Emissions from Older Vehicles*, API Research Report #053, August, 1990

14. Rhodes, Anne, *Oil & Gas Journal*, August 7, 1995, pp 60-62

CHAPTER 7

CRUDE OIL PROCESSING

In the field where crude oil is produced, water that is co-produced with the oil is separated, and the oil "stabilized" to the desired RVP by flash vaporization of light hydrocarbons. Some salt remains in this crude in the form of brine and solid particles.

Desalting of the crude oil is normally considered a part of the crude distillation unit (CDU) since heat from some of the streams in the CDU is used to heat the crude in the desalting process. First, water is mixed (as much as 7 volume percent) with the crude to dissolve salt crystals and to dilute the brine already present in the crude. In the case of some heavier crudes, naphtha is added as a diluent to reduce the viscosity of the crude. One or more chemicals are added to facilitate the separation of the aqueous phase from the crude oil. The pH of the effluent water is held between 5.5 and 6.5. This is particularly important in the case of naphthenic crudes with TAN greater than 1.0. The mix is heated to a temperature just below the BP for the pressure on the mix (usually below 325°F). The mix is then passed through an electrostatic field to further facilitate the separation of the aqueous phase. Depending on the amount of salt contained in the crude and the difficulty in removing it, a second stage of desalting may be employed to reduce the salt remaining in the crude down to a tolerable value.

The crude is further heated by exchange with other streams and finally passes through a fired heater where it is heated to the temperature desired for introduction into the atmospheric distillation column.

The reduced crude from the atmospheric tower is heated and charged to a vacuum distillation tower. In order to reduce the production of residual fuel oil and increase the yield of gas oil cracking stock, many refiners are going to lower flash zone pressures (as low as 20 mm Hg) and a flash zone temperature approaching 750°F to be able to cut between gas oil and resid at an atmospheric equivalent temperature as high as 1,100°F.

Seldom is a finished product produced by the CDU today. The basic purpose of the CDU is to separate the crude oil into fractions suitable for further processing. Even the fuel gas (essentially ethane and lighter) that is to be burned in the refinery usually must be treated or blended with sufficient "sweet" gas to be in compliance with regulations covering sulfur emissions.

The disposition of the remaining streams produced by the CDU is typically as follows:

Stream	Disposition
Light ends	Fed to a saturate gas plant (that may or may not be considered a part of the CDU) producing fuel gas, propane, butanes, light naphtha, and including amine treating to remove acid gases
Propane	Probably Merox treating then to LPG sales
Butanes	Probably to alkylation unit feed treater then to its deisobutanizer
Light naphtha	Treating then gasoline blending or isomerization unit
Heavy naphtha	Hydrotreating then catalytic reforming or possibly to jet fuel
Kerosene	Treating then sales or jet fuel
Diesel fuel	Treating then sales

Atmos. gas oil	Feed to fluid catalytic cracking unit (FCCU), possibly after hydrotreating
Vacuum gas oil	Feed to the FCCU, possibly after hydrotreating
Vacuum residue	SDA, coking, or blending to fuel oil

Figures 7–1, 7–2 and 7–3 are simplified process flow diagrams of an atmospheric CDU, a vacuum distillation unit and a saturate gas plant respectively.

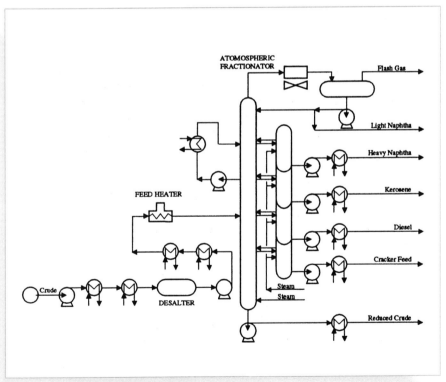

Fig. 7–1   Atmospheric Crude Distillation

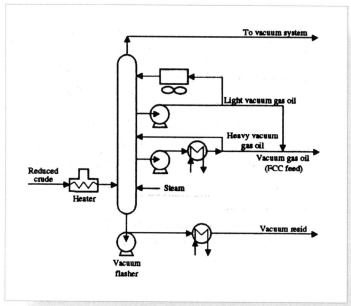

Fig. 7–2   Vacuum Distillation

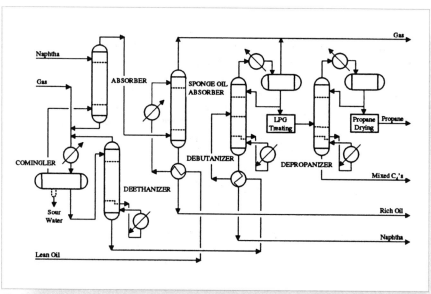

Fig. 7–3   Saturate Gas Plant

Yields

Estimating the yields of the desired fractions that might be obtained from a single crude is a fairly simple task. However, the refiner is rarely processing a single crude, but a mixture of a number of crudes. Assays are usually available for single crudes, but only for very few blends and these are unlikely to be the ones of interest.

Performing a complete assay of a crude is an expensive (and time consuming) procedure. The blend being charged to the CDU could change significantly before an assay could be completed. The refiner, therefore, must have some other means of estimating the amounts of the various streams he should realize from his current blend of crudes.

Fortunately, computer programs are available that can take crude assay data and derive from them a complex of pseudo hydrocarbon components that will satisfactorily represent the actual crude. Such pseudo assays for a number of crudes can then be blended together in the proportion desired to produce a pseudo assay for the blend. The refiner can then specify the boiling ranges (or cut temperatures) desired for the various streams and the computer program can determine not only the yields to be expected, but certain properties for these streams as well (e.g., API gravity, PONA, octane number, cetane number, etc.).

Assays

What about the individual who does not have access to such sophisticated means? In general there are three kinds of assay data available:

1. Cursory data of the type produced by the Bureau of Mines, based on a Hempel distillation at ambient pressure to a cut temperature of 527°F, continued at 40 mm Hg to 572°F.

2. Cursory data of the type published in the Oil & Gas Journal much like the Bureau of Mines reports, but based on true BP (TBP) distillations (Fig.7–4).

Oseberg premium crude is in production

Richard A. Corbett
Refining/Petrochemical Editor

Another quality crude oil is available from the North Sea. Norway's Oseberg crude oil entered the export market on Dec. 1, 1988.

This premium, low-sulfur crude oil is currently being produced at approximately 240,000 b/d from the Norwegian sector of the North Sea. Two platforms, operated by Den Norske Stats Oljeselskap A.S. (Statoil), are currently in production, with a third scheduled to go into production in 1991.

Oseberg is a typical, low-sulfur North Sea crude oil that will provide a full range of high-quality, low-sulfur products. The naphtha cuts are of intermediate to paraffinic quality, making them suitable for gasoline production.

The kerosine cut is of intermediate quality, and has good cold flow and burning properties required by jet fuels. The low-sulfur gas oil fractions have good cetane indices, and they are slightly naphthenic. FCCU feedstock will be low in sulfur and metals contents.

In September 1989, Statoil plans to blend crude oil from the Veslefrikk field with Oseberg, making the future Oseberg blend somewhat lighter and more paraffinic.

The Oseberg crude is stabilized at the offshore platform and then transported to the Sture terminal, north of Bergen, Norway. The terminal can handle crude oil tankers up to 300,000 DWT.

Oseberg, Norway

Sture terminal, north of Bergen, Norway

Whole crude:
Gravity, °API: 33.71
Sulfur, wt %: 0.31
Pour point, °C.: –6
Viscosity @ 20° C., cSt: 8.52
Concarbon residue, wt %: 2.6
Water content, vol %: 0.10
Salt as NaCl, ppm: 5.0
Toatl acid No., mg KOH/g: 0.22
Metals, V/Ni/Na, ppm: 4/<2/<2

Straight run gasoline (C_5 - 90° C.):
Yield on crude, wt %: 5.46
Yield on crude, vol %: 6.83
Density @ 15° C., kg/l: 0.68
Rvp, psi: 11.6
Mercaptan sulfur, ppm: <2
Sulfur, wt %: <0.005
RON: 71.1
MON: 70.0
Paraffins, vol %: 74.6
Naphthenes, vol%: 21.5
Aromatics, vol %: 3.9
Benzene, vol %: 3.5
Nitrogen, ppm: <5

Light naphtha (90° - 160° C.):
Yield on crude, wt %: 11.62
Yield on crude, vol %: 13.05
Density @ 15° C., kg/l: 0.76
Rvp, psi: 1.2
Mercaptan sulfur, ppm: <2
Sulfur, wt %: <0.005
Paraffins, vol %: 47.8
Naphthenes, vol %: 35.4
Aromatics, vol %: 16.8
Benzene, vol %: 0.2
Nitrogen, ppm: <5

Naphtha (90° - 180° C.):
Yield on crude, wt %: 15.09
Yield on crude, vol %: 16.82
Density @ 15° C., kg/l: 0.76
Rvp, psi: 1.0

Mercaptan sulfur, ppm: <2
Sulfur, wt %: <0.005
Paraffins, vol %: 48.2
Naphthenes, vol %: 34.9
Aromatics, vol %: 16.9
Benzene, vol %: 0.2
Nitrogen, ppm: <5

Heavy naphtha (180° - 180° C.):
Yield on crude, wt %: 3.47
Yield on crude, vol %: 3.77
Density @ 15° C., kg/l: 0.78
Aromatics, vol %: 16.5
Sulfur, wt %: <0.01
Smoke point, mm: 27
Freeze point, °C.: –56
Flash point, °C.: 42
Cetane index: 34.1
Cloud point, °C.: <30
Pour Point, °C.: <30
Viscosity @ 20° C., cSt: 1.2
Viscosity @ 50° C., cSt: 0.83
Neut. No., mg KOH/g: 0.04
Naphthalenes, vol %: 0.01
Mercaptan sulfur, ppm: 7

Light gas oil (180° - 240° C.):
Yield on crude, wt %: 14.13
Yield on crude, vol %: 14.94
Density @ 15° C., kg/l: 0.81
Aromatics, vol %: 15.7
Sulfur, wt %: 0.01
Smoke point, mm: 26
Freeze point, °C.: –56
Flash point, °C.: 57
Cetane index: 41.4
Cloud point, °C.: <30
Pour point, °C.: <30
Viscosity @ 20° C., cSt: 1.74
Viscosity @ 50° C., cSt: 1.11
Neut. No., mg KOH/g: 0.05
Naphthalenes, vol %: 2.10
Mercaptan sulfur, ppm: <2

Gas oil (180° - 240° C.):
Yield on crude, wt %: 10.66
Yield on crude, vol %: 11.17
Density @ 15° C., kg/l: 0.81
Sulfur, wt %: 0.01
Smoke point, mm: 26
Freeze point, °C.: –52
Flash point, °C.: 69
Cetane index: 43.2
Cloud point, °C.: <30
Pour point, °C.: <30
Viscosity @ 20° C., cSt: 1.96
Viscosity @ 50° C., cSt: 1.23
Neut. No., mg KOH/g: 0.05
Naphthalenes, vol %: 2.70

Mercaptan sulfur, ppm: <2
Aniline point, °C.: 59.9
Luminometer: 51

Gas oil (240° - 320° C.):
Yield on crude, wt %: 17.61
Yield on crude, vol %: 17.62
Density @ 15° C., kg/l: 0.85
Flash point, °C.: 121
Sulfur, wt %: 0.1
Viscosity @ 20° C., cSt: 5.16
Viscosity @ 50° C., cSt: 2.62
Cetane index: 48
Total nitrogen, ppm: 20
Cloud point, °C.: –23
Pour point, °C.: –24
Basic nitrogen, wt %: <0.01
Aniline point, °C.: 65.9
Neut. No., mg KOH/g: 0.10

Gas oil (320° - 375° C.):
Yield on crude, wt %: 10.34
Yield on crude, vol %: 9.96
Density @ 15° C., kg/l: 0.89
Sulfur, wt %: 0.34
Viscosity @ 20° C., cSt: 19.96
Viscosity @ 50° C., cSt: 6.89
Cetane index: 47.2
Total nitrogen, ppm: 310
Cloud point, °C.: 6
Pour point, °C.: 6
Basic nitrogen, wt %: <0.01
Aniline point, °C.: 73.3
Ramecarbon, wt %: 0.09
Concarbon, wt %: 0.02
Neut. No., mg KOH/g: 0.27
Watson K: 11.9
Wax, wt %: 6.0

Heavy gas oil (375° - 420° C.):
Yield on crude, wt %: 6.36
Yield on crude, vol %: 6.01
Density @ 15° C., kg/l: 0.9
Sulfur, wt %: 0.41
Viscosity @ 50° C., cSt: 16.82
Viscosity @ 80° C., cSt: 6.86
Cetane index: 45
Pour point, °C.: 21
Basic nitrogen, wt %: 0.02
Ramscarbon, wt %: 0.11
Concarbon, wt %: 0.03
Neut. No., mg KOH/g: 0.34
V/Ni, ppm: <2/<2

Heavy gas oil (420° - 525° C.):
Yield on crude, wt %: 16.9
Yield on crude, vol %: 15.56
Density @ 15° C., kg/l: 0.93
Sulfur, wt %: 0.51

Viscosity @ 50° C., cSt: 77.75
Viscosity @ 80° C., cSt: 21.41
Pour point, °C.: 36
Basic nitrogen, wt %: 0.05
Ramscarbon, wt %: 0.28
Concarbon, wt %: 0.15
Cetane index: 37.3
V/Ni/Na/K, ppm: <2/<2/0/0

Residuum (375° C.+):
Yield on crude, wt %: 39.28
Yield on crude, vol %: 35.25
Density @ 15° C., kg/l: 0.95
Sulfur, wt %: 0.65
Viscosity @ 50° C., cSt: 375
Viscosity @ 80° C., cSt: 75.02
Pour point, °C.: 24
Basic nitrogen, wt %: 0.06
Ramscarbon, wt %: 6.7
Concarbon, wt %: 6.0
V/Ni/Na/K, ppm: 5/5/<2/<2
Ash, wt %: <0.005
Watson K: 11.7
Asphaltenes, wt %: 0.9

Residuum (420° C.+):
Yield on crude, wt %: 32.92
Yield on crude, vol %: 29.24
Density @ 15° C., kg/l: 0.96
Sulfur, wt %: 0.9
Viscosity @ 50° C., cSt: 948.8
Viscosity @ 80° C., cSt: 151.0
Pour point, °C.: 27
Basic nitrogen, wt %: 0.1
Ramscarbon, wt %: 8.1
Concarbon, wt %: 7.2
V/Ni/Na/K, ppm: 5/5/<2/<2
Ash, wt %: 0.005
Watson K: 11.7
Asphaltenes, wt %: 1.3

Residuum (525° C.+):
Yield on crude, wt %: 16.01
Yield on crude, vol %: 13.69
Density @ 15° C., kg/l: 1.0
Sulfur, wt %: 0.9
Viscosity @ 50° C., cSt: 3,460
Viscosity @ 80° C., cSt: 827.6
Pour point, °C.: 48
Ramscarbon, wt %: 16.4
Concarbon, wt %: 17.6
V/Ni/Na/K, ppm: 13/11/<2/<2
Ash, wt %: 0.02
Watson K: 11.7
Asphaltenes, wt %: 3.13
Penetration: >300
Total nitrogen, wt %: 0.50
Pentane insolubles, wt %: 7.5
Softening point, °C.: 37.2

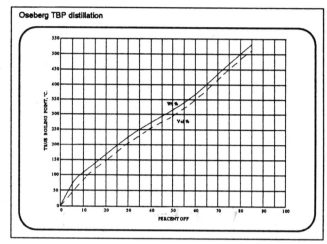

Oseberg TBP distillation

(vertical axis: TRUE BOILING POINT, °C — 0 to 550; horizontal axis: PERCENT OFF — 0 to 100; curves labeled "Wt %" and "Vol %")

Fig. 7–4   Oseburg Crude Assay

3. Comprehensive assays by the producers of the more prominent crudes. These vary considerably in detail. Some include yield and property data for catalytic reforming of naphthas and catalytic cracking of gas oils. Where appropriate, considerable detail on lubricating oil potential and/or asphalt production is given. These assays are proprietary and availability is usually limited to potential buyers/processors and sometimes to process design engineers.

Crude Break-Up Procedure

Assuming TBP data are available:

1. Establish TBP cut temperatures for the cuts desired. For example:

	°F
Light naphtha	IBP–194
Heavy naphtha	194–356
Jet fuel	356–464
Diesel fuel	464–608
Atmospheric gas oil	608–707
Light vacuum gas oil	707–788
Heavy vacuum gas oil	788–977
Vacuum residuum	977+

(These cut temperatures have been chosen to correspond with cuts in the Oil & Gas Journal assay of Oseberg Crude.)[1]

2. Tabulate volume percent and gravity of each cut and calculate the corresponding weight percents (Table 7–1).

3. Usually some minor adjustment is required to obtain a weight balance. This may be done by normalizing (distributing the necessary adjustment proportionally over all the cuts) or adjusting one or more of the largest streams (Table 7–2).

95

CUTS	°C	°F	SP GR	VOL %	WT %	% S
LSR	IBP–90	IBP–194	0.68	6.83	5.46	
MID NAPH	90–180	194–356	0.76	16.82	15.09	
GO	180–240	356–464	0.81	11.17	10.66	0.01
GO	240–320	464–608	0.85	17.62	17.61	0.10
GO	320–375	608–707	0.89	9.96	10.34	0.34
HVY GO	375–420	707–788	0.90	6.01	6.36	0.41
HVY GO	420–525	788–977	0.93	15.56	16.90	0.51
RESID	525+	977+	1.00	13.69	16.01	0.90
			TOTALS	97.66	98.43	

Table 7–1 Data on Certain Cuts from Oseberg Assay   — FIG. 7.4

				CORRECTED		
	TBP CUT	SP GR	VOL %	WT %	% S	
LSR	IBP–194	0.68	6.99	5.55		
NAPHTHA	194–356	0.76	17.22	15.33		
JET FUEL	356–464	0.81	11.44	10.83	0.01	
DIESEL	464–608	0.85	18.04	17.89	0.10	
ATM GO	608–707	0.89	10.20	10.50	0.34	
LT VAC GO	707–788	0.90	6.15	6.46	0.41	
HVY VAC GO	788–977	0.93	15.93	17.17	0.51	
RESIDUE	977+	1.00	14.02	16.27	0.90	
	TOTALS	99.99	100.00			

Table 7–2 Data on Selected Cuts after Normalizing

4. Estimate the sulfur content and the UOP K factor for each cut.

5. If the cuts are to be processed further, estimate pertinent additional properties for certain cuts, such as:

Naphthas	PONA, octanes
Jet fuel	Smoke point, percent aromatics
Diesel fuel	Pour point, CI
Gas oils	Carbon residue, nitrogen, metals
Resid	Carbon residue, nitrogen, metals

To determine the yields and properties of streams with different cut temperatures than those of a particular assay, the engineer may resort to linear regression of the assay data by a procedure described by the author.[2] It was found that a third or fourth order polynomial could be used to satisfactorily describe the yields and properties at the cut temperatures of the assay for any temperature over the range of the distillation. These equations can be used to obtain the necessary data to make blends of crudes.

Crude distillation operating requirements

The following values may be used for the operating requirements for crude distillation:

	Electric (kWh/b)	Fuel (kBtu/b)	Steam lb/b
Atmospheric	0.5	100	25
Vacuum	0.3	100	50

Crude distillation capital cost

Data on a total of 19 announced projects were scaled to 100,000 BPD and translated to the first month of 1991. The average value was $38 million. Data were found for eight vacuum units. Scaled to 60,000 BPD and the first month of 1991, the average value was $30 million.

Notes

1. Corbett, R.A., *Oil & Gas Journal*, July 24, 1989, pp. 56-57

2. Maples, R.E., *Oil & Gas Journal*, November 3, 1997, ff. 72

SECTION B:

RESIDUAL OIL PROCESSING

CHAPTER 8

SOLVENT DEASPHALTING

Further separation of vacuum residue into fractions by distillation without decomposition is very difficult and very expensive. It is not practiced commercially. Solvent extraction offers a non-destructive means of accomplishing this. In solvent extraction, separation is primarily by type of compound and only secondarily by number of carbon atoms per molecule.

Solvent deasphalting (SDA) was originally developed as a means of removing asphaltenes from lubricating oil feedstock. More recently it has become a means of obtaining additional catalytic cracking feed from residual oil. The deasphalted oil (DAO) has lower carbon residue and metals content than the untreated oil. As we shall see, the extent to which these factors are lowered is primarily a function of the amount of DAO recovered, and this can be controlled.

The carbon residue content effects the amount of coke deposited on the cat cracker catalyst and the metals decrease the activity of (or poison) the catalyst. SDA is not so effective in lowering sulfur or nitrogen content in the DAO.

SDA is sometimes combined with visbreaking or coking in a refinery process scheme. It can be employed in an indirect scheme to produce low sulfur fuel oil from resid. The DAO can be desulfurized and blended with the asphalt to produce a given sulfur content more easily and at much lower cost than by the direct desulfurization of the vacuum resid itself.[1,2]

SDA process description

The process is a typical extraction process involving partitioning of the residual oil into two or more fractions on the basis of relative solubility in a solvent. The solvent is normally a light hydrocarbon such as propane, butane, pentane, or hexane. The mixture of oil and solvent is allowed to separate into a DAO-rich fraction and an asphalt-rich fraction. Solvent is stripped from each fraction and reused. Figure 8-1 is a typical simplified process flow diagram.

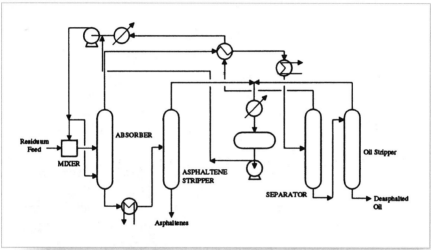

Fig. 8–1   Solvent Deasphalting Unit

A recent development in SDA technology involves operating above the critical temperature of the solvent in the DAO-solvent separator resulting in a significant reduction in energy required by the process.

Also, some processes provide for the separation of an intermediate—or resin—fraction. Whether or not this is practiced depends on the ultimate use of the asphalt, since resins have a significant effect on the properties of asphaltenes.[3]

SDA data correlation

Table 8-1 is a tabulation of the data gleaned from the references listed as their source. This was the database for the various correlations developed for reduction of the following properties in the DAO as compared with the residue feed: Conradson carbon residue, vanadium (V), nickel (Ni), sulfur (S), nitrogen (N), and API gravity.

At first glance, the size of the database is impressive, with 114 sets of data. However, on close inspection it can be seen that there are many gaps in the data. As a result, the number of complete, usable sets available to explore a particular variable is usually considerably less than this number. More sources reported volume percent of feed for DAO yield (91) than reported weight percent (34). A total of 11 sets gave both volume and weight yields.

FEED PROPERTIES						DAO – YIELDS –		DAO PROPERTIES					
API	S	N	CCR	NI	V	LV%	WT%	API	S	N	CCR	NI	V
	3.9	0.5		13	62	68.5			3.1	0.4		1.6	23.0
	1.2	0.3		14	9	86.6			0.9	0.2		4.5	10.0
	4.5	1.1	24.6	162	815	69.5			4.0		14.0	40.0	20.0
	1.1	0.5		14	11	77.6			0.6	0.4		3.5	7.0
	5.6	0.7	20.0	124	302	67.4			4.7	0.5	9.3	25.0	59.0
	0.8	0.5		15	4	79.4			0.7	0.5		5.4	
1.4	4.4	0.7	24.5			63.5	8.2	3.8	0.5		12.1		
1.4	4.4	0.7	24.5			45.5	10.6	3.7	0.4		6.9		
3.1	6.2	0.5	22.5	76	233		31.0	18.2	3.6	0.1	2.0	2.0	
3.6	5.5	0.5	25.1	58	135	47.4		20.3	3.8	0.1	6.6	2.0	
3.6	5.5	0.5	25.1	58	135	56.0		14.6	3.9	0.2		5.0	
3.6	5.5	0.5	25.1	58	135	73.8		10.8	4.4	0.3	14.2	14.0	
3.6	5.5	0.5	25.1	58	135	65.3		12.4	4.0	0.2	12.3	8.0	
4.0	5.7		24.0			68.0		4.5					
5.0	2.2		22.0	77	178		86.8	9.3	2.0		16.1	36.0	
5.0	5.9	0.8	18.0	133	1264		33.0	17.0	4.7	0.2	1.8	6.0	
5.3			21.5	28	84	35.0		19.3			4.3	0.9	
5.4	4.9		24.0	35	87	50.0		13.8	3.8		5.3	4.0	
5.4			24.0			25.0		21.2			1.3		
5.6			24.0	30	110	45.6		16.2			4.5	0.9	
5.9	2.7		22.2	120	372	51.5		16.0	1.9		6.3	7.6	
5.9	5.2		18.5	25	106	63.6	60.0	14.2	4.1		8.0	4.0	
5.9	2.7		22.2	120	372	75.0		11.0	2.3		14.3	40.0	
5.9	5.2		18.5	25	106	83.1	80.0	11.3	4.8		12.7	6.0	
6.2	3.6		24.5	37	185	76.8	73.0	13.4	2.7		12.0	4.0	
6.3			22.2	139	136	52.8		18.3			5.3	8.1	
6.5	4.1		22.0			83.0			3.5				

Table 8–1 Solvent Deasphalting Unit

FEED PROPERTIES						DAO – YIELDS –		DAO PROPERTIES					
API	S	N	CCR	NI	V	LV%	WT%	API	S	N	CCR	NI	V
6.6	4.9		20.1			31.0		19.8	2.8				
6.6	4.3		21.1	21	70	70.0	67.3	12.1	3.6		10.6	2.1	
6.6	4.9		20.1			40.5		18.0	3.2		2.3		
6.6	4.3		21.1	21	70	50.0	46.8	16.0	3.0		5.0	0.7	
6.6	4.3		21.1	21	70	85.0	82.8	10.3	3.9		14.0	7.0	
6.6	4.3		21.1	21	70	32.0	29.0	21.0	2.6		1.5	0.3	
6.8			15.0	74	365	49.8		18.1			5.9	3.5	
6.9	4.0	0.3	20.8			56.0		16.0	2.7	0.1	5.6		
6.9	4.0	0.3	20.8			78.0		12.0	3.3	0.2	10.7		
7.1	5.4	0.4	15.4	33	87		29.0	19.5	2.7	0.1	1.9	1.0	
7.3	5.1		19.1			37.5		21.3			1.6		
7.3	5.1		19.1			45.0		20.0	2.4		1.7		
7.3	5.1		19.1			44.5		19.3	2.8		2.2		
7.3	5.1		19.1			53.0		18.2	2.9		2.5		
7.5	2.3		18.0			88.0		2.0					
8.0	3.9		19.0				81.0	12.9	3.5		10.5		
8.1			19.7	30	89	54.8		17.1			5.4	0.6	
8.1	2.7	0.5	17.4	45	65	81.8		13.0	2.1	0.3	10.0	11.0	
8.1	2.7	0.5	17.4	45	65	80.0		13.2	2.0	0.3	9.6	9.0	
8.1	2.7	0.5	17.4			81.3		13.0	2.1	0.3	10.0		1.2
8.1	2.7	0.5	17.4	45	65	65.1		16.3	1.6	0.2	6.0	1.0	0.4
8.1	2.7	0.5	17.4			65.5		16.3	1.6	0.2	6.0		1.9
8.1	2.7	0.5	17.4	45	65	72.8		14.8	1.8	0.3	7.3	5.0	0.7
8.2	3.9	0.3	19.0				83.0	14.4	3.6	0.2	8.4		1.3
8.4	3.5		15.0	42	114	67.0	63.6	15.9	2.4		4.0	5.0	2.3
8.5	4.5	0.5	15.0	25	66		85.0	11.0	4.0	0.4	8.8	8.0	1.4
8.6	1.5		19.1			56.5	52.0	21.6	1.0		3.0	4.8	
8.6			16.0			36.8		24.3			0.7		
8.7			19.0			62.0		17.8			3.4		
8.7			19.0			54.0		19.5			2.6		
8.7			19.0			70.0		16.0			5.5		
9.6	4.1	0.3		19	61		70.1	16.0	3.3	0.2	5.3	2.0	
9.6			18.9	47	31	67.8		17.8			5.4	3.9	
9.6	4.1	0.3		19	61		85.5	13.8	3.7	0.2	7.9	7.0	
9.6	4.1	0.3		19	61		45.2	20.1	2.6	0.1	1.7	1.0	
9.7	3.1		17.5	30	550	75.1	72.0	15.7	2.6		8.0	5.0	
10.1			16.0	20	62	47.0		20.3			2.6	0.4	
11.3			15.0	14	23	62.9		19.0			4.2	1.1	
11.7	1.5	0.4	15.0			86.2		15.6	1.1	0.3	7.8		14.0
11.7	1.5	0.4	15.0			71.2		18.5	0.9	0.2	3.5		
12.0			12.1	16	28	66.0		19.6			2.2	1.0	
12.4	0.9	0.5	13.3	45	100	74.0	71.6	17.3	0.7	0.4	4.9	7.0	
12.6	3.0	0.3	12.9	17	26		54.0	21.5	1.8	0.1	2.0	1.0	
13.0			13.3	12	28	69.0		20.5			2.6	0.6	

Table 8–1 Solvent Deasphalting Unit cont'd

	FEED PROPERTIES					DAO – YIELDS –			DAO PROPERTIES				
API	S	N	CCR	NI	V	LV%	WT%	API	S	N	CCR	NI	V
14.3			11.5			60.0		23.0			1.5		18.0
14.9			6.7			76.3		19.4			1.7		9.0
15.5			11.7			70.0		22.7			1.4		83.0
15.5			11.9			49.0		23.4			1.3		1.1
15.5			11.7			62.0		23.7			0.9		2.5
15.5			11.9			70.0		21.2			3.9		7.0
15.5			11.9			60.0		22.4			2.4		23.0
15.5			11.7			80.0		21.2			2.7		160.0
16.7			9.1			81.5		24.3			2.3		50.0
17.5							72.9	22.4					6.1
17.5							69.6	22.0					2.3
17.5							75.8	21.8					1.7
17.5							77.0	21.5					4.5
17.5							65.5	22.5					7.5
17.8			8.8			67.0		24.5			1.5		
17.8			8.8			86.0		22.5			1.8		12.4
17.8			8.0			64.0		24.8			1.7		
19.2			4.9			76.6		23.4			1.5		
19.2			5.9			78.0		23.5			1.2		2.3
19.3			5.4			82.1		23.6			1.5		1.4
19.3			7.3			77.0		23.3			1.7		0.7
19.3			5.3			79.0		23.5			1.4		4.0
19.4			5.5			85.5		22.4			2.1		
19.4			5.8			83.0		23.4			1.5		
19.5			5.6			86.5		22.0			2.4		8.0
19.5			5.6			80.6		23.8			1.5		89.0
19.6			5.7			78.0		23.7			1.5		5.0
19.6			6.1			83.5		23.0			1.6		12.0
19.6			6.1			84.1		22.9			1.7		22.0
19.6			6.1			84.9		22.9			1.8		43.0
19.6			6.1			89.5		21.8			2.9		4.0
19.6			6.1			87.2		22.3			1.9		8.0
19.6			6.1			85.8		22.7			1.8		17.0
19.6			6.1			80.0		23.4			1.5		18.0
19.6			6.1			88.9		22.0			2.5		
19.6			6.0			89.0		22.3			2.6		1.0
19.6			6.1			89.1		21.7			2.7		1.4
19.6			6.1			89.3		22.3			2.2		2.0
19.6			6.1			89.6		21.6			2.8		4.0
19.8			5.6			86.0		22.7			1.8		10.0
19.8			5.5			81.1		23.6			1.5		1.4
20.4			5.5			82.1		23.5			1.4		2.6
27.0	1.1					69.0		37.7	0.5				15.5

Table 8–1 Solvent Deasphalting Unit cont'd

105

Data were generally given for S, Conradson carbon residue, and API. N, Ni, and V were less frequently given. Viscosity data were very sparse and were not studied.

Table 8-2 is a summary of results of solvent deasphalting correlations. The yield of DAO was chosen as the primary independent variable. Both weight percent and volume percent were tried. With one exception, API of DAO, the results were more satisfactory with weight than with volume. This is despite the fact that there were always more sets of data with volume yield.

In the case of Conradson carbon residue, S, and API, correlation results were improved by the addition of API to the feed as a second independent variable. In the case of N, V, and Ni, the square of the yield of DAO was used as a second independent variable.

Because of almost complete removal of V and Ni at DAO yields below about 50%, regressions were made using the logarithms of the DAO/feed ratios as dependent variables to reduce curvature from the resulting graphs.

DEPENDENT VARIABLE [1]	CCR	CCR	S	S	N	V	V [2]
INDEPENDENT VARIABLE							
1	FD API	FD API	FD API	WT% DAO	WT% DAO	WT% DAO	WT% DAO
2	WT% DAO	LV% DAO	WT% DAO	WT^2		WT%^2	WT%^2
3		LV%^2					
R^2	0.925	0.824	0.947	0.941	0.974	0.862	0.865
SEE [5]	0.063	0.064	0.0326	0.0293	0.0424	0.03525	0.191
S1 [6]	0.006186	0.0016	0.00038	0.000023	0.00558	0.00314	0.0144
S2 [6]	0.000731	0.0034	0.00315	0.00275	0.000047	0.000028	0.000122
S3 [6]		0.000027					
NO. OF SETS	21	85	21	21	11	22	27

NOTES:
[1] Fraction of variable in feed remaining in DAO except for API and wt% DAO
[2] Log(1,000*V in DAO/V in feed)
[3] Log(1,000*Ni in DAO/Ni in feed)
[4] Metals = V + Ni
[5] SEE denotes the standard error of the estimate of Y
[6] These are the standard errors of the corresponding coefficients

Table 8–2 Some Results of Solvent Deasphalting Correlations

In the literature, V and Ni are frequently combined as a dependent variable. To permit comparison with published correlations, regressions were made on a volume yield basis with the results shown in Table 8-2.

The correlation of the reduction of Conradson carbon residue as a function of the weight percent yield of DAO by solvent deasphalting will serve as an example of the method employed throughout this work.

Table 8-3 is a Lotus tabulation derived from the SDA database including only the sets of data germane to this correlation. Note the shrinkage from 114 sets in the database to 23 sets in Table 8-3. One of the first steps is to calculate the ratio of Conradson carbon residue in the DAO to the Conradson carbon residue in the feed. This ratio becomes the dependent variable or Y. Weight percent (WT%) DAO is the independent variable first chosen. Since this did not give a satisfactory result ($R^2 = 0.840$), the square of WT% DAO was added as a second independent variable.

The result of this regression is the one shown in Table 8-3a, with R^2 equal to 0.858. Based on the regression equation, calculated values of the

Ni	Ni[3]	API	API	WT% DAO	METALS[4]	METALS[4]
WT% DAO WT%^2	WT% DAO WT%^2	FD API LV% DAO	FD API WT% DAO	LV% DAO LV%^2	API LV% DAO LV%^2	WT% DAO WT%^2
0.925	0.884	0.963	0.953	0.999	0.897	0.7372
0.0324	0.143	0.814	0.924	0.596	0.009	0.0204
0.00261 0.000022	0.0108 0.000092	0.0195 0.00692	0.0425 0.0101	0.0854 0.000709	0.000455 0.000868 0.000008	0.00168 0.000014
22	27	88	28	11	38	20

WT%	WT^2	FD CCR	DAO CCR	DAO/FD	CALC	DIFF
29.0	841	15.4	1.9	0.12	0.03	0.09
29.0	841	21.1	1.5	0.07	0.03	0.04
31.0	961	22.5	2	0.09	0.05	0.04
33.0	1,089	18.0	1.8	0.10	0.07	0.03
45.0	2,025	16.4	1.65	0.10	0.20	−0.10
46.8	2,190	21.1	5	0.24	0.22	0.02
50.0	2,500	24.0	5.3	0.22	0.25	−0.03
52.0	2,704	19.1	3	0.16	0.27	−0.11
54.0	2,916	12.9	2	0.16	0.29	−0.14
60.0	3,600	18.5	8	0.43	0.35	0.08
63.6	4,045	15.0	4	0.27	0.39	−0.12
67.3	4,529	21.1	10.6	0.50	0.43	0.07
67.4	4,543	20.0	9.3	0.47	0.43	0.04
69.5	4,830	24.6	14	0.57	0.45	0.12
71.6	5,127	13.3	4.9	0.37	0.47	−0.10
72.0	5,184	17.5	8	0.46	0.48	−0.02
73.0	5,329	24.5	12	0.49	0.49	0.00
80.0	6,400	18.5	12.7	0.69	0.56	0.13
81.0	6,561	19.0	10.5	0.55	0.57	−0.02
82.8	6,856	21.1	14	0.66	0.59	0.07
83.0	6,889	19.0	8.4	0.44	0.59	−0.15
85.0	7,225	15.0	8.8	0.59	0.61	−0.03
86.8	7,534	22.0	16.1	0.73	0.63	0.10
						0.00

Regression Output:

Constant	−0.267247278
Std Err of Y Est	0.087901045 DAO CCR/FD CCR
R^2	0.839589190
No. of Observations	23
Degrees of Freedom	21
X Coefficient(s)	0.010344499
Std Err of Coef.	0.000986695

WT% DAO   WT% DAO^2

Table 8–3   Solvent Deasphalting—Conradson Carbon Residue Reduction

WT%	WT^2	FD CCR	DAO CCR	DAO/FD	CALC	DIFF
29.0	841	15.4	1.9	0.12	0.07	0.05
29.0	841	21.1	1.5	0.07	0.07	0.00
31.0	961	22.5	2	0.09	0.09	0.00
33.0	1,089	18.0	1.8	0.10	0.10	0.00
45.0	2,025	16.4	1.65	0.10	0.18	–0.08
46.8	2,190	21.1	5	0.24	0.19	0.04
50.0	2,500	24.0	5.3	0.22	0.22	0.00
52.0	2,704	19.1	3	0.16	0.24	–0.08
54.0	2,916	12.9	2	0.16	0.26	–0.10
60.0	3,600	18.5	8	0.43	0.32	0.11
63.6	4,045	15.0	4	0.27	0.36	–0.09
67.3	4,529	21.1	10.6	0.50	0.40	0.10
67.4	4,543	20.0	9.3	0.47	0.41	0.06
69.5	4,830	24.6	14	0.57	0.43	0.14
71.6	5,127	13.3	4.9	0.37	0.46	–0.09
72.0	5,184	17.5	8	0.46	0.46	–0.01
73.0	5,329	24.5	12	0.49	0.48	0.01
80.0	6,400	18.5	12.7	0.69	0.57	0.11
81.0	6,561	19.0	10.5	0.55	0.59	–0.04
82.8	6,856	21.1	14	0.66	0.62	0.05
83.0	6,889	19.0	8.4	0.44	0.62	–0.18
85.0	7,225	15.0	8.8	0.59	0.65	–0.06
86.8	7,534	22.0	16.1	0.73	0.68	0.06
						0.00

Regression Output:

Constant	0.006803142	
Std Err of Y Est	0.084865132	DAO CCR/FD CCR
R^2	0.8575984101	
No. of Observations	23	
Degrees of Freedom	20	
X Coefficient(s)	–0.00034467	0.000092890
Std Err of Coef.	0.006788253	0.000058407
	WT% DAO	WT% DAO^2

Table 8–3a   Solvent Deasphalting—Conradson Carbon Residue Reduction

109

WT%	WT^2	FD CCR	DAO CCR	DAO/FD	CALC	DIFF
29.0	841	15.4	1.9	0.12	0.08	0.04
29.0	841	21.1	1.5	0.07	0.08	−0.01
31.0	961	22.5	2	0.09	0.09	0.00
33.0	1,089	18.0	1.8	0.10	0.10	0.00
45.0	2,025	16.4	1.65	0.10	0.17	−0.07
46.8	2,190	21.1	5	0.24	0.19	0.05
50.0	2,500	24.0	5.3	0.22	0.22	0.00
52.0	2,704	19.1	3	0.16	0.23	−0.08
54.0	2,916	12.9	2	0.16	0.25	−0.10
60.0	3,600	18.5	8	0.43	0.32	0.11
63.6	4,045	15.0	4	0.27	0.36	−0.10
67.3	4,529	21.1	10.6	0.50	0.41	0.09
67.4	4,543	20.0	9.3	0.47	0.41	0.05
69.5	4,830	24.6	14	0.57	0.44	0.13
71.6	5,127	13.3	4.9	0.37	0.47	−0.10
72.0	5,184	17.5	8	0.46	0.47	−0.02
73.0	5,329	24.5	12	0.49	0.49	0.00
80.0	6,400	18.5	12.7	0.69	0.60	0.09
81.0	6,561	19.0	10.5	0.55	0.61	−0.06
82.8	6,856	21.1	14	0.66	0.64	0.02
85.0	7,225	15.0	8.8	0.59	0.68	−0.10
86.8	7,534	22.0	16.1	0.73	0.71	0.02
						0.00

Regression Output:

Constant		0.064053075
Std Err of Y Est		0.075378551 DAO CCR/FD CCR
R^2		0.892666386
No. of Observations		22
Degrees of Freedom		19
X Coefficient(s)	−0.00300793	0.000120877
Std Err of Coef.	0.006121349	0.000053053
	WT% DAO	WT% DAO^2

Table 8–3b Solvent Deasphalting—Conradson Carbon Residue Reduction

WT%	WT^2	FD CCR	DAO CCR	DAO/FD	CALC	DIFF
29.0	841	15.4	1.9	0.12	0.08	0.04
29.0	841	21.1	1.5	0.07	0.08	−0.01
31.0	961	22.5	2	0.09	0.09	0.00
33.0	1,089	18.0	1.8	0.10	0.10	0.00
45.0	2,025	16.4	1.65	0.10	0.17	−0.07
46.8	2,190	21.1	5	0.24	0.18	0.06
50.0	2,500	24.0	5.3	0.22	0.21	0.01
52.0	2,704	19.1	3	0.16	0.23	−0.07
54.0	2,916	12.9	2	0.16	0.24	−0.09
60.0	3,600	18.5	8	0.43	0.31	0.12
63.6	4,045	15.0	4	0.27	0.35	−0.09
67.3	4,529	21.1	10.6	0.50	0.40	0.10
67.4	4,543	20.0	9.3	0.47	0.40	0.06
71.6	5,127	13.3	4.9	0.37	0.46	−0.09
72.0	5,184	17.5	8	0.46	0.47	−0.01
73.0	5,329	24.5	12	0.49	0.48	0.01
80.0	6,400	18.5	12.7	0.69	0.59	0.09
81.0	6,561	19.0	10.5	0.55	0.61	−0.06
82.8	6,856	21.1	14	0.66	0.64	0.02
85.0	7,225	15.0	8.8	0.59	0.68	−0.09
86.8	7,534	22.0	16.1	0.73	0.71	0.02

−0.03

Regression Output:

Constant		0.107821766
Std Err of Y Est		0.070557426 DAO CCR/FD CCR
R^2		0.906857917
No. of Observations		21
Degrees of Freedom		18
X Coefficient(s)	−0.00474162	0.000134559
Std Err of Coef.	0.005800570	0.000050169
	WT% DAO	WT% DAO^2

Table 8–3c Solvent Deasphalting—Conradson Carbon Residue Reduction

111

Conradson carbon residue ratio were determined. The difference between the actual (reported) values and the calculated values were then determined. Comparing these differences with the standard error of the estimate (SEE) of 0.0828, it was seen that several values exceeded SEE significantly. Deleting the set with the largest difference (-0.18), a second regression was performed (Table 8-3b). R^2 increased to 0.8927 and SEE decreased to 0.0754. Another difference between the two tables is the sorting of the data table in terms of increasing WT% DAO to facilitate plotting results later. Again, the set giving the largest difference was deleted and a third regression was made (Table 8-3c). This deletion may be questioned, since the difference was only about 1.5 times SEE. In any case the results are as shown, some slight improvement in R^2 and SEE. A graph of the actual data points and the trace of the regression equation appears as Figure 8-2. For practical purposes, there is no difference between the traces of the three regression equations.

Figures 8-3 is a composite chart showing the fraction of Conradson carbon residue, S, N, Ni, V, and Ni + V in the feed remaining in the DAO vs. the weight percent yield of DAO. Figure 8-4 is a plot of the API of DAO in terms of the

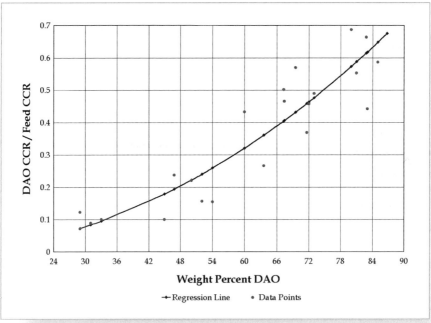

Fig. 8–2   Solvent Deasphalt CCR Reduction

112

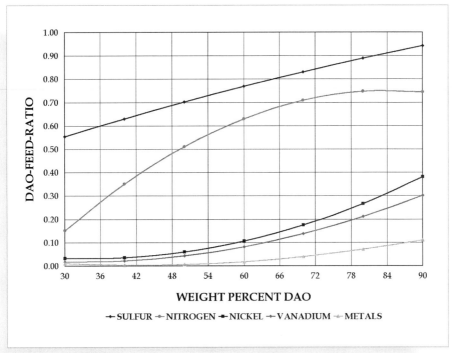

Fig. 8–3   SDA Yields

API of the feed and the liquid volume percent yield of DAO. Figure 8-5 is a plot of liquid volume percent yield of DAO vs. weight percent yield.

Comparison with other correlations

A number of correlations of SDA results have been published. All have used yield of DAO, either volume percent or weight percent, as the primary independent variable. Agreement between the author's results and some of these correlations has ranged from good to not-so-good, as illustrated by the accompanying tabulations.

Bonilla's data were on a volume yield basis,[4] so it was necessary to make some additional regressions for comparison purposes. On the whole, the author's results checked very well with Bonilla's data (Table 8-4).

Selvidge and Watkins presented data on UOP's DEMEX process for demetallizing resids.[5] The author's correlations agreed well with their API

113

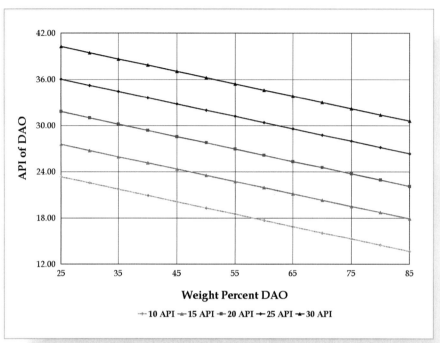

Fig. 8–4 API Gravity of DAO vs. Yield of DAO

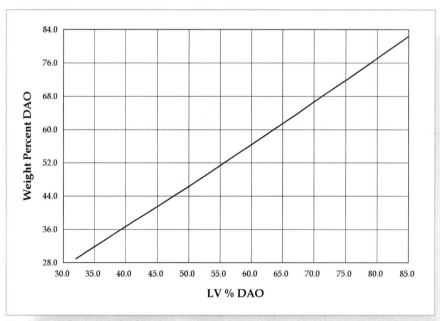

Fig. 8–5 Lv% vs. Wt% DAO

114

VARIABLE	REFERENCE	CORRELATION
API	10.6	13.5
	8.2	10.2
	16.3	15.9
	13.0	13.0
	18.5	18.2
	15.6	15.3
S	3.7	3.1
	3.8	3.6
	1.6	2.0
	2.1	2.3
	0.9	1.1
	1.1	1.3
N	0.4	0.3
	0.5	0.4
	0.2	0.2
	0.3	0.3
	0.2	0.2
	0.3	0.3
CCR	6.9	6.9
	12.1	12.1
	6.0	6.0
	10.0	10.0
	3.5	3.5
	7.8	8.0

Table 8–4 Comparison of Correlation Results with Reference[4]

and S data and fairly well on Conradson carbon residue and Ni. Results on V were not as optimistic as theirs (Table 8-5).

Viloria, *et al.*, studied Boscan crude.[6] Agreement on Conradson carbon residue and V is fair; on the others, not so good (Table 8-6).

Ditman presented data on seven oils.[2] A comparison with Ditman's average values is not good. Few of the author's values fall within the range

VARIABLE	REFERENCE	CORRELATION
API	13.4	12.4
	15.9	15.8
	15.7	15.5
	11.3	11.0
	14.2	14.2
S	2.6	3.0
	2.4	2.7
	2.6	2.4
	4.8	4.7
	4.1	4.0
CCR	12.0	10.4
	4.0	5.1
	8.0	7.9
	12.7	9.3
	8.0	5.0
Ni	4	8
	5	6
	5	6
	6	7
	4	3
V	20	34
	7	10
	59	94
	23	29
	10	7

Table 8-5 Comparison of Correlation Results with Reference[5]

of his results (Table 8-7). Ditman's data were on a volume yield basis, so the author's data were read from weight basis results after converting the volume percents to weight percents by means of the regression performed for that purpose. It is the author's opinion that this conversion had negligible effect on the outcome.

In Meyers' *Handbook of Petroleum Refining Processes*, there is a section on SDA by Bonilla, et al., of Foster Wheeler. One chart is a plot of volume yield of DAO vs. Conradson carbon residue ratio with API of the feed oil as

VARIABLE	REFERENCE	CORRELATION
API	24.7	19.3
	14.3	15.9
	14.3	15.7
	13.5	15.1
	12.3	14.8
S	43.0	66.0
	71.0	80.0
	72.5	81.0
	77.6	83.0
N	21.0	44
	33.0	64
	36.3	65
	53.0	69
	63.0	70
CCR	13.8	22.3
	45.4	46.6
	49.9	48.6
	63.5	52.6
	58.5	54.4
Ni	3.5	5.6
	35.0	20.3
	35.5	22.0
	45.7	25.9
V	2.0	2
	10.6	18
	24.0	21
	30.0	25
	22.8	27

Table 8–6 Comparison of Correlation Results with Reference[6]

a second parameter. A regression by the author on this same basis resulted in the chart shown. Agreement with Bonilla, et al., is fairly good, with the author's curves lying slightly within the area defined by their curves. The other correlations, all on a volume yield basis, were not checked.

VARIABLE	REFERENCE[2]		AUTHOR
	RANGE	AVERAGE	
S	40–57	49.0	79.0
	55–68	63.0	85.0
	75–83	80.0	92.0
N	32–38	34.0	58
	48–54	51.0	68
	70–74	72.0	78
Ni	3.2–17	7.5	16.6
	8.7–32	17.0	25.2
	28–57	41.0	35.7
V	1.1–11	5.5	11.6
	4.9–26	14.0	23.5
	22–55	36.0	40.0

Table 8–7 Comparison of Correlation Results with Reference[2]

On the whole, while it would be desirable to have exact agreement with all other sources, it is thought that the results obtained are satisfactory for general study purposes.

Operating requirements

Unlike most refinery processes, SDA operating requirements are more dependent on type of solvent and solvent-to-feed ratio than on the feed rate. However, the following set of average values can be safely used in the absence of more specific information, especially since most processes not using supercritical separation are using multi-effect evaporation. The quantities are on a per barrel of feed basis.

Electric power	2 kWh
Steam	60 pounds
Fuel	80 kBtu
Cooling water	nil

Capital costs

Five capital cost values from the literature in the past 10 years were scaled to a capacity of 30,000 BPD using a 0.7 exponent and to the first of January, 1991 using the Nelson-Farrar cost indices as shown in the following tabulation:

Bbl/Day	mm$	Year	Index	Time Factor	Size Factor	Escalated Cost, mm$
13,776	13.1	81	903.8	1.374	1.724	31.0 *
20,000	26.	90	1211	1.025	1.328	35.4 **
15,000	20.3	8/90	1226	1.013	1.624	33.4 ***
30,000	30.	89	1191	1.043	1.000	31.3 †
12,000	16.	82	976.9	1.271	1.899	38.6 ††
					Average	**33.9**

* Silkonia, et al., Oil & Gas Journal, Oct. 5, 1981, p.145
** Anon., Hydrocarbon Processing, November, 1990, p. 88
*** Ibid, p. 90
† Ibid, p. 90
†† Anon., Chemical Engineering, March 22, 1982, p. 35

An average value of $34 million for a 30,000 BPD SDA unit at the beginning of 1991 can be used as a point of departure for estimating costs of other units different in size and time in the manner shown.

Notes

1. Billon, A., Peries, J.P., Fehr, E., and Lorenz, E., *Oil & Gas Journal*, January 24, 1973, pp. 43–48

2. Ditman, J.G., *Hydrocarbon Processing*, May, 1973, pp. 110–113

3. Newcomer, R.M., and Soltau, R.C., "Successful Operation of the Rose Process," 1982 NPRA Annual Meeting, San Antonio

4. Bonilla, J.A., "Delayed Coking and Solvent Deasphalting: Options for Residue Upgrading," AIChE National Meeting, Anaheim, June 1982

5. Selvidge, C.W., and Watkins, C.H., "Demetallizing Vacuum Residuals by the Demex Process," UOP 1973 Technology Conference, Des Plaines, Illinois

6. Viloria, D.A., Krasuk, J.H., Rodriguez, O., Buenfama, H., and Lubowitz, J., *Hydrocarbon Processing*, March 1977, pp. 109–113

7. Bonilla, J.A., Feintuch, H.M., and Godino, R.L., pp. 8–19 through 8–51 in *Meyers' Handbook of Petroleum Refining Processes*, McGraw–Hill, New York, 1986

8. Silkonia et al., *Oil & Gas Journal*, October 5, 1981, p. 145

9. Anon., *Hydrocarbon Processing*, Nov. 1990, p. 88

10. Ibid, p. 90

11. Ibid, p. 90

12. Anon., *Chemical Engineering*, Mar. 22, 1982, p. 35

References

Anon., "Solvent Decarbonizing," Kelloggram, 1956 Series, No. 3

Billon, A., Peries, J.P., Fehr, E., and Lorenz, E., *Oil & Gas Journal*, January 24, 1977, pp. 43–48

Bonilla, J.A., "Delayed Coking and Solvent Deasphalting: Options for Residue Upgrading," AIChE National Meeting, Anaheim, California, June 1982

Ditman, J.G., *Heat Engineering*, Vol. XXXX, No. 5, September–October 1965

Ditman, J.G., and Zahnstecher, L.W., "Solvent Deasphalting for the Production of Catalytic Cracking—Hydrocracking Feed & Asphalt," NPRA National Meeting, San Francisco, March 1971

Ditman, J.G., *Hydrocarbon Processing*, May 1973, pp. 110–113

Ditman, J.G., *Oil & Gas Journal*, February 18, 1974, pp. 84–85

Gearhart, J.A., and Garwin, L., *Oil & Gas Journal*, June 14, 1976, pp. 63–66

Gearhart, J.A., *Hydrocarbon Processing*, May 1980, pp. 150–151

Nelson, S.R., and Roodman, R.G., *Chemical Engineering Progress*, May, 1985, p. 63

Marple, S., Jr., Train, K.E., and Foster, F.D., *Chemical Engineering Progress*, Vol. 57, No. 12, 1961, pp. 44–48

Newcomer, R.M., and Soltau, R.C., "Sucessful Operation of the Rose Process," NPRA National Meeting, San Antonio, March 1982

Nysewander, C.W., and Durland, L.V., *Oil & Gas Journal*, March 23, 1950, pp. 216–218

Penning, R.T., Vickers, A.G., and Shah, B.R., *Hydrocarbon Processing*, May 1982, pp. 145–150

Rossi, W.J., Deighton, B.S., and MacDonald, A.J., *Hydrocarbon Processing*, May, 1977, pp. 105–110

Selvidge, C.W., and Watkins, C.H., "Demetallizing Vacuum Residuals by the DEMEX Process," UOP 1973 Technology Conference, Des Plaines, Illinois

Selvidge, C.W., and Ocampo, F., "Processing High Metal Residues by the Demex Process," NPRA National Meeting, San Antonio, April 1973

Sherwood, H.D., *Oil & Gas Journal*, March 27, 1978, pp. 148–158

Sinkar, S.R., *Oil & Gas Journal*, September 30, 1974, pp. 56–64

Sprague, S.B., "How Solvent Selection Affects Extraction Performance," NPRA National Meeting, Los Angeles, March 1986

Thegze, V.B., Wall, R.J., Train, K.E., and Olney, R.B., *Oil & Gas Journal*, May 8, 1961, pp. 90–94

Viloria, D.A., Krasuk, J.H., Rodriguez, O., Buenafama, H., and Lubkowitz, J., *Hydrocarbon Processing*, March 1977, pp. 109–113

CHAPTER 9

VISBREAKING AND AQUACONVERSION

Visbreaking "is an effective and inexpensive way to produce more valuable products from heavy residues."[1] Initially it was used to reduce the viscosity and/or pour point of a fuel oil. It is currently employed to obtain additional cat cracker feed and to reduce fuel oil production. Residual fuel oil is the least valuable of the refiner's products, selling at a price below that of crude oil. Therefore, it is in the interest of the refiner to minimize its production.

Visbreaker process description

Visbreaking is a mild thermal cracking process. There are two versions practiced—the furnace process and the soaker process.

The furnace version is most widely practiced. It consists of a furnace and a fractionator. The reaction takes place in the furnace. The furnace effluent is quenched to stop further reaction prior to fractionation. Figure 9–1 is a simplified process flow diagram.

In the soaker version, a vessel is interposed between the furnace and the fractionator. This vessel is the soaker and it provides additional time for reaction. In fact, most of the reaction takes place in the soaker. The reactants are held for a longer period of time but at a lower temperature than in the case of the furnace process.

All the visbreaker products require further processing or treating with the possible exception of the resid (if it meets fuel oil specifications for viscosity, sulfur, etc.). An easy way to take care of the naph-

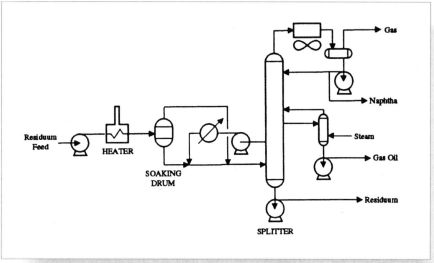

Fig. 9–1   Visbreaking Unit

tha is to feed it to the FCC fractionator. The heavy naphtha can be hydrotreated and reformed, providing its nitrogen content will permit. The gas oil can be fed to the FCC (directly or after hydrotreating) or to a hydrocracker.

Visbreaking data correlation

Early visbreaking correlations used the API of the feed as the primary parameter. This was a carryover from the thermal cracking of gas oils. The author has found, as have others, conversion to be a satisfactory parameter. Conversion is defined as the yield of gas plus gasoline. Some workers use gas plus gasoline plus distillate (also defined by cut temperature).

A major concern in visbreaking is the stability of the fuel oil resulting. It is important that the asphaltenes in the blended fuel oil remain in solution until used. Asphaltenes are soluble in aromatics; insoluble in paraffins. Therefore, the cut–back oil blended with the visbroken resid must be compatible with the resid. (Cut–back oil is a low viscosity oil blended with resid to lower its viscosity into specification.)

Allan, *et al.*,[2] report that the limiting conversion (maximum conversion for asphaltenes to remain in solution) for a resid correlates very well with its asphaltene content.

124

According to Rhoe and Bignieres,[3] conversions of 6% to 7% were the practice when viscosity reduction was the purpose; 8% to 12%, when maximizing distillate production.

Table 9–1 is the database used in correlating visbreaking yields. There were fewer sets (30) of VB data (on a weight percent basis) found in the literature than in the case of SDA. However, it will be observed that there are far fewer gaps in the VB data. The data are for visbreaking of both atmospheric and vacuum resids as is evident from the API gravity of the feed. The main difficulty experienced with these data was the frequent absence of cut temperatures defining the boiling ranges of the various cuts. This became particularly evident when distillate and gas oil data were plotted against conversion.

API	CCR	CONV	GAS	GASO	DIST	GO	RESID	GO+	DIST+	C5+
15.10	9.50	3.80	1.80	2.00	16.20		80.00	80.00	96.20	98.20
14.40	10.00	4.00	1.90	2.10	17.00		79.00	79.00	96.00	98.10
7.50	20.60	5.00	2.00	3.00	12.00		83.00	83.00	95.00	98.00
6.90	17.40	6.08	2.70	3.38	5.64	26.11	62.17	88.28	93.92	97.30
8.60		7.00	2.50	4.50	13.00		80.00	80.00	93.00	97.50
30.00	1.80	7.00	1.90	5.10	17.00		76.00	76.00	93.00	98.10
3.40	21.40	7.50	1.70	5.80	8.20		84.30	84.30	92.50	98.30
15.40		7.90	1.70	6.20	16.90		75.00	75.00	91.90	98.10
16.90	7.60	8.40	2.30	6.10	10.70	57.50	23.40	80.90	91.60	97.70
6.90	20.80	8.40	2.50	5.90			91.60	91.60	91.60	97.50
6.90	20.80	8.40	2.50	5.90	8.80	19.40	63.40	82.80	91.60	97.50
12.20	10.99	8.72	1.84	6.88	23.58	39.72	27.98	67.70	91.28	98.16
8.30	20.20	8.80	3.10	5.70	15.00	25.90	50.30	76.20	91.20	96.90
8.50	14.20	8.90	1.70	7.20	12.40		78.70	78.70	91.10	98.30
17.80	10.10	10.06	2.96	7.10	11.93	17.81	60.20	78.01	89.94	97.04
21.50	7.90	10.20	1.80	8.40	12.60		77.20	77.20	89.80	98.20
17.80	10.10	10.40	1.66	8.74	12.23	15.25	62.12	77.37	89.60	98.34
14.70	16.30	10.82	3.27	7.55	11.25	15.50	62.43	77.93	89.18	96.73
18.70		11.00	4.00	7.00	23.00		66.00	66.00	89.00	96.00
17.80	10.10	11.74	2.81	8.93	10.56	16.48	61.22	77.70	88.26	97.19
12.50	14.50	11.83	3.27	8.56	11.50	25.76	50.91	76.67	88.17	
8.40	14.00	11.88	3.01	8.87	11.27	38.61	38.24	76.85	88.12	96.99
12.20	10.99	11.92	2.96	8.96	25.46	36.89	25.73	62.62	88.08	97.04
8.40	14.00	12.56	2.78	9.78	12.71	39.05	35.68	74.73	87.44	97.22
17.80	10.10	12.84	3.49	9.35	14.06	17.52	55.50	73.02	87.08	96.43
		13.50	5.20	8.30	54.70		31.80	31.80	86.50	94.80
17.80	10.10	13.97	3.87	10.10	13.93	18.44	53.66	72.10	86.03	96.13
8.40	14.00	14.08	3.32	10.76	12.55	42.01	31.36	73.37	85.92	96.68
8.40	14.00	14.67	3.37	11.30	12.03	41.30	32.00	73.30	85.33	96.63
12.20	10.99	15.89	4.34	11.55	27.34	33.71	23.06	56.77	84.11	95.66

Table 9–1 Visbreaker Database

A summary of some of the visbreaking correlation results is presented in Table 9–2. Very satisfactory results were obtained for gas and gasoline using the single independent variable conversion.

DEPENDENT VARIABLE	GAS	GAS	GASO	GASO	GASO	DIST	DIST	400-950	RESID	RESID	C5+	C5+
INDEPENDENT VARIABLE												
1	CONV	CONV	CONV	CONV	CONV	CONV	CONV	CONV	CONV	CONV	CONV	CONV
2		CONV^2			CONV^2		CONV^2	CONV^2		END PT		CONV^2
3												
R^2	0.826	0.84	0.973	0.981	0.985	0.693	0.89	0.833	0.695	0.907	0.639	0.697
SEE	0.285	0.281	0.446	0.3828	0.354	1.078	0.675	6.74	9.94	4.82	0.447	0.419
S1	0.018	0.103	0.0085	0.0234	0.124	0.0911	0.3027	0.648	0.5989	0.7053	0.0274	0.143
S2		0.0053			0.00627		0.0166	0.0525		0.029		0.00723
NO. OF SETS	21	21	25	25	25	14	14	19	25	12	26	26

Table 9–2 Summary of Results of Visbreaker Correlations

In the case of resid, the addition of the gas oil end point as an independent variable, improved the result considerably. Distillate results were significantly improved by the addition of the square of the conversion as an independent variable. Lacking clear and precise definition and corresponding yield data for material boiling between gasoline and resid, namely, distillate and gas oil, one may resort to some indirect derivations.

By definition, the 400°F plus material is equal to 100 minus gas plus gasoline. With a reasonable correlation for resid, the yield of distillate plus gas oil can be obtained by subtracting yield of resid from yield of 400 plus. A satisfactory correlation of 400–950 material was obtained using a second order polynomial in conversion. A composite plot of the yields is shown in Figure 9–2.

Table 9–3 gives an average composition of the butanes and lighter produced in visbreaking, with H_2S excluded since it is a function of sulfur contained in the feed as well as of the extent of conversion. The few sets of data on H_2S (5) were regressed using conversion and sulfur in feed as independent variables. The results are plotted in Figure 9–3. The few sets of data and the narrow range of conversion (7.2 to 14.1) and of sulfur in feed (2 to 4) on which this plot is based, should be borne in mind by the user.

126

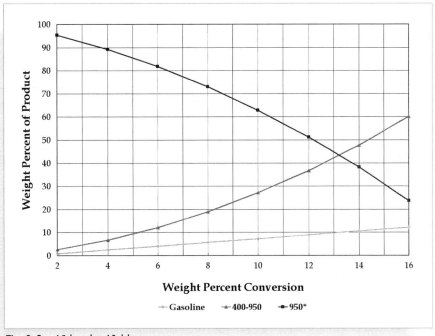

Fig. 9–2 Visbreaker Yields

AVERAGE COMPOSITION OF BUTANES AND LIGHTER
(H_2S EXCLUDED)

COMPONENT	WT% OF LT ENDS
H2	0.33
C1	16.45
C2=	12.80
C2	15.73
C3=	12.62
C3	15.03
C4=	15.13
IC4	2.20
NC4	9.71
TOTAL	100.00

Table 9–3 Visbreaker Yields

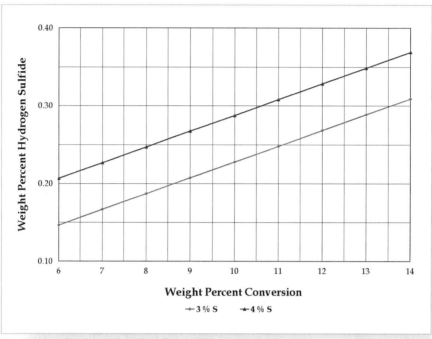

Fig. 9–3   Visbreaker H₂S Yield (Sulfur in Feed as Parameter)

The following additional correlations were developed and the results are depicted in the indicated figures: Research Octane of VB Gasoline (Fig. 9–4), Motor Octane of Visbreaker Gasoline (Fig. 9–5), Visbreaker Gasoline API Gravity (Fig. 9–6), Visbreaker Distillate API Gravity (Fig. 9–7), Visbreaker Gas Oil API Gravity (Fig. 9–8), Visbreaker Resid API Gravity (Fig. 9–9), Conradson Carbon Residue 650+ (Fig. 9–10), Conradson Carbon Residue of Visbreaker Resid (Fig. 9–11), Sulfur in Visbreaker Products (Fig. 9–12), Visbreaker Lv % vs. Wt % (9–13).

Comparison with other correlations

Beuther, *et al.*,[1] presented a set of curves with volume percent of 10 Reid vapor pressure (RVP), 300°F end point gasoline as the independent variable. By means of the volume percent vs. weight percent chart, volume percent yields were read from their curves and converted to weight percent. The results agreed very well over the conversion range of the author's correlations. Beuther, *et al.*, went to much higher conversions. They also pre-

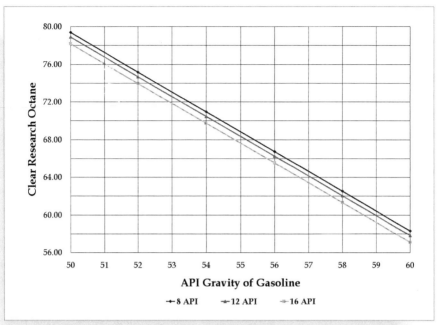

Fig. 9–4 Research Octane of Visebreaker Gasoline

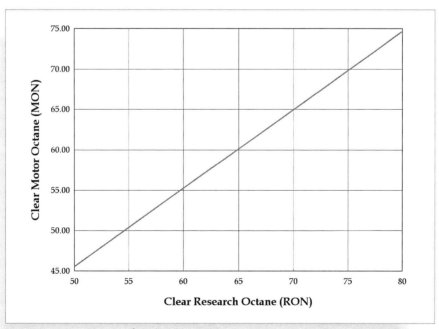

Fig. 9–5 Motor Octane of Visebreaker Gasoline

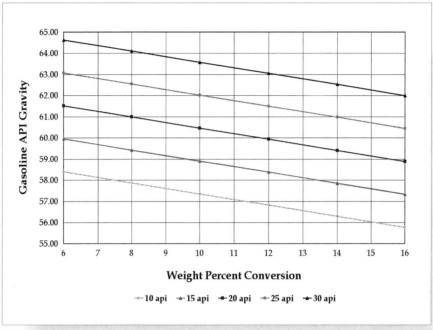

Fig. 9–6   Visebreaker Gasoline API Gravity

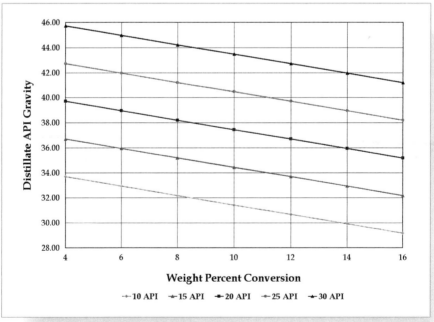

Fig. 9–7   Visebreaker Distillate API Gravity

130

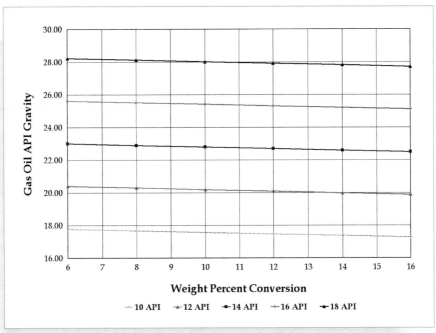

Fig. 9–8 Visebreaker Gas Oil API Gravity

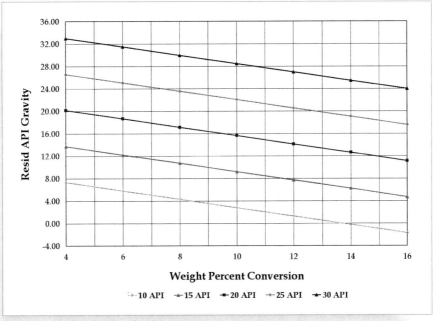

Fig. 9–9 Visebreaker Resid API Gravity

131

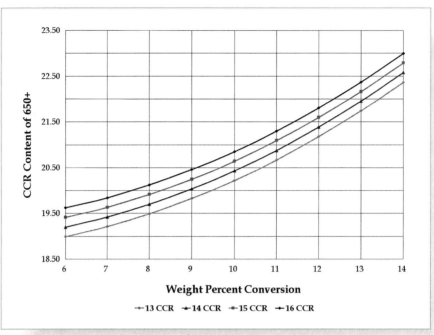

Fig. 9–10 Conradson Carbon Residue of Visbreaker 650+ (Feed CCR is Parameter)

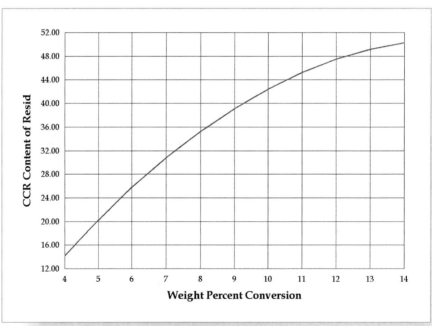

Fig. 9–11 Conradson Carbon Residue of Visbreaker Resid

132

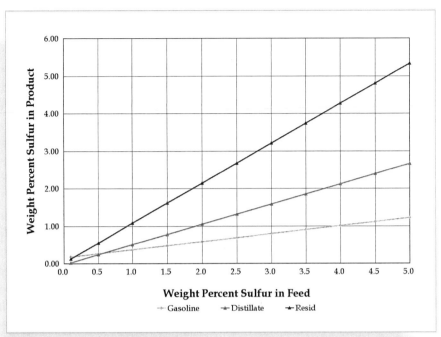

Fig. 9–12 Sulfur in Visbreaker Products

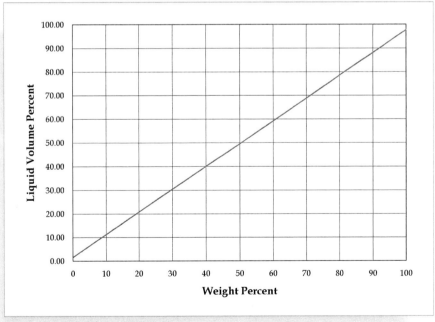

Fig. 9–13 Visebreaker Lv% vs. Wt%

133

sented gasoline and naphtha yields at maximum severity in terms of normal pentane insolubles and of ring and ball softening point of the feedstock. These yields represented lower conversions than the other set of curves. However, this second set could not be checked because the properties of the feed used were not available in the author's database.

Bakshi and Lutz[4] correlated their yields in terms of pentane insolubles also. Their curves for distribution of sulfur in the products was in terms of the sulfur content of the feed and agreed very well with the author.

Wuithier of Institut Francais du Petrole published a set of charts[5] for visbreaking based on producing a Bunker C fuel (300 SSF at 122°F). The API gravity and the K factor were used as parameters to characterize the feed. The yields were volume percent of feed. For the limited comparisons made, there was good agreement with the author's correlations.

Kuo[6] reported on pilot plant research into stability limits in visbreaking. Unfortunately, there are no numerical scales in his yield chart, so no check was possible.

Operating requirements

The following average values can be used for most purposes:

Fuel	80 kBtu per bbl
Electric power	0.5 kWh per bbl
Steam	(50) pounds per bbl (produced)
Cooling water	essentially zero, maximum air cooling

Capital cost

Of a total of 25 plant costs noted in the literature in the past 14 years, seven announcements of domestic projects were considered appropriate for consideration. Scaling these data to 25,000 BPD and January, 1991 as illustrated under SDA, a value of $24 million was calculated.

Aquaconversion

Aquaconversion is "a new hydrovisbreaking technology (that) uses a dual catalyst system to achieve higher conversion levels, lower asphaltene, and Conradson carbon contents, and more stabilized residue than conventional visbreaking technologies. It is currently being promoted as either a replacement of, or a modification to, conventional visbreaking."[7]

In the presence of the proprietary catalyst, water dissociates into hydrogen and oxygen. This nascent hydrogen reacts with aromatic free radicals and inhibits their condensing to asphaltenes.

In a commercial demonstration, conversion to material boiling lower than 662°F was as high as 31 WT% and conversion to material boiling below 330°F was 7.8 WT%.[8]

Notes

1. Beuther, H., Goldthwait, R.G., and Offutt, W.C., *Oil & Gas Journal*, Vol. 57, no. 46, pp. 78–84

2. Allan, D.E., Martinez, C.H., Eng, C.C., and Barton, W.J., *Chemical Engineering Progress*, January 1983, pp. 85–90

3. Rhoe, A., and Blignieres, C., *Hydrocarbon Processing*, January 1979, pp. 131–136

4. Bakshi, A.S., and Lutz, I.H., *Oil & Gas Journal*, July 13, 1987, pp. 84–87

5. Wuithier, P., *Revue de l'Institut Francais du Petrole*, Vol. XIV, no. 9, pp. 1,160–1,163 and 1,174–1,180

6. Kuo, C.J., *Oil & Gas Journal*, September 24, 1984, pp. 100–102

7. Marzin, R., Pereira, P., McGrath, M.J., Feintuch, H.M., and Thompson, G., *Oil & Gas Journal*, November 2, 1998, ff. 79

8. Ibid.

References

Aiba, T., and Kaji, H., *Chemical Engineering Progress*, February 1981, pp. 37–44

Akbar, M., and Geleen, H., *Hydrocarbon Processing*, May 1981, pp. 81–85

Allan, D.E., Martinez, C.H., Eng, C.C., and Barton, W.J., *Chemical Engineering Progress*, January 1983, pp. 85–90

Allen, J.G., Little, D.M., and Waddill, P.M., *Oil & Gas Journal*, June 14, 1951, pp. 78–84

Anon., *Hydrocarbon Processing*, May, 1986, pp. 42–44

Anon., *Oil & Gas Journal*, May 26, 1975, pp. 96–103

Anon., *Oil & Gas Journal*, November 6, 1978, pp. 56–59

Bakshi, A.S., and Lutz, I.H., *Oil & Gas Journal*, July 13, 1987, pp. 84–87

Beuther, H., Goldthwait, R.G., and Offutt, W.C., *Oil & Gas Journal*, Vol. 57, No. 46, 1959, pp. 151–157

Gadda, L., *Oil & Gas Journal*, October 18, 1982, pp. 120–122

Hournac, R., Kuhn, J., and Notarbartolo, M., *Hydrocarbon Processing*, December 1979, pp. 97–102

Hus, M., *Oil & Gas Journal*, April 13, 1981, pp. 109–120

Kuo, C.J., *Oil & Gas Journal*, September 24, 1984, pp. 100–102

Nelson, W.L., *Oil & Gas Journal*, February 23, 1950, p. 195

Ibid, June 29, 1950, p. 88

Ibid, February 25, 1952, p. 185

Ibid, March 10, 1952, p. 129

Ibid, March 24, 1952, pp. 208–209

Ibid, March 31, 1952, pp. 114–115

Ibid, May 26, 1952, p. 219

Ibid, September 15, 1952, p. 141

Ibid, April 13, 1953, p. 143

Ibid, April 20, 1953, p. 167

Ibid, March 14, 1960, p. 189

Ibid, November 10, 1969, p. 229

Ibid, December 7, 1970, pp. 62–64

Ibid, April 17, 1978, pp. 106–108

Notarbartolo, M., Menegazzo, C., and Kuhn, J., *Hydrocarbon Processing*, July, 1979, pp. 114–118

Rhoe, A., and de Blignieres, C., *Hydrocarbon Processing*, January, 1979, pp. 131–136

Wood, J.R., *Oil & Gas Journal*, April 22, 1985, pp. 80–84

Wuithier, P., *Revue de l'Institut Francais du Petrole*, Vol. XIV, No. 9, pp. 1,160–1,163 and 1,174–1,180

Yepsen, G.L., and Jenkins, J.H., *Hydrocarbon Processing*, September, 1981, pp. 117–120

CHAPTER 10

DELAYED COKING

Coking is by far the most widely practiced means of reducing the carbon-hydrogen ratio of residual oils. Of the two main processes— delayed coking and fluid coking—the bulk of the capacity (about 90%) is in delayed coking units.

Both processes have the dual purposes of increasing cat cracker feedstock availability and of reducing the production of residual fuel oil.

In delayed coking, carbon is removed in the form of a black solid that is referred to as green coke. It contains moisture, volatile combustible matter (VCM), sulfur, and metals concentrated from the feed. Fixed carbon content ranges from 85% to 95%. After roasting to reduce the moisture and VCM, typically in a rotary kiln, the calcined coke contains 98% to 99.5% fixed carbon.[1] The uses and specifications of the various grades of coke will not be elaborated on here.

Delayed coking process description

As shown in Figure 10-1, the fresh feed is introduced directly into the fractionator. This permits flashing of lowering boiling material not desired in the furnace feed, combining remaining fresh feed with recycle (if any), and preheating the feed to the furnace. In the furnace, the endothermic heat of the coking reaction is supplied in a way to minimize coking in the furnace itself. The furnace effluent goes to a coke chamber where coke is formed

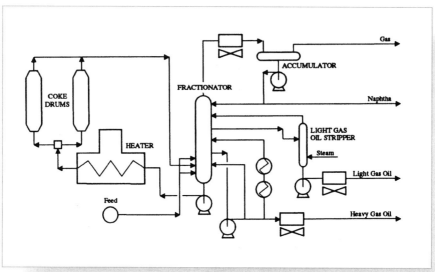

Fig. 10–1 Simplified Delayed Flow Coking Scheme

and accumulates. Overhead vapors from the coke drum enter the lower part of the fractionator to be separated into gas, naphtha, light and heavy gas oils, and recycle.

While coke is being formed in one drum, coke is being removed from another by high–pressure water jets. Currently drum cycles range from 32 to 48 hours for a complete turn–around.

Coking is another of those processes where no finished fluid product (liquid or vapor) is made. Every stream from the coker requires further processing.

Delayed coking data correlation

The database for delayed coking yield correlation is displayed in Table 10–1. Again, there are many gaps in the table. The chief problem encountered with these data was the general lack of boiling ranges of cuts. The terms distillate and gas oil are ambiguous in this respect, and gasoline can vary in end point and be butane–free or not. The assumption was made that the gasoline was butane–free (BFG) and had a 400°F end point. Further, liquid boiling above gasoline was considered gas oil (data labeled distillate and gas oil were added together in the same set).

140

FD API	FD CCR	COKE	C4-	C5-400	GAS OIL	C5+
15.85	2.84	4.36	8.39	18.23	69.02	87.25
13.20	3.96	7.86	4.04	21.76	66.34	88.10
12.80	5.20	10.00	6.20	18.50	65.30	83.80
19.30	6.49	22.00	11.10	18.20	48.70	66.90
18.90	6.71	12.60	6.40	15.80	61.40	77.20
16.90	7.60	21.00	12.30	19.90	46.80	66.70
16.90	7.60	15.40	7.00	11.80	65.80	77.60
21.50	7.90	16.70	7.80	16.90	58.60	75.50
16.80	8.00	20.80				
17.40	8.20	20.63	8.06	14.89	56.42	71.31
19.00	8.20	18.60				
14.48	8.61		7.38	38.78	48.84	87.62
15.00	9.00	22.00	7.50	17.10	53.00	70.10
15.00	9.00	11.00	6.40	10.20	72.00	82.20
15.00	9.00	22.00				
16.10	9.10	18.50	9.20	17.20	54.90	72.10
15.20	9.30	20.70	9.70	15.60	53.90	69.50
12.00	9.40	19.10				
12.00	9.40	14.10				
12.00	9.60	21.60	12.00	15.70	50.70	66.40
16.40	10.30	20.09	7.91	13.74	58.26	72.00
18.60	10.60	24.70				
18.60	10.60	18.60				
15.70	10.70	21.15	7.85	13.42	57.58	71.00
16.40	10.70	20.09	8.02	13.91	57.89	71.80
15.70	10.80	22.36	7.74	13.34	56.56	69.90
15.70	10.80	22.30				
17.10	11.10	17.70	7.40	20.40	54.50	74.90
14.80	11.10	22.50	8.13	13.43	55.94	69.37
14.80	11.10	20.30	9.00	13.30	57.50	70.80
15.20	11.20	19.60				
12.30	11.30	28.90				
12.30	11.30	21.00	6.50	16.00	56.50	72.50
15.30	11.50	22.63				
15.30	11.50	22.63	7.34	10.35	58.85	69.20
14.00	11.60	22.90	7.10	20.50	49.50	70.00
14.00	12.10	22.90				
16.33	12.85	21.42	7.41	22.28	48.89	71.17
12.30	13.00	23.70	8.00	17.40	47.90	65.30
13.00	14.10	23.60				
8.50	14.20	24.80	12.00	17.70	45.50	63.20

Table 10–1a Delayed Coking Database

Following the procedure outlined previously, some of the results obtained using first order relations are shown in Table 10–2. Coke yield correlates very well with Conradson carbon residue content of feed. Gas, gas oil, and C_5^+ correlate slightly better with coke yield than with

FD API	FD CCR	COKE	C4-	C5-400	GAS OIL	C5+
12.30	14.60	22.00	7.00	18.60	52.40	71.00
12.50	14.80	26.00	6.40	21.60	46.00	67.60
7.30	14.80	19.80				
12.50	15.00	26.00				
11.70	15.00	22.20	7.10	18.50	52.20	70.70
7.40	15.40	27.00	11.10	16.10	45.80	61.90
8.20	15.60	24.90	9.20	17.40	48.50	65.90
8.90	16.10	26.50	11.30	14.60	47.60	62.20
8.10	17.40	28.00	11.70	14.10	46.20	60.30
8.90	17.80	28.40	11.70	21.90	38.20	60.10
7.30	17.90	21.00	7.00	15.40	55.00	70.40
7.30	17.90	29.60	13.10	12.90	44.90	57.80
7.40	18.10	30.20	9.90	15.00	44.90	59.90
7.40	18.10	27.20	9.10	12.50	51.20	63.70
7.70	18.80	28.35	8.55	10.36	52.74	63.10
7.70	18.80	31.41	9.66	12.42	46.51	58.93
7.50	19.10	31.80	10.05	12.30	45.85	58.15
7.50	19.10	28.73	8.98	10.32	51.97	62.29
7.40	19.60	29.70	9.00	19.40	41.90	61.30
7.40	19.60	32.60	5.60	17.80	43.70	61.50
6.70	19.80	30.20	8.70	19.90	41.20	61.10
2.80	20.60	34.60	16.10	12.00	37.30	49.30
6.90	20.80	32.80	10.70	13.90	42.60	56.50
7.70	20.90	35.60				
7.70	20.90	35.60	10.10	17.90	36.40	54.30
3.40	21.40	31.50	9.50	16.00	43.00	59.00
3.87	21.76	20.54	11.54	14.36	53.26	67.62
4.00	22.00	35.10	10.50	21.40	33.00	54.40
7.20	22.20	33.00	12.10			
5.60	22.50	26.00	8.00			
5.60	22.50	40.20				
5.60	22.50	40.20				
5.60	22.50	40.20	9.20			
5.10	22.80	20.30	4.60	12.10	63.00	75.10
2.60	23.30	37.90	16.30	16.20	29.60	45.80
4.30	23.50	38.08	10.15	12.43	39.34	51.77
4.30	23.50	34.24	8.78	9.91	47.07	56.98
2.50	24.00	26.50	11.37	12.51	48.62	61.13
4.50	24.20	33.00	13.20	13.50	40.40	53.90
1.40	24.50	39.60	16.70	15.40	28.30	43.70
2.60	25.50	38.30	13.20	19.30	29.20	48.50

Table 10–1b Delayed Coking Database cont'd

Conradson carbon residue. The gasoline data are very scattered in either case with no strong definite trend indicated. It was decided to obtain gasoline yield by difference. Figure 10–2 is a composite plot of the yields.

DEPENDENT VARIABLE	COKE	GAS	GAS	GAS OIL	GAS OIL	C5+	C5+	H$_2$S
INDEPENDENT VARIABLE	CCR	CCR	COKE	CCR	COKE	CCR	COKE	FD S COKE
R^2	0.921	0.756	0.792	0.865	0.875	0.902	0.9662	0.974
SEE	2.29	0.877	0.654	3.29	3.4	2.76	1.61	0.06
S1	0.0515	0.025	0.0146	0.0865	0.0593	0.071	0.0303	0.0291
S2								0.0067
NO. OF SETS	66	39	38	47	56	49	53	9

Table 10–2 Some Results of Delayed Coking Yields Correlations

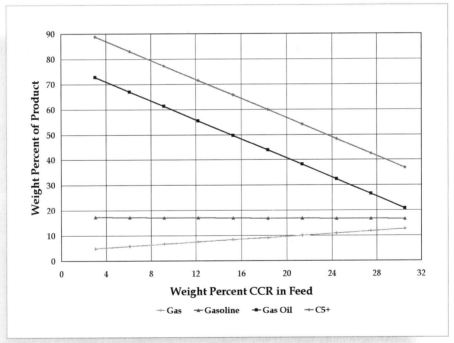

Fig. 10–2 Delayed Coker Yields

The butanes and lighter gas can be assumed to have approximately the following composition in weight percent:

Hydrogen	0.9
Methane	27.2
Ethylene	6.7
Ethane	17.8
Propylene	10.8
Propane	14.4
Butylenes	11.0
Isobutane	2.4
Normal butane	8.8
Total	100.0

In the case of H_2S, a second parameter, percent sulfur in feed, was used along with weight percent coke yield as an independent variable. The result is plotted in Figure 10–3. The weight percent contents of sulfur in coke, gasoline, and gas oil are plotted in Figure 10–4.

A number of product properties was explored, including gasoline octane, gasoline PONA, Conradson carbon residue content of gas oil, and API and sulfur content of both gasoline and gas oil. There was insufficient data to develop a relationship for gasoline octane, only to indicate a range (RON of 61 to 68) in which it might fall. Of more significance, since the gasoline requires further processing, is its PONA (average values of 40, 37, 13, 10). API gravity is important in making an overall material balance (Fig. 10–5). Conradson carbon residue is important for gas oil fed to a cat cracker (Fig. 10–6). Too few data were found on nitrogen contents to be useful. A plot of liquid volume percent vs. weight percent for coke products appears in Figure 10–7.

Comparison with other correlations

Early attempts to correlate delayed coking yields used the API of the feed as the independent variable as had been the practice with thermal cracking. The results were not very good. George Armistead presented a

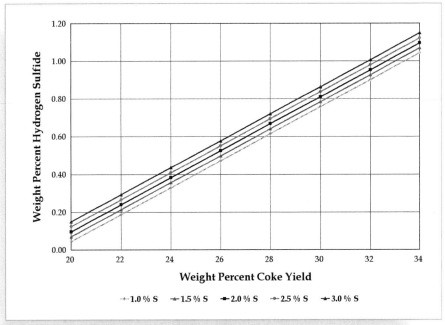

Fig. 10–3   Delayed Coker Hydrogen Sulfide Yield

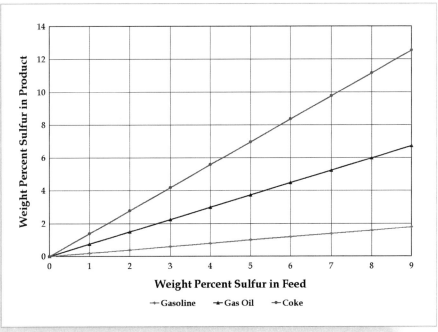

Fig. 10–4   Sulfur in Delayed Coker Products

145

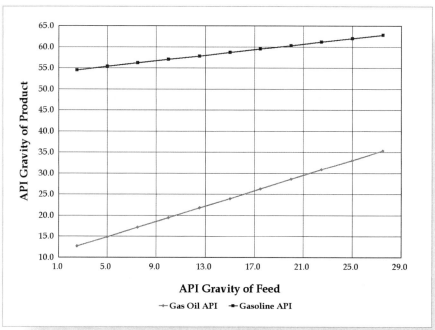

Fig. 10–5   Delayed Coker Product Gravities

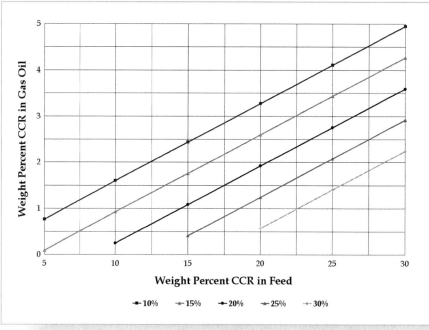

Fig. 10–6   CCR in Delayed Coker Gas Oil (Wt% Yield of Coke Plus Gas is Parameter)

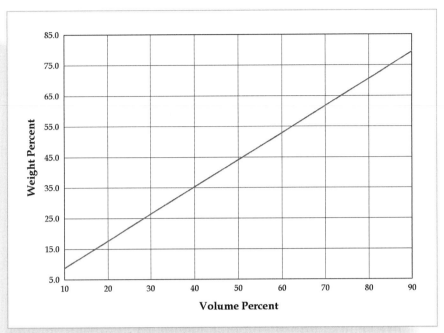

Fig. 10–7 Lv% vs. Wt% of Coker Products

plot[2] of gas, gasoline, and coke yields vs. API gravity of feedstock with a discussion of the various other variables at play in the process.

Nelson[3] developed equations relating the volume percent yield of gasoline with the API gravity of the feed and with the volume yield and API gravity of gas oil, along with tabulated yields calculated from these equations. In a later article,[4] Nelson published coke and gas yields corresponding with those gasoline yields. Still later,[5] he published yields of coke in terms of Conradson carbon residue content of feed (see comparison below).

Martin and Wills in Advances in Petroleum Chemistry and Refining,[6] gave equations for coke and gas yields in terms of Conradson carbon residue. Jakob[7] published a curve for coke yield in terms of Conradson carbon residue that appears to be a plot of a third order equation. Gary and Handwerk (G & H)[8] presented equations for coke, gas, and gasoline yields with gas oil obtained by difference.

The following tabulation shows how some of these relationships compare with those developed by the author:

147

Coke Yield

WT% Conradson carbon residue	Author	Nelson	M & W	Jakob	G&H
5	8.2	8.5	10.3	15	8
10	16.4	18	18.6	20	16
15	24.6	27.5	26.9	29	24
20	32.7	35.5	35.2	36	32
25	40.9	42	43.5	—	40

G & H state that their relationships are based on straight–run residuals as are those of the author. It is not known if this is true for the others.

Attempts at deriving a simple correlation for gasoline yields were so discouraging it was decided to obtain them by difference.

A comparison of the author's results with G & H's gas yields appears in the following tabulation:

WT% Conradson carbon residue	G&H	Author
5	8.52	5.47
10	9.24	6.87
15	9.96	8.27
20	10.68	9.67
25	11.40	11.07
30	12.12	12.47

It is assumed that the differences in results are due to differences in the data populations employed in the correlations. Such an explanation may account for the differences found between the author's results for sulfur content of coke and that of others. Though the author's regression results look good they are higher than those reported by others:

Feed S	Author	Nelson[9]	Kutler[10]	Jakob[7]
0.25	—	—	0.5	—
0.9	—	—	1.5	—

Feed S	Author	Nelson[9]	Kutler[10]	Jakob[7]
1	1.4	1.2	0.7	—
2	2.8	2.1	—	1.4–2
3	4.2	3.2	—	2.5–3.1
4	5.6	4.4	—	3.8–4.3
5	7	5.7	5	5.5–5.8
6	8.4	7.2	—	7.5

Jakob used Conradson carbon residue of feed as a second independent variable. Nelson's values are averages. He commented that he observed greater variation than Jakob. For the author's data, the improvement in adding Conradson carbon residue as a parameter was not significant. Kutler, *et al.*, point out that certain California crudes produce cokes with significantly lower sulfur content than would be expected from the feed sulfur content.[10]

G & H give percentages for the distribution of sulfur and nitrogen in the feed to the various products. As stated previously, the author did not find sufficient nitrogen data to work with. Results on sulfur are summarized here as percent of sulfur in feed appearing in product:

	G & H	Author	Average Range
Gas	30	14.9	6.3–35.7
Gasoline	5	3.3	2.3–5.5
Gas oil	35	33.5	12–42
Coke	30	29.7	15.6–50.2

It is recommended that Figures 10–3 and 10–4 be used rather than author's average values above.

Delayed coking operating requirements

The following data are for the continuous operation of the coking process. There is an increase in steam requirement during the purging of the drum before and after decoking and in electric power during decoking.

149

Electric power, kWh/b	3.6
Fuel, mBtu/b	120
Cooling water, gpm/b/h	0.6
High pressure steam, #/b	< 40 >

Delayed coker capital cost

Eight published capital costs for delayed cokers were adjusted to 20,000 BPD and January 1991 with an average result of $46 million (ranging from 40.7 to 51.3).

Notes

1. Reis, T., *Hydrocarbon Processing*, June 1975, pp. 97–104

2. Armistead, G., Jr., *Oil & Gas Journal*, March 16, 1946, pp. 103–111

3. Nelson, W.L., *Oil & Gas Journal*, July 7, 1952, p. 103

4. Ibid, February 15, 1954, p. 181

5. Ibid, January 14, 1974, p. 70

6. Martin, S.W., and Wills, L.E., *Advances in Petroleum Chemistry and Refining*, Vol. 2, Interscience Publishers Inc., New York, 1959, pp. 364–419

7. Jakob, R.R., *Hydrocarbon Processing*, September 1971, pp. 132–136

8. Gary, J.H., and Handwerk, G.E., *Petroleum Refining Technology and Economics*, Marcel Dekker, New York, 2nd ed., 1984

9. Nelson, W.L., *Oil & Gas Journal*, Oct. 9, 1978, p. 71

10. Kutler, A.A., DeBiase, R., Zahnstecher, L.W., and Godino, R.L., *Oil & Gas Journal*, April 5, 1970, pp. 92–96

References

DeBiase, R., and Elliott, J.D., *Oil & Gas Journal*, April 19, 1982, pp. 81–88

Foster, A.L., *Petroleum Engineer*, April 1951, pp. C–53 to C–62

Gibson, C.E., *Refining Engineer*, April, 1958, pp. C–46 to C–50

Heck, S.B., *Oil & Gas Journal*, July 24, 1972, pp. 46–48

Hengstebek, R.J., Petroleum Processing, McGraw–Hill Book Co., New York City, 1959, p. 142

Jakob, R.R., *Hydrocarbon Processing*, September 1971, pp. 132–136

Kutler, A.A., DeBiase, R., Zahnstecher, L.W., and Godino, R.L., *Oil & Gas Journal*, April 5, 1970, pp. 92–96

Mekler, V., and Brooks, M.E., *Petroleum Refiner*, Vol. 39, No. 2, 1960, ff. 158

Meyer, D.B., and Webb, H.C., *Petroleum Refiner*, Vol. 39, No. 2, 1960, pp. 155–158

Meyers, R.A., *Handbook of Petroleum Refining Processes*, McGraw–Hill Book Co., New York City, 1986, pp. 7–18, 7–25, 7–26

Mohammed, A–H.A.K., Abdullah, M.O., and Abdul–Ammer, A.A., *Hydrocarbon Processing*, November 1979, pp. 66–F to 66–L

Murphy, J.R., Whittington, E.L., and Chang, C.P., *Hydrocarbon Processing*, September 1979, pp. 119–122

Nelson, W.L., *Oil & Gas Journal*, July 7, 1952, p. 103

Ibid, March 9, 1953, pp. 125–126

Ibid, March 23, 1953, p. 359

Ibid, February 15, 1954, p.181

Ibid, November 28, 1955, p. 117

Ibid, February 25, 1963, p. 115

Ibid, January 14, 1974, p. 70

Ibid, October 9, 1978, p. 71

Ibid, December 18, 1978, pp. 68–69

Reis, T., *Hydrocarbon Processing*, June 1975, pp. 97–104

Rose, K.E., *Hydrocarbon Processing*, July 1971, pp. 85–92

Stolfa, F., *Hydrocarbon Processing*, May 1980, pp. 101–109

CHAPTER 11

FLUID COKING / FLEXICOKING

Flexicoking is a proprietary process of Exxon and is an outgrowth of their fluid coking technology. It produces a large volume of low Btu heating value gas by gasification of a high percentage of the coke produced by fluid coking. It is a relatively expensive process, but can be advantageous from the ecological standpoint particularly in a new grass-roots installation where heaters can be designed for the low Btu fuel. It also considerably reduces the problem of disposing the coke.

Flexicoking process description

There are three fluidized beds in the process—a reactor, a heater, and a gasifier. Fresh resid is fed into the reactor where it contacts hot, fluidized coke particles that supply heat of reaction plus sensible and latent heat. Products of reaction are separated from circulating coke by means of cyclones in the top of the reactor. The vapors rise through a scrubber where the vapors are quenched by a wash oil and any remaining coke is washed back into the reactor along with the wash oil. The scrubbed vapors are separated in a fractionator into gas, gasoline, distillate, and gas oil.

New coke formed in the reactor is deposited on the circulating particles. These particles are circulated to a heater where devolatilization of the coke occurs. The devolatilized coke passes to a gasifier where much of it reacts at elevated temperature with air (or oxygen) and steam to produce a product gas. This gas is sent to the heater where it supplies the heat required in the process. It leaves the heater and is cooled in a steam generator. Further treatment to remove last traces of coke and recover sulfur leaves a clean, low Btu fuel. A net (or purge) coke is withdrawn from the reactor to keep in bounds its metals content and particle size. The process description for fluid coking is essentially the same except for the absence of the gasifier as shown on the simplified process flow diagrams, Figures 11–1 and 11–2.

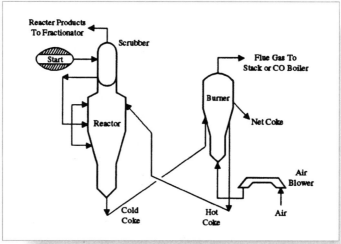

Fig. 11–1   Simplified Fluid Coking Flow Scheme

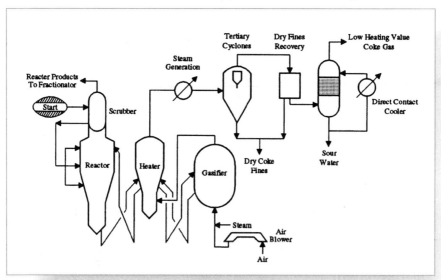

Fig. 11–2   Simplified Flexicoking Flow Scheme

154

Fluid coking/flexicoking data correlation

Though this technology is not practiced as extensively as the other resid conversion processes previously discussed, a fair amount of data has been published. Except for gasification of more coke with the attendant fuel gas production in the case of flexicoking, the yields are the same for both fluid coking and flexicoking and they have been combined in Table 11–1. Where both distillate and gas oil were reported separately, they were combined and designated as gas oil. Coke and gas are reported as weight percent of feed. All normally liquid materials are reported as volume percent of feed.

CCR WT%	COKE WT%	C1–C3 WT%	C4 LV%	GASO LV%	GAS OIL LV%
41.0	48.5	11.5	3.0	14.5	2.0
34.0		11.3	2.2	13.4	36.9
33.0	36.0	10.0	3.0	17.5	44.5
32.9	34.6	13.2	2.2	10.3	29.9
31.0	34.0	10.3	2.0	13.2	40.5
30.6	28.1	9.5	1.6	11.8	28.0
30.4	35.9	12.3	2.8	15.8	33.2
30.2	25.6	8.2	1.9	11.1	35.0
30.0	36.0	14.0	5.5	21.0	36.0
30.0	34.0	11.0	4.0	18.0	45.0
27.8				18.1	44.1
27.7		11.1		15.0	45.6
27.7				14.5	44.4
26.5		10.5	2.1	14.7	48.3
26.0	30.0	10.0	3.0	20.0	49.0
24.4	30.7				
24.2		10.5	3.8	21.1	42.5
24.0	26.5	9.5	1.9	12.5	49.6
24.0	27.5	9.5	3.5	19.5	52.0
23.2	19.0	7.0	2.5	15.0	36.5
23.2		11.1	3.8	21.3	44.3
22.8	27.8	9.8	3.6	21.1	46.8
22.5	26.0	8.0	2.2	19.3	55.0
22.3		8.1		19.5	49.3
22.2	27.8	9.5	3.5	20.7	48.3
22.0	28.7	11.6		21.2	45.6

Table 11–1a Fluid Coking/Flexicoking Yield Database

CCR WT%	COKE WT%	C1–C3 WT%	C4 LV%	GASO LV%	GAS OIL LV%
21.8	28.0	10.0	4.0	21.0	48.0
21.8		11.3	3.7	21.3	46.2
21.5	23.0	12.0	4.5	22.0	50.0
21.4		9.9	1.9	15.4	55.1
21.1	24.9	11.2		18.9	53.8
21.0	24.0	9.0	3.0	20.0	54.0
21.0	26.2	10.0		20.6	49.8
20.8		10.8	3.6	21.5	48.0
20.0		9.4		16.0	57.3
19.2	20.0	12.0	4.0	24.0	52.0
19.0	17.5	8.0	2.5	20.5	61.0
19.0	20.0	9.0	3.0	21.0	58.0
18.2	22.4	8.9	3.3	21.2	54.4
18.2	21.3	7.8		17.1	57.5
18.0		13.7		20.8	51.8
18.0		8.9	11.6	16.6	59.3
18.0			26.0	19.3	50.4
17.9		8.5		16.1	60.7
17.0	20.5	8.0	2.5	16.5	61.5
15.5	18.5	8.0	2.8	17.6	62.5
14.4				11.1	70.1
14.0	16.0	8.0	2.5	21.5	61.0
13.0	13.5	6.5	2.0	16.5	69.5
13.0				16.6	66.2
13.0	17.0	8.0	3.0	21.0	61.0
12.7				17.9	65.5
11.7	12.5	14.5		22.0	51.0
11.0	11.5	7.0	2.0	21.0	68.5
10.6				18.0	56.8
9.0	11.0	5.5	1.6	13.0	75.0
8.4	9.9	7.3		18.9	81.8
5.0	8.0	6.0	2.0	17.0	74.0

Table 11–1b Fluid Coking/Flexicoking Yield Database cont'd

Coke, gas, gas oil, and C_5 correlate well with CCR of feed. Dry gas (propane and lighter) also correlates with coke yield—butanes, with gasoline. To obtain a satisfactory direct correlation for gasoline, it was necessary to use two independent variables—CCR and gasoline end point. Figure 11–3 is a composite plot of regression lines for these products. Table 11–2 is a summary of some of the yield correlation results.

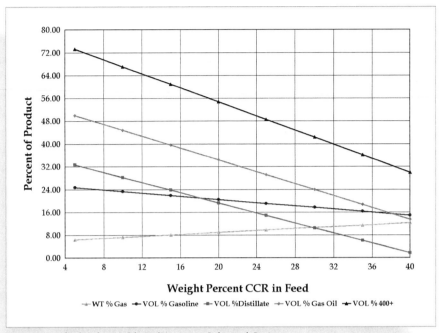

Fig. 11–3 Fluid Coker Yields (Vol% except Coke and Gas)

DEPENDENT VARIABLE	COKE	GAS	C4S	GASO	GAS OIL	400+	C5+	C5+	H₂S
INDEPENDENT VARIABLE	CCR	CCR	GASO	CCR EP	CCR	CCR	CCR GASO	CCR	FD S
R^2	0.982	0.8485	0.7701	0.8397	0.9265	0.922	0.9295	0.8227	0.915
SEE	1.35	0.5344	0.368	1.089	3.03	3.062	3.005	4.65	0.126₇
S1	0.0321	0.0126	0.0178	0.0288	0.0752	0.0775	0.0763	0.0957	0.02
S2				0.0057			0.1851		
NO. OF SETS	24	28	31	32	23	25	28	45	17

Table 11–2 Some Results of Fluid Coking/Flexicoking Yield Correlations

Figure 11–4 shows the relationship between dry gas (C_3 and lighter), C_2 and lighter, and C_4 hydrocarbons, all on a weight percent basis. The following average compositions may be used for these fractions:

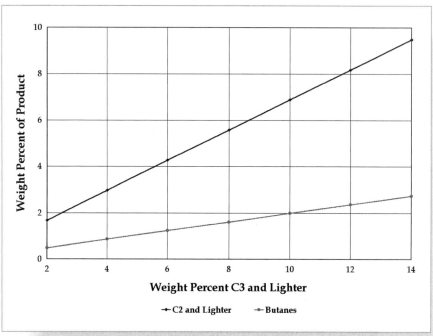

Fig. 11–4 Fluid Coker Light Ends

	C_2 and lighter	C_3's	C_4's
Hydrogen	2.4		
Methane	44.7		
Ethylene	20.0		
Ethane	32.9		
Propylene		56.0	
Propane		44.0	
Butenes			69.5
Isobutane			4.9
Normal Butane			25.6
Total, WT%	**100.0**	**100.0**	**100.0**

Figure 11–5 shows the volume percent yield of C_4's as a function of gasoline yield.

158

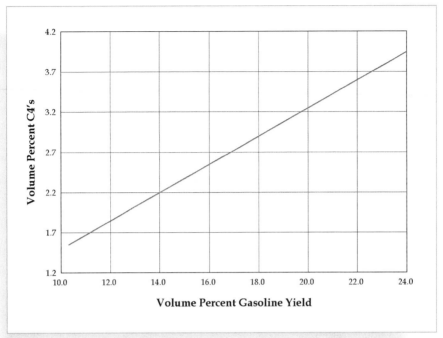

Fig. 11–5 Fluid Coker C$_4$ Yields

The distribution of sulfur in flexicoker products is displayed in Figure 11–6. Figure 11–7 gives a rough indication of the research octane to be expected of flexicoker gasoline in terms of CCR in the feed and the API gravity of the feed. It is based on very few data. It is followed by Figure 11–8 that gives the motor octane corresponding to a given research number.

Figure 11–9 shows the relationship between API gravity of gas oil in terms of feed API gravity and CCR in feed. Figure 11–10 is a plot of feed specific gravity vs. CCR in feed. Attempts to find a satisfactory relation for gasoline gravity were not successful. Figure 11–11 is a plot of volume percent vs. weight percent for flexicoker liquid products.

Comparison with other correlations

Though few correlations of fluid coking or flexicoking yields were found in the literature, some comparisons are possible. Martin and Wills[1] presented equations for yield of coke and of coke plus gas and a plot of these equations. Johnson and Wood[2] published a graph relating coke yield

159

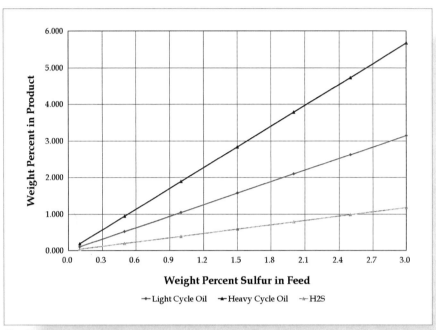

Fig. 11–6 Sulfur in Fluid Coker Products

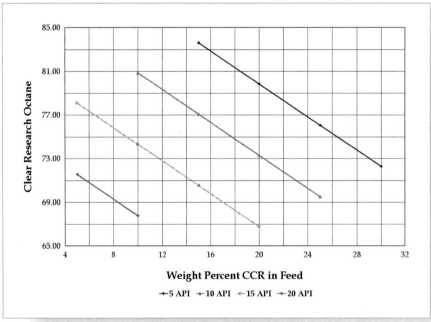

Fig. 11–7 Fluid Coker Research Octane (API of Feed is Parameter)

160

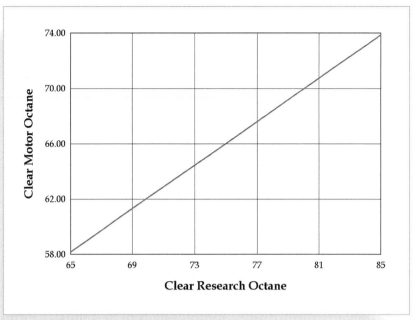

Fig. 11–8 Fluid Coker Motor Octane

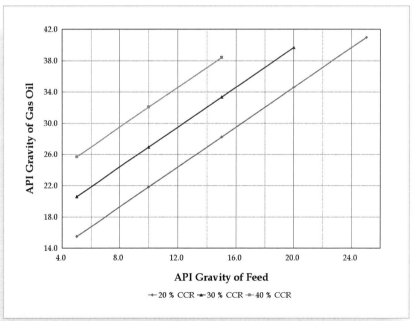

Fig. 11–9 API Gravity of Fluid Coker Gas Oil (CCR of Feed is Parameter)

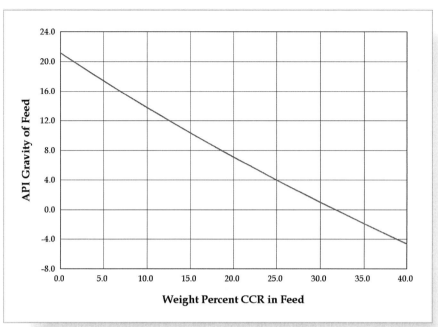

Fig. 11–10 CCR in Fluid Coker Feed

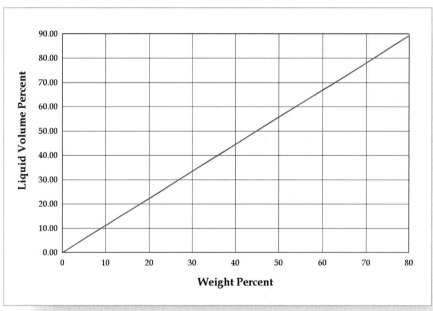

Fig. 11–11 Lv% vs. Wt%

162

to CCR of feed. Nelson[3] tabulated coke yield vs. CCR. These are compared with the author's results below:

CCR	M & W	J & W	Nelson	Author	M & W	Author
	Coke Yields				**Coke + Gas**	
5	5.8		3	6.5	11.5	12.7
10	11.5	11+	11.5	12.1	18	19.1
15	17.3		17	17.2	24.5	25.1
20	23	22.5	23	22.4	31	31.1
25	28.8		29	27.6	37.5	37.1
30	34.5	34	34.5	32.8	44	43

Allegedly the agreement is good. As additional data have been published since the earlier reports, the present work is based on a larger population of data. Supposedly the author's correlations can be relied on in the performance of the preliminary kinds of studies envisioned.

Flexicoking operating requirements

The following data have been listed for the operation of a flexicoker:

Electric power, kWh/b	13
Cooling water, gal/b	30
Low pressure steam, #/b	100
High pressure steam, #/b	< 200 >

It is assumed that the air blower drive is electric.

Flexicoker capital cost

Based on data in the literature and adjusting for size and time as before, an average cost of $46 million was calculated for a 20,000 BPD flexicoker starting operation in January, 1991.

163

Notes

1. Martin, S.W. and Wills, L.E., *Advances in Petroleum Chemistry and Refining*, vol. 2, Interscience Publishers, New York, 1959, pp. 390–421

2. Johnson, F.B., and Wood, R.C., *Oil & Gas Journal*, Nov. 29, 1954, pp. 60–62

3. Nelson, W.L., *Petroleum Refinery Engineering*, McGraw–Hill Book Co., New York, 4th ed., 1958, pp. 641–642

References

Aalund, L., *Oil & Gas Journal*, September 11, 1972

Allan, D.E., Metrailer, W.J., and King, R.C., *Chemical Engineering Progress*, December 1981, pp. 40–47

Allan, D.E., Blaser, D.E., and Lambert, M.M., *Oil & Gas Journal*, May 17, 1982, pp. 93–102

Anon., *Oil & Gas Journal*, May 22, 1978, pp. 76–77

Anon., *Hydrocarbon Processing*, September 1986, p. 96

Barr, F.T., and Jahrig, C.E., *Chemical Engineering Progress*, Vol. 51, No. 4, 1955, pp. 167–173

Blaser, D.E., "Flexicoking for Improved Utilization of Hydrocarbons," 43rd Midyear Meeting of the API Ref'g. Div., Toronto, May 1978

Busch, R.A., Kociscin, J.J., Schroeder, H.F., and Shah, G.N., Hydrocarbon Processing, Sept. 1979, pp. 136–142

Carlsmith, L.E., Haig, R.R., and Holt, P.H., *The Oil Forum*, March 1957, pp. 90–93

Jahnig. C.E., *Encyclopedia of Chemical Technology*, 3rd. ed., vol. 17, pp. 210–218

Johnson, F.B., and Wood, R.G., *Oil & Gas Journal*, November 29, 1954, pp. 60–62

Kett, T.K., Lahn, G.C., and Schuette, W.L., *Chemical Engineering*, December 23, 1974, pp. 40–41

Martin, S.W., and Wills, L.E., *Advances in Petroleum Chemistry and Refining*, vol. 2, Interscience Publishers, New York, 1959, pp. 390–421

Matula, J.P., Weinberg, H.N., and Weisman, W., *Oil & Gas Journal*, September 18,1972, pp. 67–71

McDonald, J., and Rhys, C.O., Jr., *Refining Engineer*, September 1959, pp. C–15 to C–17

Molstedt, B.V., and Moser, J.F., Jr., *Industrial and Engineering Chemistry*, Vol. 50, No. 1, 1958, pp. 21–26

Nelson, W.L., *Petroleum Refinery Engineering*, McGraw–Hill Book Co., New York, 4th Ed., l958, pp. 641–642

Voorhies, A., Jr., and Martin, H.Z., *Petroleum Engineer*, Ref. Ann. 1954, pp. C–3 to C–18

Wuithier, P., *Revue de L'Institut Francais du Petrole*, vol. 14, no. 2., pp. 1,164–1,165, 1,181–1,185

SECTION C:

HEAVY DISTILLATE PROCESSING

CHAPTER 12

FLUID CATALYTIC CRACKING

For 55 years catalytic cracking has been the workhorse of the petroleum refining industry, making small- and medium-sized molecules out of big ones (gasoline and distillate out of gas oils). In recent years, it has taken on bigger and bigger molecules.

The modern riser cracker bears little resemblance to the original fixed-bed catalytic cracking units. The changes have not been in hardware alone. The refiner today has a multitude (about 250 according to Reichle)[1] of catalysts from which to choose to meet his needs. The changes continue apace.

Recent changes have included:

- Hardware:
    Redesign feed nozzles to improve feed atomization
    Redesign stripper baffles to reduce coke production
    Install closed cyclones to reduce post-riser thermal cracking
    Inject feed radially to improve oil/catalyst mixing
    Two-step catalyst regeneration to lower coke on regenerated catalyst

- Catalyst/Additives:
    To reduce SO_x
    Propylene selective catalyst
    Passivators

- Process Changes:

    Naphtha cracking

    Deep catalytic cracking (DCC)

    Pretreating feed—hydrotreating, desalting

    Naphtha post-treatment

    Oxygen enrichment of air to regenerator

As pointed out in an earlier chapter, catalytic gasoline is the largest volume component in the current gasoline pool of the average refiner. It is also the largest source of sulfur in gasoline. With pressure on to reduce sulfur in gasoline, much effort is being expended developing means of accomplishing this—from pretreatment of FCC feed, to hydrotreatment of the produced naphtha.

The cat cracker is the major source of olefin feed for alkylation units. In fact, in studying the economics of an FCC operation, the alkylation unit is frequently included as part of the complex.[2]

The FCC unit converts about 40% of the sulfur in the feed to H_2S, which is easily removed.[3]

FCC provides a means of reducing the carbon-to-hydrogen ratio by depositing coke on the circulating catalyst. This coke is removed more or less completely in the regenerator.

FCC process description

Fresh feed plus recycle (if any) is introduced into a vertical pipe (riser reactor) where it is combined with and propelled by fluidized, regenerated catalyst into a large diameter vessel where catalyst (assisted by cyclones) is disengaged from the hydrocarbon vapors. The vapors are separated into gas, gasoline, distillates, and recycle in a fractionator.

The catalyst flows from the disengager to the regenerator where its activity is partially restored by the combustion of the coke deposited on it. Flue gases are separated from the catalyst by means of cyclones and are then utilized in waste heat recovery or power recovery. Figure 12–1 is a simplified process flow diagram (PFD).

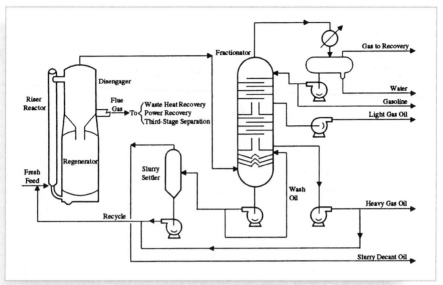

Fig. 12-1 Fluid Catalytic Cracking Unit

Since the FCC unit is the major source of olefins in the refinery, an unsaturates gas plant is generally considered a part of it. Here fuel gas (C_2 and lighter) is separated by means of an absorber-deethanizer. Figure 12–2 is a simplified PFD of an unsaturate gas plant.

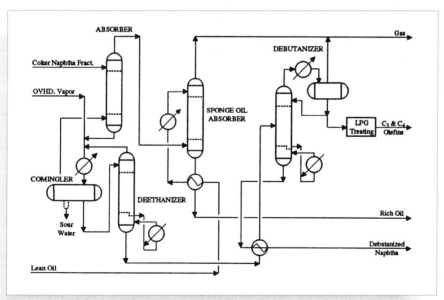

Fig. 12-2 Unsaturate Gas Plant

FCC data correlation

Because of its importance, it is not surprising that the catalytic cracking literature is quite voluminous. A sizable table of data has been compiled (Table 12–1).

As a first step, the data for each product were plotted against the corresponding conversion level. (Here conversion is defined as 100 minus the volume percent yield of liquid boiling higher than gasoline.) The plots for coke and gas appear to be made up of two different families. (This is particularly apparent in the case of gas.) In part, this reflects the progress in catalyst type from silica-alumina to zeolite. Consideration here is restricted largely to results with zeolite catalysts.

A fair correlation was obtained for coke in terms of conversion and API gravity of feed. Results for gas were less than satisfactory for all combinations tried. Propane was another product that did not respond readily. A part of the problem here is that it was not always clear in the literature when or how H_2S was accounted for, if at all.

Table 12–2 is a summary of some of the yield correlation results. Not all of the results are as good as would be desired, but they are satisfactory for their intended purpose—preliminary studies where a macroscopic view is taken. For a more detailed study—a microscopic view—one would conduct pilot plant studies and/or obtain information from licensors of the technologies of interest.

Figure 12–3 is a composite plot of yields, based on obtaining the yield of heavy cycle oil by difference (100 minus conversion minus yield of light cycle oil). Figure 12–4 is a plot of coke yield in terms of conversion and API gravity of feed. Figure 12–5 shows the yields of propane and propylene in terms of conversion. Figure 12–6 shows yields of the C_4 hydrocarbons in terms of conversion. Figure 12–7 shows the yield of gasoline vs. conversion.

Other figures (Fig. 12–8 through 12–16) show properties of products.

FD K	LV% CONV	FD API	WT% FD S	CCR	COKE	GAS	C4	GASO	LCO	LV% HCO	C3	C3=	I-C4	N-C4	C4=
	93.0	34.0	0.1		4.8	2.5	22.5	73.0	5.0	2.0	3.3	8.7	8.0	2.5	12.0
	93.0	29.6	0.4	0.1	9.7	13.7	24.4	61.6	1.1	5.9	3.2	5.2	12.4	3.8	8.2
12.2	90.6	28.5			6.5			70.6							
	90.6	28.5						70.6							
	90.0	27.5	0.5		8.5	4.9	32.2	55.7	7.0	3.0	5.6	12.0	14.0	3.2	15.2
	90.0	27.5	0.5		7.5	1.8	10.2	47.0	39.0	3.0	1.1	3.9	4.4	0.8	5.0
	90.0	27.5	0.5		7.6	3.0	20.5	71.1	7.0	3.0	2.6	7.0	8.1	1.4	11.0
12.1	89.8	28.5	0.4	0.5	8.2			71.5	4.6	5.6					
	89.6				5.7	3.0		70.2				9.5			9.5
	89.4							64.5				11.5			10.6
	88.9	25.9	0.7	0.3	5.4			57.8	23.0	4.7					
	88.3				7.5			70.3							
	88.3	25.9	0.7	0.3	5.6			57.6	19.9	6.1					
12.1	88.2	28.2			6.5			69.5							
	88.2	28.2						69.5							
	87.5							68.9				10.6			8.5
12.1	86.3	28.5	0.4	0.5	8.8			67.9	8.5	5.2					
	85.7							71.4				7.8			7.4
	85.5				7.1	5.9		54.5			3.2	14.9	7.1	1.6	11.4
	85.5	27.1	0.8	0.4	3.5	1.9	19.6	60.7		14.5	1.9	10.2	7.6	1.3	10.7
11.3	85.2	27.1	0.8	0.4	3.4	1.9	19.0	61.5		14.8	1.9	10.2	7.2	1.3	10.5
11.6	85.2	27.1	0.8	0.4	3.6	2.0	19.8	60.7		14.8	1.9	10.4	7.6	1.4	10.8
	85.0	27.0			5.4	2.8	19.0	70.0	10.0	5.0					
	85.0				7.2	4.3		62.5	10.0	5.0					
	85.0				5.3	2.8	16.4	69.5	10.0	5.0	3.0	7.0			7.6
	85.0	34.0	0.5				18.5	68.0							
11.9	85.0	27.1	0.8	0.4	3.3	2.0	19.8	59.9		15.0	1.9	10.6	7.5	1.5	10.8
	85.0	28.7	0.4		5.4	2.8	19.0	70.0	10.0	5.0	2.4	6.1	7.3	1.9	9.8
	84.8	27.1	0.8	0.4	3.4	1.9	19.4	60.6		15.7	1.8	10.6	7.4	1.2	10.8
	84.8	29.6	0.4	0.1	7.8	11.8	22.4	57.6	7.2	8.0	2.9	4.5	11.4	3.5	7.5
	84.7	27.1	0.8	0.4	3.2	1.9	19.3	61.5		15.3	1.8	10.2	7.1	1.3	10.9
	84.5	27.1	0.8	0.4	3.4	1.9	19.3	60.8		15.5	1.8	10.2	7.2	1.3	10.8
	84.2		0.1	2.9	6.8			62.5	13.8						
	84.0						16.5	67.3	12.8	3.2	2.8	7.3			8.2
	84.0				6.4	2.4		62.8	13.8	2.2					
	84.0	22.5	2.5		7.2			54.8	10.0	6.0					
12.2	84.0	22.5	2.5		7.2	6.4		54.8	10.0	6.0					
	83.7				4.9	2.4		68.3	12.7	3.6	3.0	6.4	7.6	2.7	7.7
	83.6				6.5	2.3		61.6	14.4	2.0					
12.0	83.5	26.2	0.7	1.0	5.0			66.7							
12.2	83.5	27.4	0.6	0.2	7.2			64.1	9.7	6.8					
12.2	83.5	27.4	0.3	0.6	3.5	2.0	15.0	68.0		16.5					
	83.3				5.4			70.0							
12.0	83.0	27.4	0.6	0.2	8.7			59.0	12.1	4.9					
	82.9	27.4	0.6	0.2	7.5			61.0	10.3	7.8					
	82.7							53.2				11.2			11.5
	82.5				4.9			65.1	12.2	10.3					
	82.5							63.0	13.0	4.5	3.5	8.5	7.8	2.1	9.6
	82.5						19.0	64.5	14.0	3.5	3.0	8.0	7.5	2.0	9.5
	82.5	26.9			4.9	2.4		70.1	13.5	4.0	1.9	7.8	7.4	1.7	6.3
	82.2				8.6			47.0							
	82.0				5.1		12.2	69.5							
	82.0				6.7	3.5		59.5			2.3	11.8	6.8	1.3	7.9
	82.0				8.0	7.2		48.6			4.3	11.4	6.3	2.1	12.5

Table 12–1a Fluid Catalytic Cracking Database

LV%		WT%									LV%				
FD K	CONV	FD API	FD S	CCR	COKE	GAS	C4	GASO	LCO	HCO	C3	C3=	I-C4	N-C4	C4=
12.0	81.9				7.5			61.0	10.3	7.8					
	81.9	26.9			4.8	2.6	15.4	68.6	13.6	4.5	3.4	7.3	8.3	2.1	5.0
	81.8	26.9			4.5	2.2		70.4	16.2	2.0	2.8	6.7	8.0	2.2	5.2
	81.7				8.0			48.6							
12.0	81.5	26.9			4.9	2.3	14.8	69.7	14.3	4.3	2.6	6.5	7.3	1.7	5.9
12.0	81.3	28.8	0.4	0.3	5.8			60.3	15.8	2.9					
	81.0				4.3	1.2		64.3	13.9	5.2					
	80.7				3.4			60.2	12.3	12.1					
	80.7				4.7	1.5		68.6				6.5			7.8
	80.6				2.6	1.9	17.6	66.4	14.2	5.2	2.0	8.1	6.1	1.7	9.8
11.4	80.3				5.2			62.1	15.6	4.1	2.2	7.5	9.9	2.8	8.4
11.4	80.3	28.4	0.2		4.5	2.4	15.5	66.3	14.7	5.0	2.6	6.2	5.5	1.8	8.2
12.0	80.3				5.1			64.7	15.7	4.0	2.0	6.3	8.7	2.7	7.5
	80.2				9.1	6.8		59.1	12.1	7.9					
	80.2				4.9	1.9		63.0				6.5			5.7
	80.1							56.4				9.5			10.2
	80.0	23.3	0.3		5.7		18.1	63.0							
	80.0	27.0	1.0				17.0	65.0							
	80.0						16.1	58.5	15.0	5.0	4.5	8.0			6.0
12.0	80.0	26.7	0.2		2.4		17.6	63.8	15.0	5.0	3.1	7.2	6.6	2.2	8.8
12.1	80.0	23.8	0.3		6.7	3.1	17.0	63.8	16.0	4.0	3.0	7.1	6.3	2.1	8.6
12.1	79.7		0.3	3.6	8.1			57.9	13.2	2.0					
	79.5				6.4	2.3		62.0			1.7	10.0	7.1	1.3	5.8
	79.5						20.1	55.0	14.2	6.3	4.8	8.6	8.4	2.3	9.4
	79.4	28.4			4.8	1.2	15.1	69.1	13.7	6.9	2.2	4.6	8.1	1.9	5.1
12.0	79.4	28.8	0.4	0.3	5.8			56.6	18.0	2.6					
11.8	79.3	22.0	2.2					55.5							
	79.3	25.9	0.8				18.5	64.2	13.0	6.0					10.6
	79.0				5.3			55.4	18.7	2.2	3.9	10.2	7.9	2.4	9.2
	78.9				8.0		116.1	55.2	14.8	6.3					
11.8	78.9	22.0	2.2					55.2							
11.8	78.9	23.7	2.5					57.6							
	78.8				8.4	6.6		61.8	13.6	7.6					
11.8	78.8	24.6			7.9			59.6							
	78.8	24.6						59.6							
12.2	78.7	28.1	0.3	0.3	5.5			59.6	17.0	4.3					
11.8	78.6	25.9	0.7	0.3	6.3			60.3	16.9	4.5					
	78.5				4.7	2.1		65.7	16.4	6.0					
	78.2				4.2			67.1							
12.0	78.2	28.8	0.4	0.3	5.8			57.6	19.6	2.2					
	78.2	30.2	1.2		5.5	10.4	17.8	55.8	8.7	13.1	4.8	5.2	8.5	2.5	6.8
	78.1				3.1			62.3	13.5	8.4	3.6	7.7	7.4	2.5	7.7
	78.1				5.7	4.5		53.3	20.9	1.0	3.9	10.4	7.2	2.2	9.8
	78.0	22.5			6.2	2.7	16.4	62.8	17.0	5.0	2.8	6.7	5.9	2.3	8.2
	78.0	22.5	2.5		5.5			59.0	15.0	7.0					
	78.0	22.5	2.5		5.5	4.7		59.0	15.0	7.0					
	77.9				4.2			60.8							
	77.6						14.4	57.3	15.9	6.5	2.5				7.9
	77.6	23.8	0.5	0.5	6.1	1.7		55.4	17.6	4.8	2.4	11.0	6.6	1.4	11.7
	77.6	23.8	0.5	0.5	5.9	1.8		59.5	17.9	4.5	2.1	8.6	6.3	1.3	9.8
	77.6	23.8	0.5	0.5	6.2	1.6		52.0	17.1	5.3	2.4	11.9	6.9	1.5	14.7
	77.6	23.8	0.5	0.5	6.1	1.6		53.2	17.4	5.0	2.4	11.6	6.7	1.5	13.3
	77.5				4.5	1.1		67.1				5.9			6.8
	77.3				6.3			54.4							

Table 12-1b Fluid Catalytic Cracking Database cont'd

	LV%				WT%						LV%				
FD K	CONV	FD API	FD S	CCR	COKE	GAS	C4	GASO	LCO	HCO	C3	C3=	I-C4	N-C4	C4=
	77.3	26.9			3.1	1.9	13.4	68.3	16.6	6.1	1.7	7.1	6.3	1.4	5.8
	77.2						14.8	58.7	17.8	5.0	2.4				8.9
	77.0				4.7		11.1	66.0	15.7						
	77.0	27.7			3.8	1.4	10.4	61.0	11.0	12.0	2.0	6.4	4.9	1.0	4.5
	77.0	27.7			3.3	1.3	8.8	62.5	10.3	12.7	1.4	6.9	4.1	0.5	4.2
	77.0	27.7			3.8	1.8	9.0	69.5	14.3	8.7	1.1	7.6	4.6	0.6	3.8
	77.0	27.7			4.0	1.5	10.1	60.5	11.0	7.1	1.8	6.6	5.0	0.9	4.2
	77.0	27.7			3.8	2.0	11.2	65.0	10.0	13.0	1.2	10.8	5.8	0.6	4.8
	77.0	27.7			3.8	1.6	10.0	60.5	10.5	12.5	1.9	6.8	5.0	0.8	4.2
11.8	76.5	23.7						55.4							
	76.5				3.5	1.9		65.8	17.7	5.8					
	76.5	28.0	0.6		4.9	2.5		63.8	19.4	4.1					
	76.1				5.0	3.1		58.2	15.0	8.9	3.7	6.8	6.6	2.1	8.2
	76.0	24.0	1.8	0.3	5.2	6.0		55.6	14.5	9.5					
	76.0				3.1		21.5	57.0	15.0	9.0			7.2		13.6
	76.0	22.6	0.8	3.8	3.1		21.5	57.0	15.0						
	75.9	24.6	0.3		5.0		17.7	60.8							
	75.8	24.9	0.1	0.1	5.0	3.4		63.3	18.3	5.9					
	75.7				12.4			51.2							
	75.5				5.3		9.9	64.5	17.1						
	75.5				3.0			60.4	17.7	6.8	3.0	6.7	5.9	2.0	7.4
	75.5	20.6			5.8	2.9	16.2	61.3	19.0	5.5	2.8	6.5	5.8	2.3	8.1
	75.0				3.9	1.5		62.5	19.5	5.5	1.2	9.0	6.0	1.2	3.2
	75.0				6.5	2.8		58.5	20.0	5.0					
	75.0	24.8	2.2	0.2	7.2			57.2	14.5	10.5	2.4	8.0	6.3	1.2	8.9
	75.0	24.8	2.2	0.2	6.4			56.1	14.5	10.5	2.7	8.9	5.5	1.3	9.6
	75.0	23.8	0.3		5.5	2.5	16.2	60.8	20.0	5.0	2.8	6.7	6.0	2.0	8.2
11.8	75.0	19.4	0.3		6.7	2.9	16.5	60.7	20.0	5.0	2.9	5.5	6.2	2.1	8.2
11.8	75.0	24.8	2.2	0.2	8.7			52.4	14.5	10.5	2.6	9.1	5.5	1.3	9.9
12.0	75.0	24.8	2.2	0.2	9.9			53.3	14.5	10.5	2.4	8.2	6.3	1.2	9.3
12.0	75.0	35.3			2.5	9.8	22.0	52.0	8.0	17.0		10.2	7.5	1.8	12.7
	75.0				3.0	1.6		67.0	10.5	14.5	1.2	7.3	5.0	0.8	3.5
	74.9	25.2	0.2	0.1	5.0	3.6		61.2	18.2	6.9					
12.0	74.9	26.9			4.0	1.9	11.6	66.6	18.1	6.9	1.5	6.1	5.3	1.1	5.2
	74.7	23.4	2.3	0.3	5.3	5.9		55.3	13.3	12.0					
	74.6				3.9	1.3	12.9	64.7			5.9	4.7	5.0	6.9	
	74.1				3.9	1.4	13.6	63.4			7.1	5.9	5.1	7.6	
	74.0				4.4	1.1	10.5	64.5	8.5						
11.8	74.0	25.9	0.7		5.6			57.6	19.9	6.1					
	74.0				4.9	1.1		64.5	26.0				5.7	0.9	3.9
12.2	74.0	28.1	0.3	0.3	5.4			59.2	22.7	3.3					
	73.7				4.5	1.8		59.5	19.0	7.3	3.1	5.4	8.0	4.7	7.4
12.1	73.6	29.6	0.4	0.1	5.9	9.8	19.4	51.6	15.4	11.0	2.5	3.8	9.9	3.0	6.5
	73.4				4.5			59.5	19.0	7.3	3.1	5.4	8.1	4.7	7.4
12.0	73.4	23.1	1.1		5.4	8.9	13.4	56.9	12.9	13.7	3.8	4.1	6.5	2.1	4.8
12.0	73.2				6.7	2.6		51.6	26.5	0.3	2.3	7.0	5.4	1.4	9.0
12.0	73.2				6.7	2.4		53.9	26.5	0.3	2.3	6.2	5.1	1.4	7.9
12.0	73.0	19.1					15.7	59.8	21.0	6.0	2.7	6.3	5.6	2.3	7.8
	73.0				5.2	1.5	9.7	61.0			2.4	6.4	4.6	1.1	4.0
	73.0				2.8	1.6		65.0	11.5	15.5	1.2	7.2	4.6	0.7	3.2
12.0	72.8	26.9			2.2	1.7		64.5	21.9	5.3	1.7	6.2	6.1	1.4	6.8
	72.8	26.1	0.9		4.3	2.3	17.4	58.3	22.2	5.0	2.1	6.1	5.9	1.6	9.9
11.8	72.8	24.9	0.6	2.2	5.2	2.7		58.9			2.7	7.0	4.8	1.2	7.3
	72.7	24.9	0.6	2.2	5.2	2.7	13.3	58.9	21.0	6.2	2.7	7.0	4.8	1.2	7.3

Table 12–1c Fluid Catalytic Cracking Database cont'd

FD K	LV% CONV	FD API	FD S	CCR	COKE	GAS	C4	GASO	LCO	HCO	C3	C3=	I-C4	N-C4	C4=
	72.7	30.0			4.7			60.3	17.0	10.3					
	72.5				4.2	1.0	10.1	63.0	9.8	17.7	1.0	6.0	4.6	0.4	
12.2	72.5	27.0	0.3		4.6	1.5	15.8	60.0	8.9		1.8	6.9	7.8		
11.8	72.3	25.9	0.7		5.4			57.8	23.0	4.7					
	72.2						12.6	55.3							
	72.0				3.1			59.7	19.9	8.1	4.5	7.4	6.3	3.0	7.5
	72.0				3.1	1.9		59.6	19.9	8.1	4.5	7.4	6.3	3.0	7.5
	72.0				4.4	1.2		63.1	28.0		1.5	5.9	5.3	0.9	3.6
12.1	72.0	28.0	0.6		5.2	2.2	13.3	59.3	24.3	3.7	2.4	5.4	5.3	1.4	6.6
	71.4				4.9			59.7	20.7	12.8	4.2	6.9	6.3	1.5	4.4
	71.4				6.2	2.5		58.3							
	71.0				5.2	1.1	9.1	61.0	10.2	18.8	1.8	5.5	4.4	3.8	
	71.0				2.6			61.3					3.0	0.3	5.6
11.8	71.0	27.0	0.3		3.6	1.2	11.8	61.5	9.3		1.7	5.7	6.0		
	70.5				3.7	1.5		59.0	26.0	3.5	0.7	8.6	5.6	0.9	4.0
	70.5				2.9			63.5							
	70.0	20.0			6.3	3.0	16.8	54.2	20.0	10.0					
	70.0	20.5	0.3				13.7	61.0							
	70.0	14.3	3.6				14.2	58.5							
	70.0	20.0	1.8				14.8	54.2							
	70.4				9.6			52.6							
	70.2				4.7	3.3		57.6							
11.5	70.0	22.0	0.1	0.0	4.7	2.3	12.8	58.2	19.4	10.6	1.9	7.6	4.8	1.0	7.0
11.5	70.0	22.0	0.1	0.0	6.0	2.5	13.3	55.5	16.8	13.2	2.3	7.3	6.0	1.3	6.0
11.5	70.0	22.0	0.1	0.0	6.0	2.5	13.4	56.3	16.8	13.2	2.5	7.0	6.3	1.4	5.7
11.5	70.0	22.0	0.1	0.0	5.5		12.5	57.7	17.8	12.2	2.2	6.8	5.5	1.2	5.8
11.5	70.0	22.0	0.1	0.0	4.1		11.5	60.4	19.2	10.8	1.6	6.7	4.4	0.9	6.2
11.5	70.0	22.0	0.1	0.0	4.7		12.0	59.0	19.4	10.6	1.8	7.2	4.5	0.9	6.6
11.5	70.0	22.0	0.1	0.0	4.1		12.2	59.0	19.2	10.8	2.0	6.9	4.2	1.0	7.0
11.5	70.0	28.4	0.2		3.6	1.8	13.5	58.2	25.0	5.0	2.4	5.6	4.7	1.6	7.2
12.1	70.0	20.0	0.9		6.3	3.0	16.8	54.2	20.0	10.0	2.0	5.5	5.9	1.4	9.5
12.1	70.0	27.4	0.3	0.6	4.0	2.5	10.9	58.0	8.5	21.5	1.9	10.0	3.9	0.5	6.5
12.2	69.8	28.1	0.3	0.3	5.2			55.9	26.4	3.8					
	69.5				4.1	1.1		60.2	30.5		1.4	5.7	5.0	0.8	3.4
	69.3				7.4			53.1							
12.1	69.3	22.5	3.0		5.8	2.3	12.7	56.5	25.7	5.0	2.2	5.0	4.9	1.6	6.2
	69.1	27.6	0.1		4.8										
	69.0				2.7			60.5					3.0	0.4	6.0
12.1	69.0	26.9			2.8	1.6		61.1	21.5	9.5	1.0	6.4	4.7	0.9	7.0
12.1	68.9	31.6		0.0	2.0	0.9		53.1	26.6						
	68.7	24.2		0.6	5.8	2.1		46.4	17.3						
	68.5				5.5	1.1		58.5			1.4	5.3	6.0	0.9	4.1
	68.0	27.0	0.7		3.9	2.3		57.6	18.0	14.0	2.0	6.8	4.2	1.1	6.1
	68.0	27.0	0.7		3.8	2.4		56.8	17.9	14.1	1.9	7.4	4.0	0.9	6.9
	68.0	27.0	0.7		3.6	2.8		56.8	17.6	14.4	2.0	7.2	3.9	0.9	7.1
	68.0	27.0	0.7		4.2	2.8		56.2	19.9	12.1	2.1	7.0	3.4	0.9	7.9
	68.0	19.4	0.3		6.1	2.5	14.5	55.6	27.0	5.0	2.6	5.9	5.5	1.8	7.2
	68.0				2.3	1.6		61.5	14.0	18.0	0.9	7.0	3.6	0.6	2.9
11.5	67.9	22.2			7.5			51.4							
11.8	67.9	23.7	2.5					51.6							
	67.9	22.2						51.4							
	67.4	30.7	0.5		3.7	9.8	16.0	52.9	24.9	7.7	3.2	4.8	8.2	2.2	5.5
	67.2	22.6	0.8	3.8	13.6	5.5		49.5	21.3		1.3	3.3	0.3	0.8	5.3
	67.0				3.8	2.9		53.8	27.6	5.4			5.3	1.5	6.4

Table 12–1d Fluid Catalytic Cracking Database cont'd

LV%		WT%									LV%				
FD K	CONV	FD API	FD S	CCR	COKE	GAS	C4	GASO	LCO	HCO	C3	C3=	I-C4	N-C4	C4=
	67.0				4.4	2.9		53.1	20.7	5.3			4.6	1.3	7.7
	67.0				1.8	9.1	20.6	50.0	9.0	26.0		9.5	7.0	1.6	12.0
	67.0	24.5	0.4	0.2	7.0	11.1	9.7	54.0	19.0	14.0		4.6	1.7	1.3	6.7
	66.9	30.0			4.1			59.8	22.5	10.6					
	66.8				4.6			59.6	18.5	19.3	3.5	4.5	5.3	1.1	2.9
11.8	66.6	24.9	0.6	2.2	4.9	3.0	12.2	50.5	21.7	11.7	2.9	6.6	4.7	1.4	6.2
	66.5	26.8	0.6	0.2	6.4	7.7	16.5	47.9	15.5	18.0	2.3	2.7	9.0	2.6	4.9
	66.2	22.0	0.1	0.0	4.1	2.4	11.3	56.4	21.2	12.6	1.7	6.4	4.0	0.9	6.4
	66.0	27.4	0.3	0.6	3.5	2.0	8.0	57.0		34.0					
	66.0	31.4	0.5	0.7	5.3	15.4	15.1	46.5	31.5	2.5	3.1	8.6	5.9	2.0	7.2
	65.2						15.5	51.3							
	65.2	29.6	0.4	0.1	5.1	8.3	17.1	46.6	27.8	7.0	2.9	3.3	8.8	2.6	5.7
	65.0				2.4	1.3		58.7			0.7	3.2	4.0	0.7	6.0
	65.0				2.7	1.4		57.5			0.6	3.7	3.4	0.6	8.6
	65.0	22.5	1.2	0.4	2.8			58.2					3.0	0.4	7.0
	65.0				3.3		19.0	46.5	17.5	17.5			5.2		13.6
	65.0				1.5	1.4		57.2	17.9	18.9	1.1	5.5	5.2	0.9	7.8
	65.0				1.7	1.3		56.8	18.3	18.5	1.0	5.8	4.9	0.8	8.5
	65.0				1.6	1.4		57.4	18.2	18.6	1.0	5.5	4.8	0.8	7.8
	65.0	24.8	2.2	0.2	5.4			51.9	13.1	21.9	1.8	5.5	6.3	1.1	7.8
	65.0	24.8	2.2	0.2	4.7			51.6	16.2	18.8	1.8	6.8	3.9	1.0	8.3
	65.0	24.8	2.2	0.2	6.5			48.9	16.2	18.8	1.7	6.9	3.9	1.0	8.6
	65.0	24.8	2.2	0.2	6.5			45.7	13.1	21.9	2.6	7.9	5.4	1.2	9.5
	65.0	24.8	2.2	0.2	7.3			49.5	16.2	18.8	1.6	6.3	4.6	0.9	8.2
	65.0	24.8	2.2	0.2	4.7			51.3	13.1	21.9	2.0	6.1	5.4	1.2	8.4
	65.0	23.8	2.5		5.0	2.1	13.7	51.6	22.0	13.0	2.4	5.6	5.3	1.6	6.8
	65.0	24.8	2.2	0.2	7.3			46.5	13.1	21.9	2.3	7.1	6.3	1.1	8.8
	65.0	24.8	2.2	0.2	5.4			52.3	16.2	18.8	1.6	6.2	4.6	0.9	7.9
	65.0	23.2			3.3		19.0	46.5	17.5						
	64.9	28.2			6.1	7.1	13.9	49.5	26.1	9.0	1.5	2.8	7.2	2.3	4.4
	64.9	23.3	1.3		5.0	1.9	12.3	53.5	30.1	5.0	1.1	5.2	4.1	0.7	7.5
	64.8				4.4		9.6	50.4							
	64.7	24.6			4.7	2.0	13.4	51.9	30.3	5.0	2.3	5.5	5.3	1.3	6.8
	64.7	26.1		0.3	4.6	1.4		48.2	20.7						
	64.2	24.6			4.7	1.8	13.5	51.5	30.8	5.0	2.3	5.5	5.4	1.3	6.8
	64.0				2.9	1.6		52.0	32.0	4.0	1.2	8.5	5.9	0.4	4.0
	64.0	27.4	0.3	0.6	4.0	3.0	14.7	48.0	9.7	26.3	3.2	9.1	5.6	1.0	8.1
	63.2						15.4	49.8							
11.8	63.0	23.7	2.5					45.1							
	63.0	31.4	0.5	0.7	3.9	9.1	15.8	47.6	27.1	9.9	3.1	4.3	9.2	1.8	4.8
	62.7	31.4	0.5	0.7	4.1	11.3	15.1	47.0	35.0	2.3	3.4	5.5	7.6	2.1	5.4
	62.6	22.0	0.1	0.0	4.1	2.3	10.5	53.0	21.2	16.2	1.7	5.7	4.0	1.0	5.5
	62.5	25.1	1.6		4.2	1.3	13.0	50.7	32.5	5.0	2.2	4.9	5.4	1.2	6.4
	62.3						12.3	52.0							
	62.0				3.3	1.0		53.5	38.0		1.1	4.6	4.7	0.8	3.2
	61.6	22.0	0.1	0.0	4.1	2.3	11.6	51.1	20.6	17.8	1.8	5.7	4.6	1.1	5.9
	61.2	30.2	1.2		3.4	8.0	13.7	45.0	25.0	13.8	3.2	5.0	5.5	1.7	6.5
	61.0	23.7	1.6		4.8	1.9	14.4	48.3	36.0	3.0	2.8	5.2	6.4	1.5	6.5
	60.6	26.3	0.5		5.2	2.3	14.5	46.1	36.4	3.0	2.3	5.5	5.9	1.3	7.4
	60.0	22.5	2.6	0.3	2.2	1.8	8.6	42.4	23.8	16.2	0.7	4.3	2.4	0.5	5.6
	60.0	22.5	2.6	0.3	2.2	2.0	11.8	34.9	23.3	16.7	0.9	8.2	3.1	0.6	8.1
	60.0	22.5	2.6	0.3	1.9	1.8	12.2	37.1	18.5	21.5	0.9	6.1	2.5	0.6	9.1
	60.0				6.5	1.3		45.9			1.4	5.7	5.2	1.0	5.4
	60.0				7.2	1.5		40.2			2.0	8.7	6.9	1.3	7.0

Table 12–1e Fluid Catalytic Cracking Database cont'd

FD K	LV% CONV	FD API	FD S	CCR	WT% COKE	GAS	C4	GASO	LCO	HCO	LV% C3	C3=	I-C4	N-C4	C4=
	60.0				7.9	1.7		33.8			2.3	11.2	7.8	1.4	10.1
	60.0				1.8	1.4		55.5	15.0	25.0	0.6	5.5	2.8	0.4	2.7
	60.0	29.0	0.6		2.9		12.5	48.8							
	60.0				2.3		13.1	47.5							
	60.0				5.9		11.9	43.2							
	60.0				5.4	2.4	13.1	47.1	35.0	5.0	2.0	6.0			7.9
	60.0	30.5			2.9	0.8	8.1	53.0				4.0	3.9		3.6
	60.0	30.5			2.3	0.7	9.1	53.5				4.4	3.9		4.6
	60.0	28.7			4.4	1.1	8.5	54.0				4.9	4.0		3.9
	60.0	26.2	0.7		3.7	1.1	7.7	53.5				3.8	4.0		3.0
	60.0	26.2	0.7		3.5	1.0	8.3	54.0				4.9	4.0		3.6
	60.0	28.9			4.8	8.9	16.8	42.3	19.9	19.1	2.1	3.9	8.7	2.5	5.6
	60.0	28.7			4.8	1.1	7.7	53.5				4.0	4.0		3.0
	59.4	26.8	0.6	0.2	5.1	6.7	14.5	44.0	16.1	24.4	3.0	2.5	7.7	2.2	4.6
	59.2	30.0			3.6			48.8	29.8	11.0					
	59.1	30.7	0.8	0.3	3.9	7.3	16.7	41.9	23.9	17.0	2.1	3.1	9.5	2.2	5.0
	59.0				2.0		13.6	49.2	27.0	14.0			4.9		7.9
	59.0	23.2			2.0		13.6	49.2	27.0						
	59.0	29.3	0.3		3.0	6.2	12.1	46.0	17.0	24.0		5.2	3.9	1.0	7.2
	59.0	21.4	0.5		6.0	5.6	7.6	47.8	21.4	19.6		3.5	2.5	1.1	4.0
	58.9	23.4	2.2		9.1	4.0		38.1	35.8	5.3					
	58.8	27.8	1.4		4.3	4.2	9.4	42.2	37.4	3.8	4.2	3.3	4.2	1.4	3.8
	58.4	32.5			3.5	5.6	11.1	46.0	32.4	9.2	1.4	2.0	4.5	2.0	4.6
	58.0	25.1	1.6		4.7	1.4	11.9	46.7	37.0	5.0	2.1	4.7	4.8	1.1	6.0
	58.0	29.8	1.4		4.5	4.2	8.2	41.6	38.6	3.4	4.8	3.5	3.4	1.1	3.7
	57.6						13.8	45.7							
	57.0	12.9	0.8	2.5	10.0	5.8	8.9	41.0	17.0	26.0		3.0	2.3	1.2	5.4
	56.7	30.7	0.5		3.9	9.2	10.7	44.5	30.5	12.8	3.1	4.3	3.2	2.5	5.0
	56.0				5.0	5.4	7.0	49.0	23.0	19.0		3.9	1.0	0.8	5.2
	56.0	29.6	0.4	0.1	3.9	8.2	14.4	39.1	25.1	18.9	2.1	3.1	7.1	1.9	5.4
	55.9	28.5	0.4	0.2	3.9	6.3	14.4	40.8	25.4	18.7	2.0	2.4	8.0	2.2	4.2
	55.7	25.1	0.34	0.2	4.0	8.3	10.6	44.8	33.9	10.4	2.7	3.8	5.9	1.4	3.3
	55.5	29.6	0.4	0.1	3.9	6.8	14.3	40.5	26.4	18.1	1.9	2.6	7.5	2.1	4.7
	55.5	21.4	0.2	0.1	3.7	6.3	13.5	43.0	37.9	6.6	2.1	2.2	8.9	2.6	2.0
	55.5	23.5	0.6	3.8	10.8	3.8		41.8	18.0		1.3	2.5	0.8	1.1	5.1
	55.4				6.4		12.0	39.5	37.7	6.9					
	55.1	29.6	0.4	0.1	3.9	6.0	13.8	41.3	26.4	18.5	1.8	2.3	7.8	2.0	4.1
	55.0	24.8	2.2	0.2	3.5			45.8	17.5	27.5	1.1	5.2	2.6	0.8	6.9
	55.0	24.8	2.2	0.2	3.5			45.5	14.5	30.5	1.3	4.7	3.7	1.0	7.1
	55.0	24.8	2.2	0.2	5.4			44.2	17.5	27.5	1.0	4.8	3.1	0.7	6.8
	55.0	24.8	2.2	0.2	4.0			45.9	14.5	30.5	1.2	4.2	4.2	0.9	6.7
	55.0	24.8	2.2	0.2	4.8			43.8	17.5	27.5	1.1	5.3	2.7	0.8	7.1
	55.0	24.8	2.2	0.2	4.0			46.3	17.5	27.5	1.0	4.7	3.1	0.7	6.6
	55.0	24.8	2.2	0.2	4.8			41.4	14.5	30.5	1.7	6.0	3.7	1.0	7.9
	55.0	24.8	2.2	0.2	5.4			42.0	14.5	30.5	1.6	5.4	4.2	0.9	7.5
	54.8				5.2		6.7	45.7							
	54.8				4.0			37.6							
	54.5	31.2			3.1	5.5	13.3	40.9	30.5	15.0	1.5	2.3	6.7	1.8	4.8
	54.5	25.1	0.3	0.2	2.9	8.0	11.4	43.5	34.7	10.8	2.8	3.6	6.1	1.5	3.8
	53.7						11.1	44.1							
12.1	53.6	28.8	0.6		2.3		10.7	43.6							
	53.5	23.4	0.9		4.8	1.3	9.9	44.8	41.5	5.0	1.8	3.9	3.9	1.2	4.8
	53.0				1.5	1.1		50.0	17.0	30.0	0.5	4.2	2.2	0.3	2.4
	52.3	30.5	1.3	0.1	3.9	6.7	13.2	37.2	38.0	9.7	1.8	2.3	6.7	2.0	4.5

Table 12–1f Fluid Catalytic Cracking Database cont'd

178

	LV%			WT%					LV%						
FD K	CONV	FD API	FD S	CCR	COKE	GAS	C4	GASO	LCO	HCO	C3	C3=	I-C4	N-C4	C4=
11.1	52.1	24.4			7.3			41.5							
12.0	51.8	26.6	0.5		4.4	1.3	12.0	42.3	45.3	3.0	2.1	5.0	4.4	1.5	6.1
12.0	51.6	26.8	0.6	0.2	3.8	6.0	12.5	38.8	17.5	30.9	1.7	2.4	6.4	1.8	4.3
	51.3	22.4	0.8		7.5										
12.0	51.3	30.7			2.0	4.9	11.0	39.7	41.2	7.5	1.2	2.2	5.8	1.2	4.0
	50.6	22.9	0.6	4.2	11.7	3.1		39.3	20.4		0.6	1.9	0.5	0.4	4.1
	49.9	26.2	2.3	0.3	3.8	6.9	13.6	35.2	31.8	18.3	1.6	2.7	6.8	1.8	5.0
	49.5	23.1	1.1		3.4	6.6	8.5	38.3	29.2	21.3	2.4	3.7	2.9	1.1	4.5
	49.0	18.0	0.5		5.5	5.6	9.3	36.9	23.0	28.0		1.7	3.1	1.0	5.2
	48.6	28.8	1.5		3.6	4.0	6.6	34.0	42.1	9.3	4.5	2.5	1.8	0.8	4.0
	48.4	23.4	2.2		7.2	2.9		33.2	23.8	27.8					
	48.0				4.9	4.3	5.3	41.7	23.7	28.3		2.8	1.7	0.7	2.9
	48.0	28.0	0.2		3.2	5.9	10.9	34.1	33.1	18.9		4.8	4.7	1.0	5.2
	48.0				2.0	4.7	8.8	39.0	20.0	32.0		3.9	3.1	0.7	5.0
	47.0				2.1		11.9	37.3	26.0	27.0			4.3		7.4
	47.0	23.2			2.1		11.9	37.3	14.0						
	46.0				7.8	4.0	7.1	35.0	20.0	34.0		2.0	1.9	1.0	4.2
	46.0	22.5	2.5		5.1			32.0	46.0	8.0					
	46.0	22.5	2.5		5.1	3.6		32.0	46.0	8.0					
	45.2						11.1	35.1							
	45.2	25.1	0.4	0.2	3.6	8.1	9.5	33.8	37.9	16.9	2.3	3.6	4.2	1.1	4.2
	45.0	24.8	2.2	0.2	3.0			39.1	15.0	40.0	0.8	3.2	2.6	0.6	5.2
	45.0	24.8	2.2	0.2	3.6			35.9	15.0	40.0	1.1	4.6	2.2	0.7	6.0
	45.0	24.8	2.2	0.2	2.6			38.8	15.0	40.0	0.9	3.6	2.2	0.7	5.5
	45.0	24.8	2.2	0.2	4.0			36.3	15.0	40.0	1.0	4.2	2.6	0.6	5.8
12.1	42.5	28.8	0.6		2.3		10.0	31.8							
	40.0	17.9	0.7	3.5	8.5	5.8	6.8	27.4	29.0	31.0		4.2	1.3	1.0	4.5
	38.0	19.8	0.6	3.3	9.0	4.0	6.9	26.0	32.0	30.0		3.4	1.6	0.7	3.6
	38.0				2.7	3.8	5.6	31.0	25.0	37.0		1.5	1.2	0.5	3.9
	38.0	22.5	0.9		4.5	5.1	9.6	27.4	30.6	31.4		3.2	4.3	1.3	4.0
	38.0	29.0	6.0		3.7	5.1	10.0	27.9	30.6	31.4		3.2	4.3	1.3	4.4
	37.0				1.8	4.4	7.5	29.0	37.9	25.1		3.0	3.1	0.6	3.9
12.0	36.1	30.0			4.1	2.4		17.0	60.3	4.7					
	36.0	30.4	0.4		3.9	4.7	7.2	23.0	55.0	9.0		3.9	3.0	0.7	3.5
	35.2	23.4	2.2		7.8	0.5		30.0	58.9	5.9	0.8	1.6	2.2	0.3	3.2
	35.0	39.4	0.3		1.6	4.5	8.2	25.6	64.0	1.0		3.3	3.8	0.8	4.6
	32.0	29.5	0.4		2.4	4.8	8.4	22.5	68.0			3.8	4.1	0.7	3.6
	31.0				6.0	3.8	5.2	23.0	30.0	36.0		2.6	0.8	0.6	3.8
11.3	30.3	20.3	2.7		3.0		5.8	23.6							
	29.6	30.0			4.1	2.3		22.5	10.6	59.8					
12.1	29.4	30.0			4.7	2.4		17.0	10.3	60.3					
	29.0	31.0	0.6		1.3	3.4	6.8	22.6	71.0			2.7	2.7	0.6	3.5
	29.0	28.7	0.4	0.1	1.9	4.2	7.4	21.0	71.0			3.4	3.5	0.6	3.3
	29.0	22.5	0.7		3.3	3.7	7.2	30.0	34.0	37.0		2.4	3.0	0.8	3.4
	27.0				2.9	3.4	6.2	21.1	34.8	38.2		2.2	2.7	0.7	2.8
	27.0				5.8	2.5	3.2	21.0	37.0	36.0		2.0	0.5	0.3	2.4
	27.0				2.1	3.4	6.6	21.1	34.8	38.2		2.2	2.7	0.7	3.2
	25.0				1.9	2.7	4.6	16.5	65.0	10.0		2.4	1.5	0.6	2.5
11.3	24.5	20.3	2.7		3.0		4.6	18.2							
	24.0				0.8	2.7	4.9	20.0	73.4	2.6		2.1	1.9	0.4	2.6
	21.0				1.1	2.8	5.0	15.5	79.0			2.2	2.1	0.4	2.5
	18.0				2.6	2.1	3.9	12.7	38.0	44.0		1.4	1.4	0.3	2.2
	18.0				0.6	2.0	2.6	16.1	82.0			1.4	0.9	0.3	1.4
	18.0				0.9	2.3	4.1	13.4	82.0			1.8	1.6	0.3	2.2

Table 12–1g Fluid Catalytic Cracking Database cont'd

179

DEPENDENT VARIABLE	COKE	GAS	PROPANE	PROPENE	C4S	IC4	NC4	C4=	GASOLINE	LCO
INDEPENDENT VARIABLE	CONV CONV^2 FD API	COKE CONV FD API	CONV CONV^2	CONV CONV^2	CONV CONV^2	CONV CONV^2	CONV CONV^2	CONV CONV^2	CONV CONV^2	CONV CONV^2
R^2	0.735	0.665	0.648	0.844	0.762	0.832	0.456	0.739	0.917	0.767
SEE	0.589	0.506	0.397	0.67	2.248	0.793	0.322	0.96	3.63	2.49
S1	0.0247	0.0328	0.0398	0.0206	0.061	0.0249	0.0084	0.00813	0.0186	0.1274
S2	0.00002	0.00384	0.00031	0.0002	0.00054	0.00023	0.00007	0.00012	0.00026	
S3	0.0184	0.0192								
NO.	104	87	56	112	119	83	173	95	339	86

Table **12-2** Some Results of FCC Yield Correlations

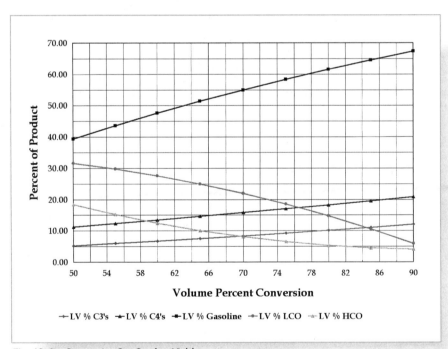

Fig. 12-3   Composite Cat Cracker Yields

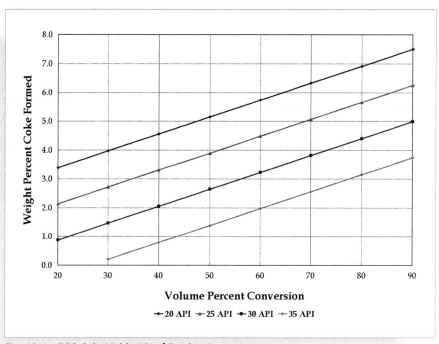

Fig. 12–4 FCC Coke Yield (API of Feed as Parameter)

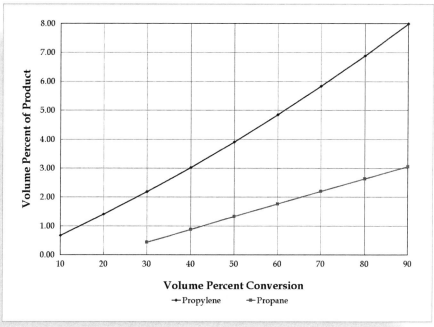

Fig. 12–5 Cat Cracker C₃ Yields

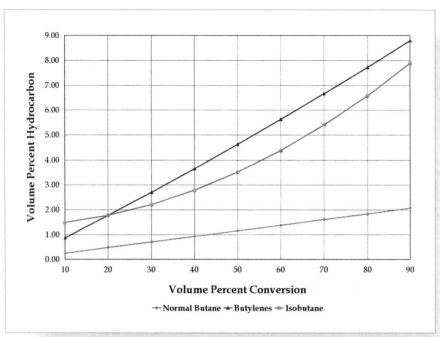

Fig. 12–6   Cat Cracker C$_4$ Yields

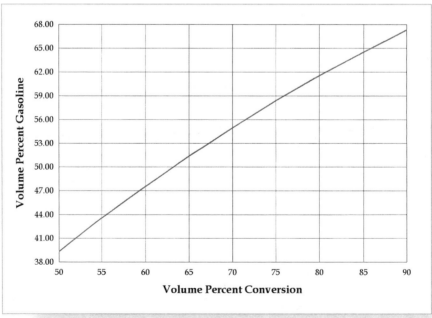

Fig. 12–7   FCC Gasoline Yield

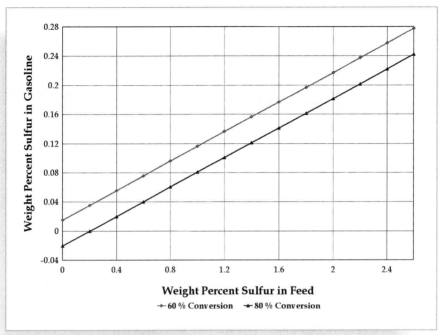

Fig. 12–8 Sulfur in FCC Gasoline

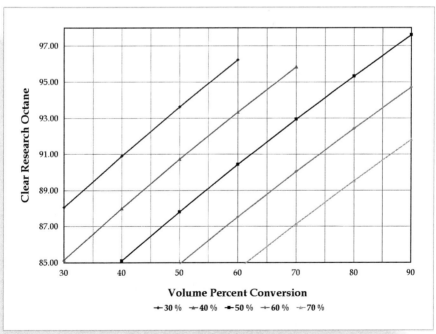

Fig. 12–9 Clear Research of FCC Gasoline (Gasoline Yield as Parameter)

183

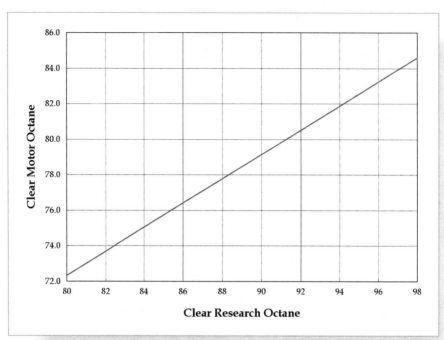

Fig. 12–10 Motor Octane of FCC Gasoline

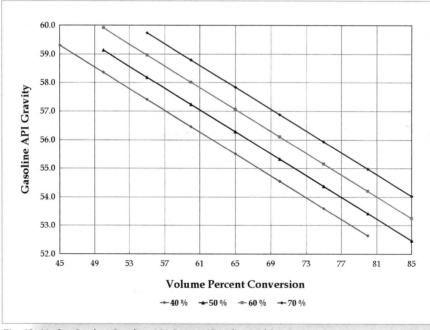

Fig. 12–11 Cat Cracker Gasoline API Gravity (Gasoline Yield as Parameter)

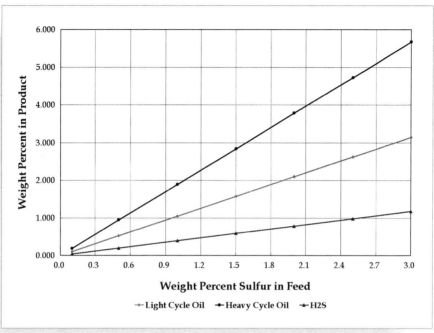

Fig. 12–12 Sulfur in Cat Cracker Products

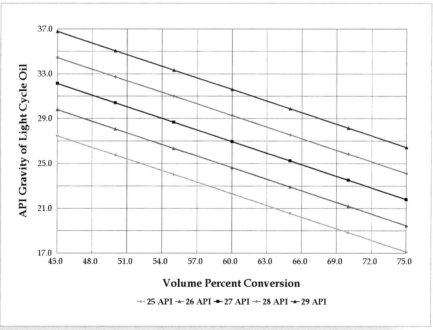

Fig. 12–13 API Gravity of Light Cycle Oil (API Gravity of Feed as Parameter)

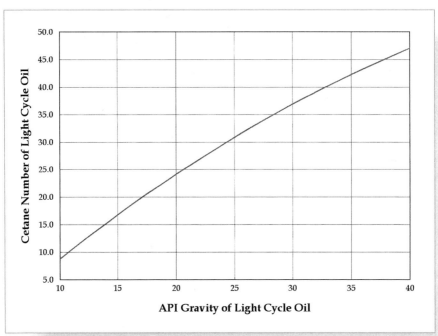

Fig. 12–14 Cetane Number of Light Cycle Oil

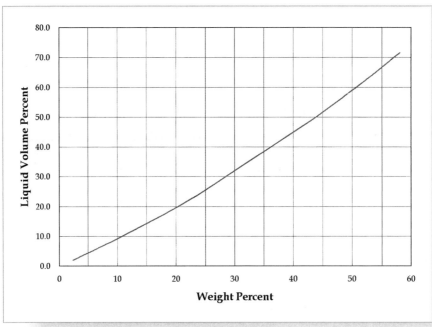

Fig. 12–15 Lv% vs. Wt% for FCC C$_{5+}$ Liquids

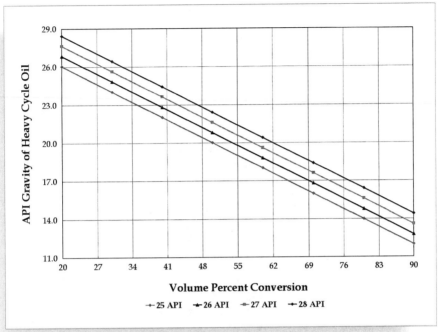

Fig. 12–16 API Gravity of FCC Heavy Cycle Oil (API Gravity of Feed as Parameter)

Comparison with other correlations

Though much has been published on cat cracker yields, very few usable, complete correlations are found. Castiglioni[4] published a means of predicting FCC yields employing a correlating factor calculated from the VABP, specific gravity, aniline point, and sulfur content of the feed. Unfortunately, the VABP and aniline point are generally not included in the published data. Ewell and Gadmer[5] classify feeds as poor, medium, or good in some undefined manner. Gary and Handwerk[6] present correlations requiring only the API gravity of the feed in addition to the conversion assumed by the user. In the case of silica-alumina catalyst, they use K factor as a parameter.

Data have been read from small-scale graphs appearing in the literature on the FCC process. Tabulations of these data along with values read from the author's graphs appear in the following Table 12–3.

COKE YIELDS

REFERENCES	A	B	C*	E	F	G	H	I	J	K*
CONVERSION										
30									1.6	
35										
40					1.0			2.3	1.8	2.2
45										
50				1.9	1.9			2.5	2.1	2.8
55	2.5		3.8			2.0				
60	3.0		4.0	2.2	2.5	2.5	1.8	2.8	3.1	3.3
65	3.4		4.4			3.0	2.0			
70	4.0		4.7	3.0	4.0	4.0	2.3	4.0	4.2	3.8
75		3.5	5.1			5.0	3.0			
80		4.0	5.6	4.5	5.8				7.0	4.4
85		4.9	6.4							
90		5.9								

YIELD OF C4S

REFERENCES	A	C**	F	H	I	K
CONVERSION						
30					6.0	
35						
40			6.5		6.0	8.0
45						
50			9.0		7.6	10.0
55	4.7					
60	6.0	11.4	12.0	12.0	10.1	12.0
65	7.7			15.0		
70	10.3	14.4	14.0	17.5	14.5	15.0
75				20.5		
80		17.4	17.0		20.0	17.0
85						
90			20.0			20.0

Table 12–3a Comparison of Correlations cont'd

Cat cracker operating requirements

The following average values may be used for preliminary studies. Actual requirements vary with such things as type of driver for air blower, presence of power recovery, use of maximum air cooling, etc.

REFERENCES	B	C*	D	E	F	G	H	I	K
CONVERSION						**GASOLINE YIELD**			
30									24.0
35						32.5	41.0		
40					41.0	37.0		50.0	32.0
45						40.5	58.0		
50				45.0	50.0	44.9		63.0	40.0
55						49.0	68.0		
60				53.0	58.0	53.0	52.5	71.0	48.0
65						56.5	54.9		
70		56.0	63.0	60.0	63.0	58.3	57.0		55.0
75	59.0		65.8			60.2	57.5		
80	62.0	62.0	67.3		68.0				63.0
85	65.0		67.4						
90	66.0		64.4		71.0				70.0

A. Baker, R.W., Blazek, J.J., Maher, P.K., Ciapetta, F.G., and Evans, R.E., *Oil & Gas Journal*, May 4, 1964, pp. 78–84

B. Blazek, J.J., *Oil & Gas Journal*, November 8, 1971, pp. 66–73

C. Gary, J.H. and Handwerk, G.E., *Petroleum Refining Technology and Economics*, Marcel Dekker, Inc., New York, 2nd ed., 1984.

D. McDonald, G.W.G., *Oil & Gas Journal*, July 31, 1989, pp. 80–83

E. Mills, G.A., Ashwill, R.E., and Gresham, T.L., *Hydrocarbon Processing*, May, 1967, pp. 121–126

F. Nelson, W.L., *Oil & Gas Journal*, September 17, 1956, p. 289

G. Stormont, D.H., *Oil & Gas Journal*, November 23, 1964, pp. 50–53

H. Voorhies, A., Jr., Kimberlin, C.N., Jr., and Smith, W.M., *Oil & Gas Journal*, May 18, 1964, ff. 107

I. Wachtel, S.J., Bailie, L.A., Foster, R.L., and Jacobs, H.E., *Oil & Gas Journal*, April 10,1972, pp. 104–107

J. Wollaston, E.G., Haflin, W.J., Ford, W.D., and D'Souza, G.J., *Hydrocarbon Processing*, September , 1975, pp. 93–100

K. The present work.

Table 12–3b Comparison of Correlations cont'd

Fuel, mBtu/b	80
Electric power, kW/b	1
Steam, #/b	< 20 >
Cooling water, gal/b	400
Catalyst, #/b	0.3

Cat cracker capital cost

A total of 18 cat cracker capital costs published since 1973 were adjusted to 50,000 BPD capacity and January, 1991 for an average of $86 million. The range was from $66 to $117 million (or plus 36%; minus 23%).

Notes

1. Reichle, A.D., Schuette, W.L., Pine, L.A., and Smith, T.E., "A New Cat Cracking Catalyst for Higher Octanes," NPRA Annual Meeting, San Antonio, March, 1981

2. McDonald, G.W.G., *Oil & Gas Journal*, Apr. 1, 1985, pp. 111–115

3. Huling, G.P., McKinney, J.D., and Readal, T.C., *Oil & Gas Journal*, May 19, 1975, pp. 73–79

4. Castiglioni, B.P., *Hydrocarbon Processing*, February, 1983, pp. 35–38

5. Ewell, R.B., and Gadmer, G., *Hydrocarbon Processing*, April, 1978, pp. 125–134

6. Gary, J.H., and Handwerk, G.E., *Petroleum Refining Technology and Economics*, 2nd ed., Marcel Dekker, New York, 1984

References

Anon., *Oil & Gas Journal*, May 26, 1975, pp. 94–95

Ibid, December 21, 1981, pp. 88–91

Ibid, April 27, 1987, pp. 67–74

Ashwill, R.E., Cross. W.J., and Schwint, I.A., *Oil & Gas Journal*, July 4, 1966, pp. 114–119

Avidan, A.A., Edwards, M., and Owen, H., *Oil & Gas Journal*, January 8, 1990, pp. 33–58

Baker, R.W., Blazek, J.J., Maher, P.K., Ciapetta, F.G., and Evans, R.E., *Oil & Gas Journal*, May 4, 1964, pp. 78–84

Baker, R.W., Blazek, J.J., and Maher, P.K., *Oil & Gas Journal*, April 1, 1968, pp. 110–113

Blazek, J.J., *Oil & Gas Journal*, November 8, 1971, pp. 66–73

Blazek, J.J., *Oil & Gas Journal*, October 8, 1973, pp. 65–70

Bryson, M.C., Huling, G.P., and Glausser, W.E., *The Gulf FCC Process*, API Div. of Ref'g., New York, May, 1972

Bunn, D.P.,Jr., Gruenke, G.F., Jones, H.B., Luessenhop, D.C., and Youngblood, D.J., *Chemical Engineering Progress*, June, 1969, pp. 88–93

Cimbalo, R.N., Foster, R.L., and Wachtel, S.J., *Metal Poisoning of Zeolite Cracking Catalysts*, API Div. of Ref'g., New York, May, 1972

Corbett, R.A., *Oil & Gas Journal*, November 18, 1985, pp. 127–132

Dale, G.H., and McKay, D.L., *Hydrocarbon Processing*, September 1977, pp. 97–102

Desai, P.H., and Haseltine, R.P., *Oil & Gas Journal*, October 23, 1989, pp. 68–72

Duval, C.A., and Holmes, L.S., *Petroleum Refiner*, August, 1952, pp. 109–111

Eastwood, S.C., Drew, R.D., and Hartzell, F.D., *Oil & Gas Journal*, October 29, 1962, pp. 152–158

Ebel, R.H., *Oil & Gas Journal*, April 1, 1968, pp. 116–118

Edgecumbe, C.D., and Valentine, S., III, *Oil & Gas Journal*, August 10, 1964, pp. 89–92

Elliott, K.M., and Eastwood, S.C., *Oil & Gas Journal*, June 4, 1962, pp. 142–144

Elvin, F.J., and Milne, R., *Oil & Gas Journal*, February 28, 1983, pp. 84–86

Evans, L.P., Hart, J.A., Johnson, E.L., and Malin, R.T., *Oil & Gas Journal*, September 9, 1963, pp. 106–114

Flanders, R.L., Girard, J.G., and Laughlin, B.D,, *Oil & Gas Journal*, Mar. 7, 1960, pp. 98–102

Gary, J.H. and Handwerk, G.E., *Petroleum Refining Technology and Economics*, Marcel Dekker, Inc., New York, 2nd ed., 1984.

Gilbert, R.J.H., and Wright, W.N.N., *Oil & Gas Journal*, February 29, 1960, pp. 70–75

Gussow, S., Higginson, G.W., and Schwint, I.A., *Oil & Gas Journal*, June 19, 1972, pp. 71–75

Hamilton, W.W., Eastwood, S.C., Potas, A.E., and Schraishuhn, E.A., *Petroleum Refiner*, August, 1952, pp. 71–78

Hemler, C.L., "Developments in Fluid Catalytic Cracking," Scientific Petroleum Council, Yugoslav Academy of Sciences and Arts, 1975

Hemler, C.L., and Vermilion, "Developments in Fluid Catalytic Cracking," UOP 1973 Technology Conference, 28

Humphries, A., and Wilcox, J.R., *Oil & Gas Journal*, February 6, 1989, pp. 45–51

Jones, H.B., *Oil & Gas Journal*, December 22, 1969, pp. 50–53

Keyworth, D.A., Yatsu, C.A., and Reid, T.A., *Oil & Gas Journal*, August 22, 1988, pp. 51–56

Knowlton, H.E., *Oil & Gas Journal*, September 18, 1967, pp. 80–83

Knowlton, H.E., Beck, R.R., and Melnyk, J.J., *Oil & Gas Journal*, November 9, 1970, pp. 57–61

Knowlton, H.E., Melnyk, J.J., and Lodge, J.C., *Oil & Gas Journal*, October 25, 1971, pp. 76–80

Letzsch, W.S., Magee, J.S., Upson, L.L., and Valeri, F., *Oil & Gas Journal*, October 31, 1988, pp. 57–63

Leuenberger, E.L., and Wilbert, L.J., *Oil & Gas Journal*, May 25, 1987, pp. 38–44

Leuenberger, E.L., *Oil & Gas Journal*, March 21, 1988, pp. 45–50

Lipinski, J.J., and Wilcox, J.R., *Oil & Gas Journal*, November 24, 1986, pp. 80–84

Macerato, F., and Anderson, S., *Oil & Gas Journal*, March 2, 1981, pp. 101–102

Magee, J.S., Ritter, R.E., and Rheaume, L., *Hydrocarbon Processing*, September, 1979, pp. 123–130

Magee, J.S., Blazek, J.J., and Ritter, R.E., *Oil & Gas Journal*, July 23, 1973, pp. 48–58

Magee, J.S., Ritter, R.E., Wallace, D.N., and Blazek, J.J., *Oil & Gas Journal*, August 4, 1980, pp. 63–67

McDonald, G.W.G., *Oil & Gas Journal*, July 31, 1989, pp. 80–83

Mills, G.A., Ashwill, R.E., and Gresham, T.L., *Hydrocarbon Processing*, May, 1967, pp. 121–126

Montgomery, J.A., *Oil & Gas Journal*, December 11, 1972, pp. 81–86

Mott, R.W., *Oil & Gas Journal*, January 25, 1987, pp.73–78

Murcia, A.A., Soudek, M., Quinn, G.P., and D'Souza, G.J., *Hydrocarbon Processing*, September , 1979, pp. 131–135

Nelson, W.L., *Oil & Gas Journal*, September 17, 1956, p. 289

Ibid, Feb. 3, 1958, p. 113

Ibid, October 23, 1961, pp. 143–144

Ibid, January 15, 1962, p. 92

Ibid, June 11, 1962, p. 161

Ibid, March 22, 1965, pp. 98–99

Ibid, February 19, 1973, pp. 86–87

Ibid, September 3, 1979, p. 107

NPRA Question and Answer *Oil & Gas Journal*, April 27, 1987, pp. 67–74

Oden, E.C., and Perry, J.J., *Petroleum Refiner*, March, 1954, pp. 191–193

Padgett, K.W., "Updating Existing FCC Units," 27th Canadian Chemical Engineering Conference, Calgary

Pierce, W.L., Ryan, D.F., Souther, R.P., and Kaufmann, T.G., *Innovations in Flexicracking*, API Div. of Ref'g., New York, May, 1972

Pohlenz, J.B., *Oil & Gas Journal*, August 10, 1970, pp. 158–165

Reif, H.E., Kress, R.F., and Smith, J.S., *Petroleum Refiner*, May, 1961, pp. 237–244

Rheaume, L., Ritter, R.E., Blazek, J.J., and Montgomery, J.A., *Oil & Gas Journal*, May 24, 1976, pp. 66–70

Ritter, R.E., *Oil & Gas Journal*, September 8, 1975, pp. 41–43

Ritter, R.E., and Creighton, J.E., *Oil & Gas Journal*, May 28, 1984, pp. 71–79

Russell, J.A., Lyke, S.E., Young, J.K., and Eberhardt, J.J., *Hydrocarbon Processing*, July, 1986, pp. 65–68

Schall, J.W., Dart, J.C., and Kirkbride, C.G., *Chemical Engineering Progress*, December, 1949, pp. 746–754

Sebulsky, R.T., Kline, R.E., Gianetti, J.P., and Henke, A.M., "Olefins from Catalytic Cracking of Naphthas," API Div. of Ref'g., Philadelphia, May, 1968

194

Stokes, G.M., Wear, C.C., Suarez, W., and Young, G.W., *Oil & Gas Journal*, July 2, 1990, pp. 58–63

Stormont, D.H., *Oil & Gas Journal*, November 23, 1964, pp. 50–53

Ibid, April 5, 1965, pp. 180–184

Ibid, April 1, 1968, pp. 103–109

Upson, L.L., Lawson, R.J., Cormier, W.E., and Baars, F.J., *Oil & Gas Journal*, October 1, 1990, pp. 64–74

Van Keulen, B., *Oil & Gas Journal*, September 26, 1983, pp. 102–105

Venuto, P.B., and Habib, E.T., *Fluid Catalytic Cracking with Zeolite Catalysts*, Marcel Dekker, Inc., New York, 1979

Vermilion, W.L., "Modern Catalytic Cracking: Production of High Octane Gasoline and Olefins," 1971 UOP Technical Seminar, Arlington Heights, Illinois

Voorhies, A., Jr., Kimberlin, C.N., Jr., and Smith, W.M., *Oil & Gas Journal*, May 18, 1964, ff. 107

Wachtel, S.J., Bailie, L.A., Foster, R.L., and Jacobs, H.E., *Oil & Gas Journal*, April 10,1972, pp. 104–107

White, P.J., *Oil & Gas Journal*, May 20, 1968, pp. 112–116

Wollaston, E.G., Forsythe, W.L., and Vasalos, I.A., "Sulfur Distribution in FCU Products," 36th Midyear Meeting, API Div. of Ref'g., San Fransisco, May 12, 1971

Wollaston, E.G., Haflin, W.J., Ford, W.D., and D'Souza, G.J., *Hydrocarbon Processing*, September , 1975, pp. 93–100

Yanik, S.J., Demmel, E.J., Humphries, A.P., and Campagna, R.J., *Oil & Gas Journal*, May 13, 1985, pp. 108–117

Yatsu, C.A., and Keyworth, D.A., *Oil & Gas Journal*, March 26, 1990, pp. 64–74

Yen, L.C., Wrench, R.E., and Ong, A.S., *Oil & Gas Journal*, January 11, 1988, pp. 67–70

Young, G.W., Suarez, W., Roberie, T.G., and Cheng. W.C, "Reformulated Gasoline: The Role of Current and Future FCC Catalysts," NPRA Annual Meeting, San Antonio, March, 1991

CHAPTER 13

HEAVY OIL CRACKER

The heavy oil cracker (HOC) is essentially the same as the FCCU except for having additional means of removing heat from the regenerator. This may be either internal steam coils (primarily in new units) or an external catalyst cooler (primarily on FCCUs revamped for HOC). This is needed because of the higher amounts of coke deposited on the catalyst when residual oil (atmospheric or vacuum) is added to the usual gas oil charge. (This is a rather ambiguous distinction since cat cracker gas oil feeds vary so widely in quality.) Each FCCU has finite coke burning capacity to remain in heat balance and not exceed temperature limits on materials, etc. The ability to remove additional heat gives the refiner a good bit more flexibility in operating his refinery in response to changes in crude supply and to changes in product demand.

HOC process description

Except for the added heat removal provision that can produce large quantities of steam for export, the HOC process flow sheet is the same as for the FCCU.

HOC yield correlation

Table 13–1 is the database for HOC developed from the literature, a much smaller table than the one for FCCU.

Yields of all products except coke and dry gas correlate well against a second degree function of conversion (where conversion

197

FD K	LV% CONV	FD API	FD S	CCR	COKE	GAS	GASO	LCO	HCO	C3	C3=	I-C4	N-C4	C4=
					WT%					LV%				
	86.6	24.2	0.9	1.0	5.2	4.2	58.6	6.4	7.0					
	83.6	23.4	0.8		12.9	4.2	63.5	11.4	5.0					
11.7	81.7	18.8	2.3	6.0		4.2	59.7	13.3	5.0	2.9	6.6	5.4	1.3	9.4
	80.3	15.1	3.0	8.0	18.6		48.0	6.9	12.8	2.7	9.5	4.5	1.5	10.9
	80.2	29.0	0.7	0.3	3.7	2.7	62.3	14.8	5.0	1.9	9.7	5.0	1.2	12.1
	80.1	22.8	0.3	3.7	9.0		65.0	8.9	11.0	1.5	6.3	4.0	1.0	9.2
	80.0	20.7	1.1	5.5	15.7	5.1	58.6	16.0	4.0	1.5	6.2	1.9	0.8	7.7
	79.9	26.0		2.2	7.2	4.5	59.3	16.3	3.8					
	79.1	23.3		3.9	8.0		63.6	15.1	5.8					
	78.8	28.6		0.9	7.6	4.3	57.3	16.6	3.1					
	78.5	21.3		3.3	6.4	5.4	61.2	12.5	9.0					
	78.1					7.0	56.9			2.6	10.0			10.6
	78.0	26.0		2.2	8.8	5.0	55.5	16.0	4.6					
	77.8	24.4		0.9	4.4	2.9	62.4	20.2	2.0					
11.7	77.8	14.8		10.9	15.1		50.4							
	77.7					8.5	53.9			2.7	9.6			10.5
	77.7	27.1	1.0	1.8	5.3	3.1	59.9	15.4	6.9	1.7	9.3	4.6	1.1	11.0
	77.5	22.8	0.9	3.8	8.4	3.4	59.1	15.0	7.5					
	77.5	24.4	1.3	3.7	7.1	3.3	58.9	15.8	6.7	1.7	9.3	4.4	1.1	10.8
	77.3	27.1	1.0	1.8	4.7	3.1	60.4	16.1	6.6	1.7	8.8	3.9	1.0	11.2
	77.0	16.1	3.5	8.6	12.4	4.2	52.6							
	77.0	23.3	0.3	4.2	8.0	1.4	65.0							
11.9	76.8	23.9	0.7	3.6	6.9	3.7	60.0	16.4	6.8					
11.8	76.6	21.3		4.5	8.4	3.3	57.8	15.0	8.4	1.8	8.7			10.1
	76.4	22.2		4.4	10.1		59.2	15.8	7.8					
	76.0	23.0		4.0	7.5	3.5	61.0	19.0	5.0					
12.0	76.0	22.1	0.3	3.2	9.6	5.8	52.4	14.2	9.8					
	76.0	21.3	1.5	5.0	11.0	3.0	57.8	15.0	9.0			3.1	0.9	8.0
	75.8	24.4	1.3	3.7	6.1	3.2	58.6	16.2	8.1	1.6	8.4	3.7	0.9	10.6
	75.8	22.2		3.0	11.0		56.8	16.3	7.9					
	75.6	21.3	1.1	5.0	9.1	3.2	56.6	14.2	10.2					
	75.6					9.7	49.8			2.9	10.3			9.2
	74.7	24.7		0.3	5.0	5.4	56.6	17.2	8.1					
	74.4	21.3		5.1	11.7		56.5	16.0	9.6					
11.7	74.3	17.6	2.2	6.7			55.9	12.9	12.8	2.1	6.9			8.4
11.8	74.1	19.2	1.2	6.9	10.8	4.1	55.6	15.0	10.9	1.8	8.3			9.9
11.7	73.3	19.1	2.1	5.6			55.6	15.3	11.4	2.2	7.0			8.3
	72.6	24.4	1.3	3.6	5.8	3.4	57.2	18.4	9.0	1.4	7.8	3.2	0.8	9.9
	72.2	19.5	2.6	2.6	6.5	6.2	51.6	16.2	11.6					
	70.3	23.7		1.6	4.9	2.5	58.6	23.3	6.4					
	70.0	24.3		0.8	4.9	4.6	55.0	19.3	10.7					
12.4	70.0	20.8	0.3		4.5	3.0	61.0	24.0	6.0	2.1	5.8	4.8	1.6	7.3
11.9	70.0	14.1	3.6		7.0	3.0	58.5	24.0	6.0	2.3	5.6	5.4	1.8	7.0
12.0	68.8	18.5	0.3	5.0	10.5	3.9	49.8	14.1	17.1					
11.5	68.4	19.0	2.1	5.6	10.6	4.8	46.8	16.5	15.1					
	67.9	25.1		1.1	4.9	2.3	59.4	25.6	6.5					
	67.2	22.6	0.8	3.9	13.6	5.5	49.5	21.3	11.5	1.3	3.8	0.3	0.8	5.3
	66.8	24.6	0.7	0.4	5.0	4.3	59.2	16.0	17.2					
	66.4	24.5		0.6	5.1	2.1	55.6	27.0	6.6					
11.7	66.0	19.5	2.7	4.8	7.6	3.4	52.2	19.9	14.1	1.7	5.5	4.3	1.1	7.2
	65.3	24.7		0.1	3.8	2.0	54.9	28.7	6.0					
	63.5	26.1		0.6	5.2	1.4	54.2	24.1	12.4					
	63.2	23.3		3.9	7.0		51.2	20.5	16.3					
	60.9	20.2		3.4	6.5	4.2	46.9	30.0	9.1					
	56.2	23.3		3.9	7.1		48.4	32.1	11.7					
	55.7	26.2		2.2	5.5	4.2	52.6	37.1	7.2					
	55.5	23.5	0.6	3.9	10.8	3.8	41.8	18.0	26.5	1.3	2.5	0.8	1.1	5.1
	50.6	22.9	0.6	4.2	11.7	3.1	39.3	20.4	29.0	0.6	1.9	0.5	0.4	4.1
	47.1	16.1	2.6	9.4	15.5	3.3	27.6	22.7	30.2	1.8	4.1	0.6	0.6	4.6

Table 13-1 Heavy Oil Cracker Database

198

is 100 minus volume percent of light plus heavy cycle oils) as shown in Table 13–2. It was found that coke had a strong correlation with Conradson carbon residue content of the feed. Numerous combinations were tried for dry gas without success, but it did correlate well against coke yield. Reasonably good results were obtained for C_3's and propylene, but not for propane. Thus, propane yield should be obtained by difference. Also, light cycle oil gave better results than heavy cycle oil. So heavy cycle oil should be obtained by difference (conversion minus light cycle oil).

DEPENDENT VARIABLE	COKE	GAS	C3S	PROPENE	C4S	IC4	NC4	C4=	GASOLINE	LCO
INDEPENDENT VARIABLE	FD CCR	COKE	CONV CONV^2	CONV CONV^2	CONV CONV^2	CONV CONV^2	CONV CONV^2	CONV CONV^2	CONV CONV^2	CONV CONV^2
R^2	0.923	0.808	0.91	0.963	0.964	0.956	0.888	0.946	0.861	0.941
SEE	0.696	0.3164	0.906	0.493	0.893	0.388	0.0936	0.5903	1.974	0.6307
S1	0.0494	0.0204	0.319	0.2375	0.323	0.1907	0.0413	0.226	0.0041	0.2132
S2			0.0025	0.0018	0.0025	0.0015	0.00032	0.0018	0.5674	0.0016
NO.	35	25	20	16	14	13	12	19	41	21

Table 13–2 Heavy Oil Cracking Yields

The literature contained a significant amount of yield data on a weight basis (primarily from sources outside the U.S.) sufficient for the development of product yield relationships on that basis. No attempt was made to do that at this time.

Figure 13–1 is a composite plot of liquid yields. Figure 13–2 is a plot of coke formed on catalyst. Figure 13–3 is a plot for C_4 and lighter.

The following additional charts were developed: Heavy Oil Cracker C_4 Yields (Fig. 13–4), Heavy Oil Cracker Gasoline Yield (Fig. 13–5), HOC Gasoline Octane (Fig. 13–6), Motor Octane of HOC Gasoline (Fig. 13–7), Heavy Oil Cracker Gasoline API (Fig. 13–8), API Gravity of Heavy Oil Cracker LCO (Fig. 13–9), Cetane Number of LCO from HOC (Fig. 13–10), API Gravity of HOC Heavy Cycle Oil (13–11), and Lv% vs. Wt% for Heavy Oil Cracker (13–12).

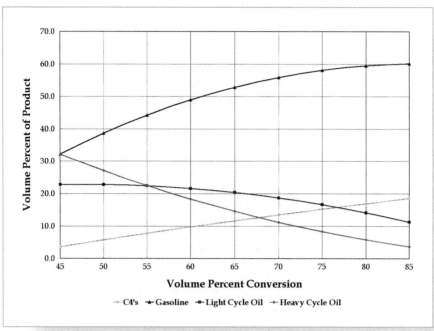

Fig. 13–1 Heavy Oil Cracker Liquid Yields

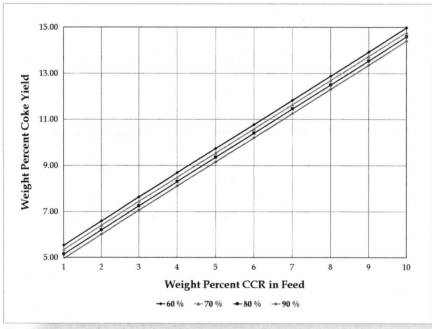

Fig. 13–2 Heavy Oil Cracker Coke Yield (Conversion as Parameter)

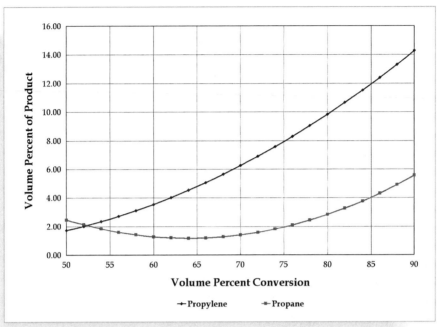

Fig. 13–3   Heavy Oil Cracker C$_3$ Yields

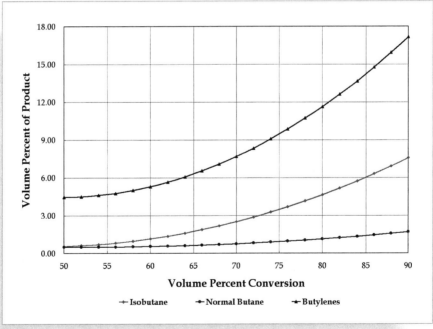

Fig. 13–4   Heavy Oil Cracker C$_4$ Yields

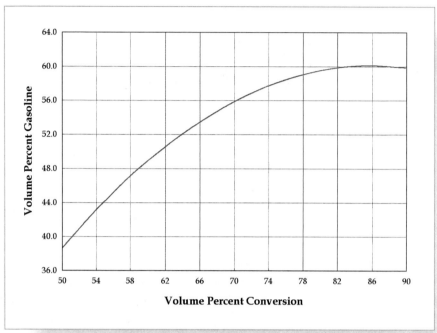

Fig. 13–5   Heavy Oil Cracker Gasoline Yield

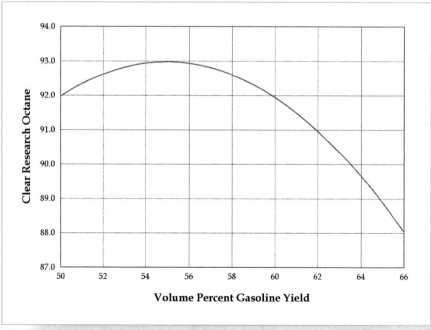

Fig. 13–6   HOC Gasoline Octane

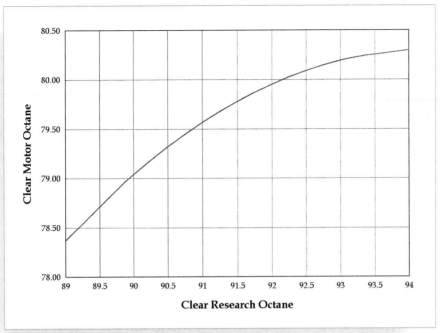

Fig. 13–7 Motor Octane of HOC Gasoline

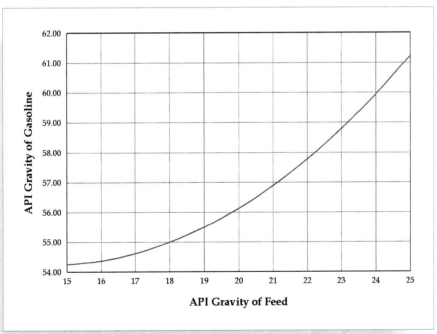

Fig. 13–8 Heavy Oil Cracker Gasoline API

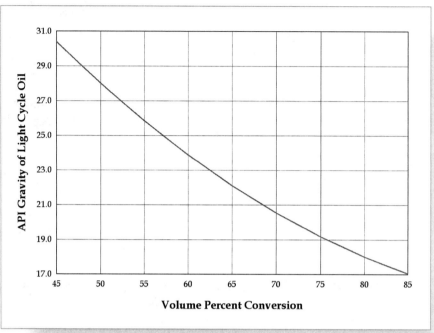

Fig. 13–9   API Gravity of Heavy Oil Cracker LCO

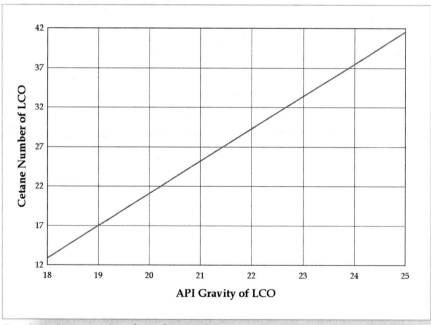

Fig. 13–10 Cetane number of LCO from HOC

204

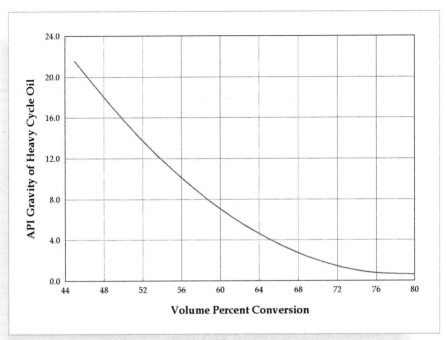

Fig. 13–11 API Gravity of HOC Heavy Cycle Oil

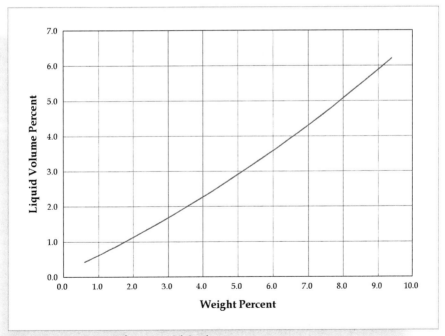

Fig. 13–12 Lv% vs. Wt% for Heavy Oil Cracker

205

Comparison with other correlations

Only one set of yield curves on a volume–percent–conversion basis was found in the literature. This was an article by Busch, *et. al.*,[1] based on seven sets of data. A comparison of readings from the charts in the article with the author's results are shown in Table 13–3. The agreement is very good.

| | —C3S— | | —C4S— | | GASOLINE | | LIGHT CYCLE | |
CONV	A	B	A	B	A	B	A	B
71	8.0	9.8	13.9	14.1	56.4	54.0	18.4	17.6
72	8.4	10.0	14.2	14.3	56.9	54.6	18.0	16.9
73	8.8	10.1	14.6	14.5	57.3	55.3	17.5	16.3
74	9.3	10.3	14.9	14.7	57.8	56.0	17.1	15.6
75	9.8	10.4	15.3	14.9	58.1	56.7	16.7	15.0
76	10.3	10.6	15.6	15.1	58.5	57.3	16.2	14.3
77	10.8	10.7	16.0	15.3	58.8	58.0	15.7	13.7
78	11.4	10.9	16.3	15.5	59.1	58.7	15.2	13.0
79	12.0	11.0	16.6	15.7	59.3	59.3	14.7	12.4

Table 13–3 Comparison with Other Correlations

Busch, *et. al.*, plotted coke yield against Ramsbottom carbon residue, whereas the author used Conradson carbon residue. Nevertheless, the agreement on coke yield is good:

WT% Carbon Residue	WT% coke Busch, et. al.	Author
4	8.4	8.5
5	9.2	9.5
6	10.+	10.6
7	11.–	11.6

HOC operating requirements

Requirements on a per barrel of feed basis can be assumed to be as follows:

Electric power, kW/b	0.25
High pressure steam production, #/b	80
Catalyst consumption, #/b	0.25

HOC capital cost

Of the few data found in the literature, it is thought that the lower values do not include fractionation or an unsaturate gas plant. When these are included, it appears that a 30,000 BPD HOC unit in early 1991 would have cost about $93 million.

Notes

1. Busch, L.E., Hettinger, W.P., Jr. and Krock, R.P., *Oil & Gas Journal*, December 10, 1984, pp. 79–84

References

Busch, L.E., Hettinger, W.P., Jr. and Krock, R.P., *Oil & Gas Journal*, December 24, 1984, pp. 54–56

Campagna, R.J., Krishna, A.S., and Yenik, S.J., *Oil & Gas Journal*, October 31, 1983, pp. 128–134

Dale, G.H., and McKay, D.L., *Hydrocarbon Processing*, September, 1977, pp. 97–102

Edison, R.R., Siemssen, J.D., and Masologites, G.P., *Oil & Gas Journal*, December 20, 1976, pp. 54–60

Elvin, F.J., *Oil & Gas Journal*, May 9, 1983, pp. 100–112

Finneran, J.A., Murphy, J.R., and Whittington, E.L., *Oil & Gas Journal*, January 14, 1974

Finneran, J.A., Murphy, J.R., and Schneider, L.W., "Application of Heavy Oil Cracking in a Fuels Refinery," 74th National Meeting of AIChE, New Orleans, March, 1973

Hansen, T.S., *Oil & Gas Journal*, August 15, 1983, pp. 47–52

Heite, R.S., English, A.R., and Smith, G.A., *Oil & Gas Journal*, June 4, 1990, pp. 81–87

Hemler, C.L., Lomas, D.A., and Tajbl, D.G., *Oil & Gas Journal*, May 28, 1984, pp. 79–88

Hemler, C.L., "Developments in Fluid Catalytic Cracking," Scientific Petroleum Council, Yugoslav Academy of Sciences and Arts, June 13, 1975

Knaus, J.A., Schwarzenbek, E.F,, Atteridg, P.T., and McMahon, J.F., *Chemical Engineering Progress*, December, 1961, pp 37–43

Krishna, A.S., "Advances in Resid Cracking Technology," Ketjen FCC Seminar, Philadelphia, 1984

Louder, K.E., Kulapaditharom, L., and Juno, E.J., *Hydrocarbon Processing*, September, 1985, pp. 80–84

McKenna G.E., Owen, C.H., and Hettick, G.R., *Oil & Gas Journal*, May 18, 1964, pp. 106–107

McKenna, G.E., Owen, C.H., and Hettick, G.R., *Heavy Oil Catalytic Cracking—A Key to Refinery Modernization*, API Div. of Ref'g., St. Louis, May, 1964

Mott, R.W., "New Technologies for FCC Resid Processing," NPRA Annual Meeting, San Antonio, March, 1991

Murphy, J.R., and Soudek, M., *Oil & Gas Journal*, January 17, 1977, pp. 70–76

Murphy, J.R., *Oil & Gas Journal*, September 1, 1980, pp. 108–110

Nieskens, M.J.P.C., Khow, F.H.H., Borley, M.J.H., and Roebschlaeger, K.W., *Oil & Gas Journal*, June 11, 1990, pp. 37–44

NPRA Questions and Answers, *Oil & Gas Journal*, March 28, 1983, pp. 108–123

Ritter, R.E., Rheume, L., Welsh, W.A. and Magee, J.S., *Oil & Gas Journal*, July 6, 1981, pp. 103–110

208

Rush, J.B. and Steed, P.V., *Oil & Gas Journal*, May 28, 1984, pp. 96–103

Rush, J.B., *Chemical Engineering Progress*, December, 1981, pp. 29–32

Shaffer, A.G., Jr., and Hemler, C.L., *Oil & Gas Journal*, May 28, 1990, pp. 62–70

Tolen, D.F., *Oil & Gas Journal*, March 30, 1981, pp. 90–109

Torgaard, H., *Oil & Gas Journal*, January 10, 1983, pp. 100–103

Upson, L.L., and Jaras, S.G., "Metal Resistant Catalyst for Heavy Oil Cracking," NPRA National Meeting, San Antonio, March, 1982

Walliser, L., *Oil & Gas Journal*, March 24, 1980, pp. 118–127

Wilson, J.W., Wrench, R.E., and Yen, L.C., *Chemical Engineering Progress*, July, 1985, pp. 33–40

Wrench, R.E., and Wilson, J.W., *Oil & Gas Journal*, October 6, 1986, pp. 53–56

CHAPTER 14

HYDROCRACKING

Hydrocracking is a catalytic cracking process conducted with a high (relative to hydrodesulfurization processes) hydrogen partial pressure. It can produce a higher conversion of refractory (resistant to cracking) stocks to lower molecular weight products than FCCUs can. Pressures employed can reach 3,000 psi and temperatures are in the range of 600°F to 800°F. Hydrogen consumption is generally in the range of 1,000 to 2,000 standard cubic feet per barrel (SCFB) for jet fuel or diesel fuel production and 1,500 to 2,500 for complete conversion of feed to gasoline and lighter.

Hydrocracker process description

The hydrocracker consists of one or two reactor stages, depending on the quality of the feed and the processing objective, followed by high- and low-pressure separators, a stabilizer, and a product fractionator as shown in Figure 14–1. Most hydrocrackers have fixed bed reactors, but there are a few installations with moving (ebbulated) catalyst beds. A simplified flow diagram of an ebbulated bed process appears as Figure 14–2. Both types appear to produce essentially the same results.

Hydrocracking is a very versatile process, but it is relatively expensive due to its high operating pressure and high hydrogen consumption. In contrast with the coking and deasphalting processes, hydrocracking decreases the carbon-to-hydrogen ratio by the addition of hydrogen rather than the removal of carbon.

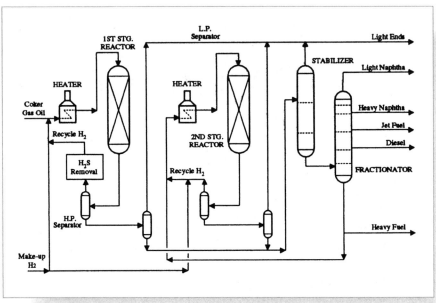

Fig. 14–1 Hydrocracker

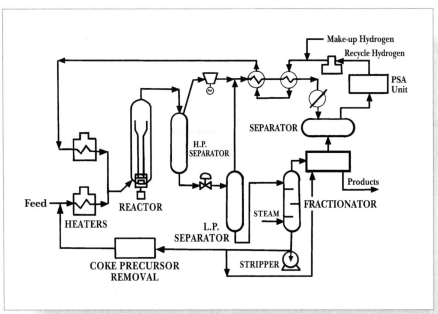

Fig. 14–2 Ebullated Bed Resid Hydrocracking Unit

212

Installations now being designed to produce the low sulfur diesel fuel soon to be required border on hydrocracking conditions and when the lower aromatic content for diesel becomes required, hydrocracking will be needed.

Hydrocracking data correlation

An extensive database of hydrocracking yields was developed from the literature in Table 14–1.

Plots of the raw data for butanes (C_4s), light and heavy hydrocrackate, distillate (jet fuel or diesel fuel) and hydrogen required were made against the yield of total gasoline (light plus heavy hydrocrackate). Light hydrocrackate is nominally C_5 to 180°F. Heavy hydrocrackate is usually 180 to 400°F unless jet fuel or diesel fuel is being co-produced in which case the end point would be significantly lower.

Table 14–2 is a summary of some of the regression results. Figure 14–3 is a composite plot of regression lines for the liquid products as second–degree functions of total gasoline. Figure 14–4 is a plot of dry gas vs. C_4s. Figure 14–5 is a plot of the components of dry gas—C_1, C_2, and C_3. It may be assumed that isobutane is 67% of total C_4s.

Because of its importance, special effort was devoted to developing the relationship of hydrogen to the hydrocracking process. Figure 14–6 is a general plot of hydrogen requirement vs. total gasoline product with API gravity of the feed as a second independent variable. In an effort to develop more definitive relations, the available data were grouped by origin of feed into virgin gas oil, cat cycle oil, and coker gas oil. These much smaller families of data were regressed and the results can be seen in Figures 14–7, 14–8, and 14–9. These more specific plots do fit their particular data better than the general plot in Figure 14–6. In all of these plots, a second–degree function of gasoline was used.

The following additional figures (14–10 through 14–19) were developed for properties of the various products:

Figures 14–10 and 14–12 illustrate that octane numbers for both light and heavy hydrocrackate are low at low yields, increase to a maximum at around 25% for light and 50% for heavy, then decrease as yield continues to increase. End point of heavy hydrocrackate was not found to have a great influence on its octane.

FD API	FD K	SCFB H2	C4	LT HC	HY HC	JET	DIESEL	RESID
6.4	10.40	3,700	3.70		40.10		77.80	
18.0	10.70	3,650	31.60	37.30	56.30			
18.0	10.70	3,380	19.90	29.00	83.40			
18.0	10.70	3,350	23.30	33.70	70.40			
19.0	11.08	3,200	15.30	33.90	81.80			
22.2	11.22	2,950	11.20	23.70	89.90			
18.5	10.60	2,715	18.30	29.60	80.40			
17.1		2,705	16.50	27.10	84.00			
20.4	10.94	2,690	18.20	29.60	79.30			
22.5	11.40	2,630	16.80	32.50	17.00	62.30		
21.2	11.77	2,590	22.10	38.70	70.00			
29.7	11.20	2,570	15.70	30.00	85.60			
17.0	11.50	2,543		28.50	87.90			
28.9	11.35	2,500	19.00	34.10	73.30			
21.9	10.70	2,460	20.00	36.60	65.00			
19.0	11.10	2,440	4.50		48.90		65.90	
21.2	11.77	2,437	18.40	31.50	78.80			
20.3	11.37	2,430	15.80	29.10	83.30			
22.5	11.40	2,425	13.70	21.50	14.90	74.40		
21.4		2,410	15.90	25.70	82.20			
22.8		2,410	14.30	26.70	84.70			
25.8	11.23	2,400	15.90	34.10	76.70			
20.0	11.30	2,400	19.90	34.20	75.30			
25.8	11.30	2,400	15.70	34.10	76.70			
23.4	10.83	2,400	2.20	3.60	8.80	98.00		
23.3	11.00	2,380	23.80	37.60	55.60			
22.8		2,370	15.00	29.20	81.90			
		2,350	14.70	39.70	70.20			
25.8	11.23	2,320	12.50	24.80	38.60	47.00		
23.3	11.00	2,310	19.40	31.80	73.50			
22.3	11.75	2,310	19.40	31.80	73.50			
19.2		2,305	13.50	21.60	88.90			
24.4		2,300	16.20	29.80	77.50			
21.9	10.70	2,267	16.30	29.60	74.50			
22.8		2,250	20.40	32.00	71.30			
23.2	10.80	2,240	20.90	31.50	69.50			
25.9	10.96	2,240	12.90	23.00	71.70		10.70	
22.8		2,225	20.00	36.60	72.90			
26.5	11.44	2,213	16.90	28.30	81.60			
22.8	11.75	2,180	17.90	33.50	67.20			

Table 14–1a Hydrocracker Database

214

FD API	FD K	SCFB H2	C4	LT HC	HY HC	JET	DIESEL	RESID
22.8		2,165	13.50	26.70	83.70			
23.0		2,160	18.40	35.50	79.40			
24.2	11.00	2,160	18.70	29.20	75.20			
22.3	11.75	2,150	15.60	31.10	79.70			
29.0	11.33	2,150	15.80	28.90	80.00			
27.8	11.28	2,150	16.00	33.00	75.00			
22.3	11.85	2,150	15.60	31.10	79.70			
24.1	11.10	2,143	28.10	37.30	54.60			
24.0	11.00	2,130	27.90	36.90	55.30			
27.9	11.15	2,130	19.50	31.20	70.00			
22.6	11.85	2,120	14.10	46.90	65.40			
20.3	11.37	2,110	8.60	16.50	34.40	61.30		
5.8		2,105	2.88		23.12		35.00	30.08
25.4	11.30	2,100	17.70	27.60	81.50			
29.6	11.50	2,100	25.80	39.50	56.80			
29.8	11.35	2,100	24.90	40.00	56.20			
27.1	11.25	2,100	9.00	18.00	95.00			
21.4	11.30	2,100	20.00	20.00	32.00			40.00
21.4	10.97	2,090	8.70	18.30	69.40		20.00	
26.8		2,090	15.30	28.60	78.10			
22.6	11.45	2,090	16.50	23.40	86.20			
29.3	11.95	2,080	10.30	25.90	25.90	58.90		
28.2		2,070	15.80	30.80	76.20			
27.6	11.85	2,070	10.70	44.90	68.40			
24.2		2,060	18.70	29.20	75.20			
27.8	11.30	2,050	16.30	25.30	79.00			
24.1	11.10	2,036	20.10	33.10	69.00			
18.8	11.40	2,020	4.20	7.20	16.20	87.80		
19.2	11.45	2,020	4.10		39.70		73.20	
19.7		2,000	5.70	13.40	24.80		76.60	
22.3	11.75	2,000	14.50	31.70	78.90			
27.6	11.30	1,950	17.30	28.60	76.30			
20.3	11.37	1,950	8.20	18.30	34.10	61.10		
24.0	11.00	1,950	16.50	27.20	76.00			
		1,950	14.60	28.40	81.50			
23.3	12.70	1,950	17.10	32.20	74.70			
22.3	11.85	1,930	5.70	12.90	24.30	75.50		
22.3	11.74	1,930	5.70	12.90	24.30	75.50		
19.7	11.80	1,920	5.40	13.00	23.30	78.00		
25.8	12.08	1,900	20.00	36.10	71.50			

Table 14–1b Hydrocracker Database cont'd

215

FD API	FD K	SCFB H2	C4	LT HC	HY HC	JET	DIESEL	RESID
29.6	11.50	1,900	21.60	36.40	63.30			
26.4	11.59	1,900	8.50	14.10	29.70	64.50		
17.7	10.45	1,900	7.00	11.00	58.00			50.00
25.8	12.10	1,900	20.00	36.10	71.50			
23.3	12.70	1,900	12.40	23.80	44.10	43.00		
22.4	11.93	1,890	6.60	16.00	14.20	85.30		
17.0	11.50	1,876		10.00	26.40		75.80	
32.8	11.67	1,876	26.90	52.80	44.10			
21.2	11.77	1,833	13.80	21.70	29.70	56.90		
30.1	11.66	1,820	22.10	38.50	61.30			
27.4	11.68	1,815	17.20	28.10	80.30			
32.4	11.65	1,800	17.00	30.00	72.00			
27.1	11.25	1,800	5.00	9.00	62.00			40.00
27.1	11.25	1,800	5.00	7.00	64.00			40.00
32.2	11.80	1,800	13.60	41.30	66.60			
29.6	11.50	1,800	19.30	32.50	68.00			
20.0	11.30	1,800	5.30	8.00	12.30	90.80		
23.3	12.70	1,800	9.30	16.60	43.90		48.60	
25.5		1,790	16.80	29.40	75.90			
		1,785		4.90	28.59		32.55	29.68
19.7	11.80	1,780	4.30	9.80	13.60			89.00
21.2	11.77	1,769	11.70	21.20	20.90	66.70		
27.3	11.85	1,760	11.70		59.10	52.80		
32.2	11.80	1,750	5.70		31.60		80.60	
20.3	11.37	1,750	12.10	23.50	40.60	45.00		
29.7	11.28	1,730	14.80	26.00	76.20			
21.2	11.77	1,705	4.90	10.40	15.60	85.70		
32.8	11.69	1,701	22.30	40.70	58.70			
24.3	11.83	1,700	6.30	15.70	34.60	63.70		
30.1	11.66	1,700	16.80	30.40	72.70			
22.4	11.93	1,675	4.00	9.30	7.40	54.80		41.50
22.3	11.75	1,660	8.30	17.20	28.00	64.40		
		1,660		31.67	70.19			
		1,650		32.67	68.84			
21.1		1,650			26.90		46.80	34.40
18.8	11.40	1,648	4.20	7.20	16.10	86.20		
21.1	11.80	1,640	2.90		48.60		65.30	

Table 14–1c Hydrocracker Database cont'd

FD API	FD K	SCFB H2	C4	LT HC	HY HC	JET	DIESEL	RESID
29.2	11.25	1,640	15.90	27.70	57.80		15.00	
31.8	12.16	1,640	13.40	34.50	71.50			
22.3	11.85	1,630	6.80	18.20	30.90	63.10		
29.4	11.69	1,630	20.20	32.70	69.40			
29.2	11.25	1,620	13.70	24.50	62.50		15.00	
22.3	11.75	1,600	5.20	13.20	26.30	70.80		
23.3	12.70	1,600	3.50	8.00	15.10		85.50	
27.5	11.45	1,581	14.20	23.50	82.40			
25.8	11.65	1,580	2.50		32.00		78.60	
29.2	11.25	1,570	13.50	23.30	63.20		15.00	
27.6		1,550	17.30	29.60	78.40			
22.4	11.93	1,550	2.80	5.90	4.60	47.10		50.70
21.2	11.77	1,541	7.50	11.40	32.70		65.60	
5.8		1,530	1.69		13.81		21.20	56.73
27.3	11.85	1,530	3.20		23.40		87.70	
32.8	11.69	1,526	18.40	32.10	69.10			
29.2		1,520	12.40	24.50	79.70			
29.7	11.28	1,510	11.50	21.10	67.70		14.00	
37.0	11.86	1,500	23.80	36.90	57.00			
20.0	11.30	1,500	3.30	6.00	11.80	92.00		
19.3		1,500			12.70	42.00	16.00	38.00
28.7	11.75	1,500	14.00	19.00	50.00			50.00
20.8		1,490	6.30	13.80	13.40		80.90	
22.3	11.75	1,480	4.90	11.50	19.60		81.10	
22.3	11.74	1,450	3.80	9.00	14.00		87.00	
22.3	11.75	1,450	3.80	9.00	14.00		87.00	
32.2	11.80	1,410	7.00		55.90	52.80		
22.3	11.74	1,400	4.10	9.30	18.00	40.00	17.00	25.00
27.1	11.25	1,400	8.00	13.00	50.00			40.00
22.3	11.75	1,400	4.10	9.30	18.00	40.00	17.00	
29.7	11.28	1,390	9.40	17.40	61.40		24.90	
28.6		1,360	7.00	12.80	23.90	71.00		
21.8	11.85	1,350		2.70	10.60		24.20	68.40
22.3	11.85	1,350	3.70	11.80	13.60		87.40	
		1,350		16.00	36.50		23.50	30.00
		1,350		16.00	23.00		35.50	31.50
29.5	11.12	1,340	16.80	26.80	73.70			

Table 14–1d Hydrocracker Database cont'd

217

FD API	FD K	SCFB H2	C4	LV%		JET	DIESEL	RESID
				LT HC	HY HC			
5.8		1,335	1.27		10.36		16.07	66.55
27.6		1,310	9.00	17.50	43.00	49.00		
29.8	11.35	1,300	8.00	13.00	55.00			50.00
28.1		1,300			8.80		88.60	15.80
29.2			6.00	9.30	17.50	76.70		
29.7	11.44		22.70	47.40	56.90			
22.2			9.00	15.90	39.10		51.90	
22.3			8.00	17.00	28.00	64.00		
29.2			12.40	24.50	79.70			
8.4	10.70		5.20	8.80	31.80		33.80	35.00
19.2			16.00	27.00	80.30			
30.1	11.42		20.80	38.10	66.30			
28.6	12.33		6.30		22.30	88.40		
22.0			4.30	9.80	14.50		86.70	
31.1	11.47		18.40	37.20	68.50			
30.7	11.42		19.40	50.40	55.40			
28.6			20.70	34.60	65.50			
29.7	11.44		21.40	43.10	59.90			
27.1			8.00	13.00	63.00			
20.0			5.30		20.70		85.80	
34.8	11.62		24.40	38.70	59.30			
20.0			18.30	33.60	76.20			
29.5			8.00	13.00	55.00			
21.7	11.85				2.40	6.80	19.20	72.00
18.6	10.50		13.50	25.60	84.30			
25.5		2,170	10.00	52.00	61.00			
23		2,260	17.00	50.00	54.00			
23.3		2,250	18.00	53.00	49.00			
		1,340	17.80	33.20	66.10			
		1,200	14.60	31.00	71.80			
		3,025	15.40	32.70	79.10			
		2,525	15.20	26.60	82.70			
23		2,200	24.00	53.00	47.00			
22.8		2170	13.50	26.70	83.70			
25.5		1,790	16.80	29.40	75.90			
28.6	11.30	1,840	20.70	34.60	65.50			
19.2	10.85	2,380	16.00	27.00	80.30			
22.2	11.20	1,760	9.00	15.90	39.10		51.90	
15.5			26.40	27.60	86.70			

Table 14–1e Hydrocracker Database cont'd

FD API	FD K	SCFB H2	C4	LT HC	HY HC	LV% JET	DIESEL	RESID
20.5			17.50	33.70	77.70			
27.7			15.00	31.30	77.70			
35.6			20.70	42.80	59.60			
33.0			18.40	20.30	79.40			
20.3			18.30	33.60	76.20			
22.3			14.50	31.70	78.90			
22.3			8.30	17.20	28.00	64.40		
22.3			4.90	11.50	19.60		81.10	
26.8	1,565		18.40	27.60	77.10			
33.2	2,020		16.80	34.40	73.60			
24.0			5.50	8.50	44.50	51.00		
23.9			5.20	7.50	27.30	41.90	27.10	
29.3	1,600		13.90	19.50	41.40			40.00
29.3	1,600		10.30	16.10	49.50			40.00
29.3	1,600		13.60	19.30	40.80			40.00
17.7	2,050		6.12	13.89	19.86	38.81	37.34	3.00
20.4	1,100		2.60		12.70	33.90	22.30	41.00
22.8	1,050		2.30		7.50	20.90	16.30	62.80
22.7	1,860		6.30	12.90	11.00	89.00		
22.7	1,550		3.80	7.90	9.40		94.10	
8.4	2,500		5.20	8.80	31.80		33.80	
25.8	2,050		17.30	27.10	77.50			
19.2	2,050		16.70	24.00	88.70			
23.2	1,950		17.10	24.40	83.80			
38.8	1,250		19.50	31.40	65.00			
26.9	1,900		11.30	21.20	59.00			
26.9	1,950		15.80	25.50	49.60			
26.9	1,760		9.48	17.35	32.53	29.94	30.00	
23.4	2,250		19.60	37.30	69.80			
23.4	2,300		24.80	39.80	60.00			
23.4	1,500		4.70	14.40	30.50	40.50	25.90	
23.4	1,550		8.30	17.40	32.10	36.00	21.80	
19.5	2,008		18.00	33.20	78.90			
19.5	2,847		25.10	37.80	65.70			
20.3	1,954		8.60	16.20	24.60			
20.3	1,989		17.10	23.30	24.90	56.10		
20.3	1,774		5.90	12.60	22.80	63.40		
20.3	1,813		16.80	23.10	23.50	41.30		

Table 14–1f Hydrocracker Database cont'd

DEPENDENT VARIABLE	DRY GAS	C4S	LT HC	HY HC	DIST	H2
INDEPENDENT VARIABLE	C4S C4S^2	BFG BFG^2	BFG BFG^2	BFG BFG^2	BFG BFG^2	BFG BFG^2 BFG^3
R^2	0.895	0.92	0.95	0.945	0.992	0.93
SEE	0.637	1.665	2.282	6.144	2.29	111.0
S1	0.0112	0.0247	0.0119	0.0371	0.0426	3.69
S2	0.00042	0.00018	0.00012	0.00037	0.000419	0.074
S3						0.00042
NO.	52	133	163	103	73	100

Table 14–2 Hydrocracker Product Yields

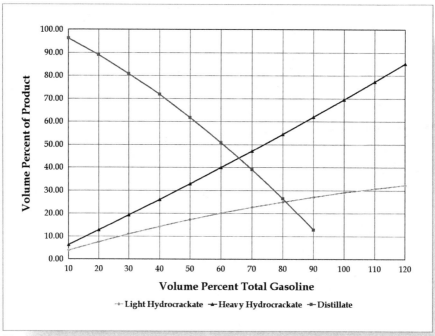

Fig. 14–3 Composite Hydrocracker Yields

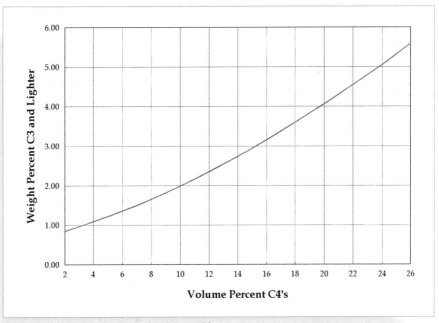

Fig. 14–4 Hydrocracker C_3 and Lighter Yield

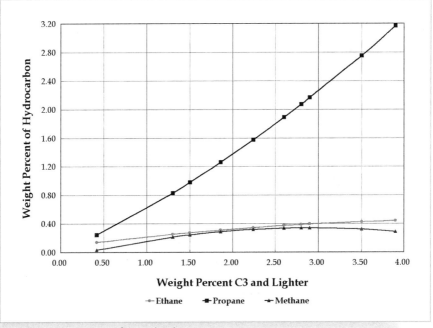

Fig. 14–5 Composition of C_3 and Lighter

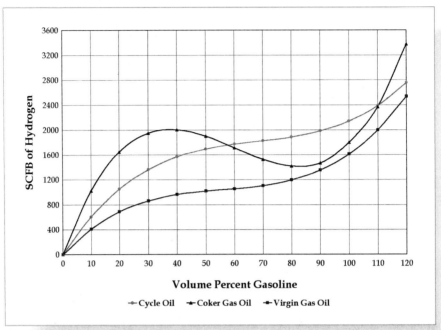

Fig. 14–6   Hydrogen Required in Hydrocracking

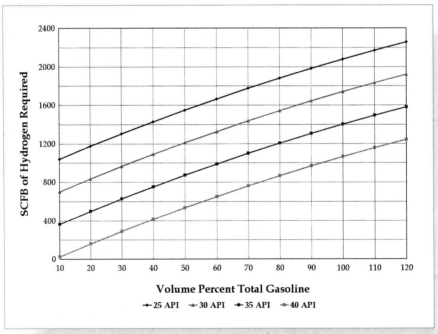

Fig. 14–7   Virgin Gas Oil H₂ Requirements (API Gravity of Feed as Parameter)

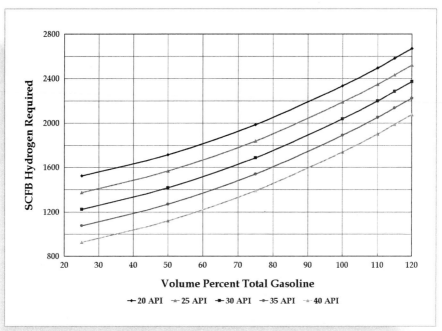

Fig. 14–8 Hydrogen Required by Cat Cycle Oil (API Gravity of Feed as Parameter)

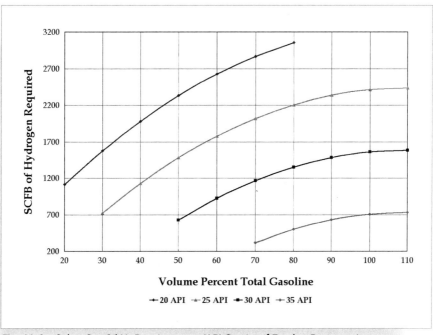

Fig. 14–9 Coker Gas Oil H_2 Requirements (API Gravity of Feed as Parameter)

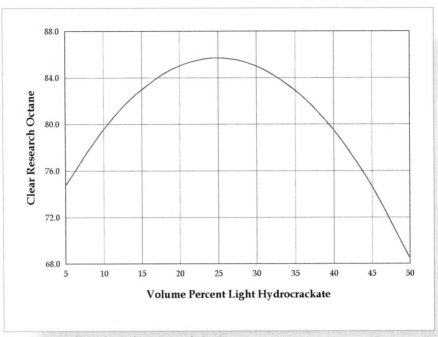

Fig. 14–10   Light Hydrocrackate Research Octane

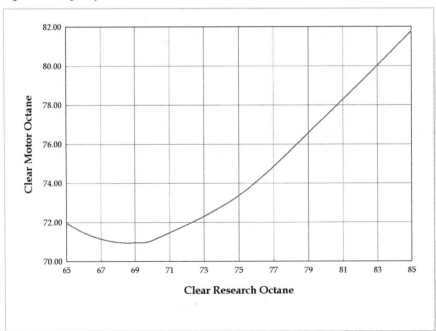

Fig. 14–11   Light Hydrocrackate Motor Octane

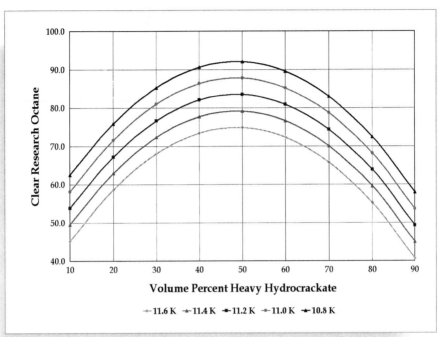

Fig. 14–12 Heavy Hydrocrackate Research Octane (K Factor of Feed as Parameter)

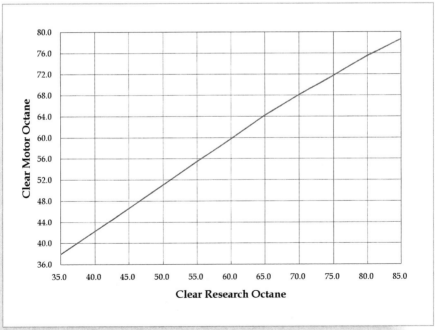

Fig. 14–13 Heavy Hydrocrackate Motor Octane

225

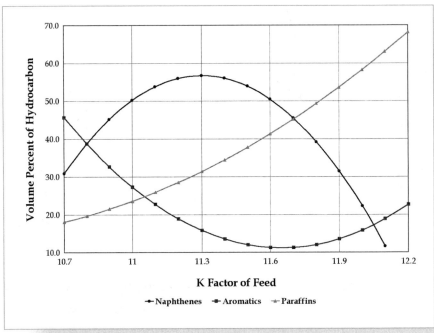

Fig. 14–14 PONA Analysis of Heavy Hydrocrakate

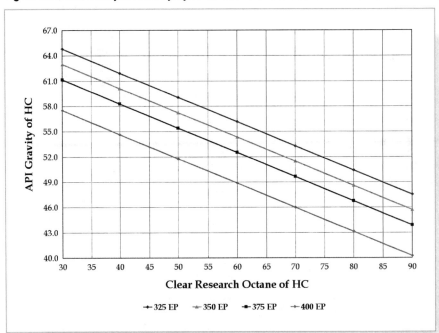

Fig. 14–15 API Gravity of Heavy Hydrocrackate (Endpoint as Parameter)

226

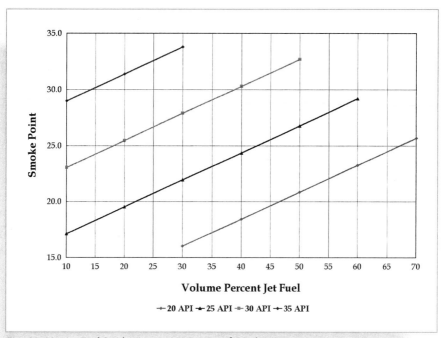

Fig. 14–16 Jet Fuel Smoke Point (API Gravity of Feed as Parameter)

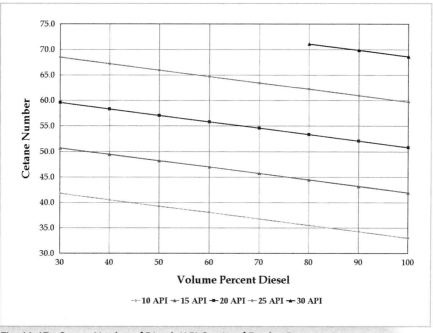

Fig. 14–17 Cetane Number of Diesel (API Gravity of Feed as Parameter)

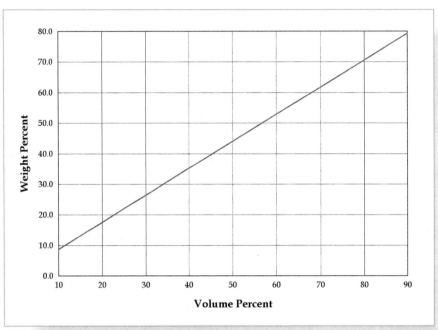

Fig. 14–18 Volume Percent vs. Weight Percent

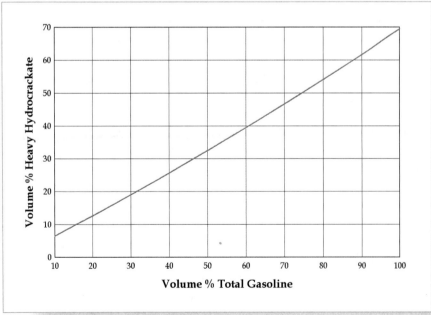

Fig. 14–19 Yield of Heavy Hydrocrackate

The PONA analysis of heavy hydrocrackate was studied because it is generally reformed to boost its octane. We will see in a subsequent chapter the importance of feed composition in catalytic reforming.

Comparison with other correlations

A comprehensive study of hydrocracking was made and published by Nelson in a series of issues of the Oil & Gas Journal.[1] Table 14–3 compares Nelson's yields and those of the author. The comparison is good in the mid–range of gasoline yield, but not as good at the extremes.

TOTAL GASOLINE	LIGHT HYDROCRACKATE		HEAVY HYDROCRACKATE		BUTANES		400+	
	WLN	REM	WLN	REM	WLN	REM	WLN	REM
10	2.5	3.9	7.5	6.3	1.1	1.6	93.5	96.2
20	5.0	7.6	15.0	12.7	1.4	3.2	85.8	89.0
30	7.3	11.0	22.7	19.3	1.7	4.7	77.8	80.8
40	10.2	14.2	29.8	26.1	2.0	6.3	69.7	71.8
50	12.8	17.3	37.2	33.0	2.2	7.9	61.6	61.8
60	15.8	20.0	44.2	40.0	2.4	9.5	53.3	50.9
70	18.8	22.6	51.2	47.2	2.6	11.1	45.0	39.1
80	22.0	25.0	58.0	54.5	2.8	12.7	36.3	26.4
90	25.2	27.1	64.8	62.0	3.1	14.2	28.1	12.8
100	29.5	29.0	70.5	69.6	3.4	15.8	20.0	
110	35.5	30.8	74.5	77.4	3.9	17.4	12.0	
120	47.5	32.2	72.5	85.3	5.6	19.0	5.6	

Table 14–3 Comparison with Other Correlations

Coonradt, et al.,[2] presented results of research on hydrocracking by Socony Mobil. Table 14–4 compares some of their results (read by the author from small scale charts), with the author's findings. Their basis is 390 end point gasoline rather than the nominal 400 end point in the author's case. There were two charts, one for light gas oil and another for heavy gas oil. The agreement is better than with Nelson, particularly in the case of butane.

Eckhouse[3] included a small–scale plot of conversion to 400°F and lower (which is equal to 100 minus distillate yield) vs. gasoline yield. Table 14–5 shows a good comparison with the author's results except for the very high conversions.

LIGHT GAS OIL

	—BUTANES—		—LT HC—		—HY HC—	
DIST	IEC	REM	IEC	REM	IEC	REM
70	5	7	8	14	31	25
60	6	8	10	17	40	33
50	7	10	12	21	50	41
40	9	12	15	23	57	49
30	10	13	19	25	63	56
20	13	14	22	27	68	62
10	16	16	29	29	70	69

HEAVY GAS OIL

80	2	5	5	10	23	18
70	3	6	6	13	31	24
60	5	8	8	17	40	32
50	6	9	10	19	47	38
40	7.5	11	12	22	55	45
30	9	12	16	24	61	52
20	10.5	14	19	26	68	60
10	14	15	23	28	72	66

Table 14–4   Comparison with Coonradt, et al.

	LV% GASOLINE	
LV% DIST YIELD	JGE	REM
80	28	30
60	52	52
40	73	70
20	91	85
0	108	99

Table 14–5   Comparison with Eckhouse

One of Nelson's yield curves where yield of heavy hydrocrackate is in terms of yield of light hydrocrackate, shows the heavy increasing with light up to a maximum at about 75% and then declining as yield of light continues to increase. Figure 14–19 is the author's confirmation of this phenomenon. This is not evident when both light and heavy are plotted against total gasoline (the sum of light and heavy).

Hydrocracking operating requirements

As one would expect, the consumption of fuel and electricity in hydrocracking vary with the amount of hydrogen being consumed as shown in Table 14–6.

SCFB	ELEC (kW)	FUEL (kBtu)
1,000	8.4	92.7
1,200	9.3	116.9
1,400	10.3	141.2
1,600	11.2	165.4
1,800	12.2	189.7
2,000	13.1	213.9
2,200	14.1	238.2
2,400	15.0	262.5
2,600	16.0	286.7
2,800	16.9	311.0
3,000	17.9	335.2

Table 14–6 Hydrocracker Utilities

Hydrocracker capital cost

The published capital costs for hydrocracking units since 1960 were scaled to 30,000 BPD of feed and translated to the first of 1991. The results can be divided into three groups that are more or less distinct. From 1960 through 1977 costs were reasonably steady with an average of $60 million for our plant. Between 1980 and 1985, the average jumped to $175 million. Between 1987 and 1991 the values deemed usable averaged about $87 million. It is very likely that many of the projects announced in the early 1980s were never built. It is thought that $90 to $100 million represents a realistic range of cost for our plant.

Notes

1. Nelson, W.L., numerous articles appearing in the *Oil & Gas Journal*. The following are particularly pertinent to this study: December 23, 1963, p. 66.; November 22, 1965, pp. 83–84.; March 14, 1966, pp. 127–128.; April 10, 1967, pp. 100–101.; May 8, 1967, p. 134.; June 26, 1967, pp. 84–85.; July 10, 1967, p. 200.; September 22, 1969, pp.153–154

2. Coonradt, H.L., Ciapetta, F.G., Garwood, W.E., Leaman, W.K., and Miale, J.N., *Industrial and Engineering Chemistry*, September, 1961, pp. 727–732

3. Eckhouse, J.G., *Oil & Gas Journal*, November 6, 1961, pp. 117–119

References

Anon., *Oil & Gas Journal*, April 18, 1960, pp. 104–106

Ibid, March 26, 1962, pp. 184–185

Ibid, June 16, 1969, pp. 74–75

Ibid, December 21, 1981, p. 88

Baral, W.J., and Miller, J.W., "Hydrocracking—A Route to Superior Distillate Products from Heavy Oils," Kellogg Symposium, Nice, France, September, 1982

Barnet, W.I., Duir, J.H., Hansford, R.C., and Tulleners, A.J., *Petroleum Refiner*, April, 1961, pp. 131–136

Cheadle, G.D., Welsh, C.J., Lappin, T.A., and Dana, P.M., *Oil & Gas Journal*, July 18, 1965, pp. 76–82

Craig, R.G., White, E.A., Henke, A.M., and Kwolek, S.J., *Hydrocarbon Processing*, May, 1966, pp. 159–164

Duir, J.H., *Oil & Gas Journal*, August 21, 1967, pp. 74–79

Hansford, R.C., Reeg, C.P., Wood, F.C., and Vaell, R., *Petroleum Refiner*, June, 1960, pp. 169–176

Huffman, H.C., Helfrey, P.F., Draeger, K.E., and Reichle, A.D., *Hydrocarbon Processing*, June, 1964, pp. 181–186

Jakobs, W.L., and Thornton, D.P., Jr., *Chemical Engineering*, November 16, 1970, pp. 79–81

Light, S.D., Bertram, R.V., and Ward, J.W., "New Zeolite Technology for Maximum Midbarrel Production," NPRA National Meeting, San Antonio, March, 1981

Mathews, J.W., Robbins, L.V., Jr., and Sosnowski, J., *Chemical Engineering Progress*, May, 1967, pp. 56–59

Michaelian, M.S., Shlegeris, R.J., and Haritatos, N.J., *Oil & Gas Journal*, May 18, 1970, pp. 72–81

Peralta, B., Reeg, C.P., Vaell, R.P., and Hansford, R.C., *Chemical Engineering Progress*, April, 1962, pp. 41–46

Read, D., Sterba, M.J., and Watkins, C.H., *Oil & Gas Journal*, May 20, 1963, pp. 110–116

Reichle, A.D., and Weller, N.O., *Oil & Gas Journal*, June 24, 1968, pp. 79–81

Reichle, A.D., and Wilkinson, H.F., *Oil & Gas Journal*, May 31, 1971, pp. 70–73

Robbers, J.A., Paterson, N.J., and Lane, W.T., "Commercial Isocracking of Heavier Gas Oils," 49th Annual Meeting, Western Petroleum Refiners Association, San Antonio, April, 1961

Rossi, W.J., Mayer, J.F., and Powell, B.E., *Hydrocarbon Processing*, May, 1978

Rossi, W.J., *Hydrocarbon Processing*, December, 1965, pp. 109–114

Scott, J.W., Robbers, J.A., Mason, H.F., Patterson, N.J., and Kozlowski, R.H., *Hydrocarbon Processing*, July, 1963, pp. 131–136

Sikonia, J.G., Jacobs, W.L., and Gambicki, S.A., *Hydrocarbon Processing*, May, 1978

Stormont, D.H., *Oil & Gas Journal*, April 25, 1966, pp. 146–167

Unzelman, G.H., and Gerber, N.H., *Petro/Chem Engineer*, October 1965, pp. 32–52

Vaell, R.P., Lafferty, J.L., and Sosnowski, J., "Produce Quality and Quantity Jet Fuel with Unicracking—JHC," 35th Midyear Meeting of API Div. of Refining, Houston, May, 1970

Van Driesen, R.P., and Fornoff, L.L., *Energy Progress*, January, 1982, pp. 47–52

Ward, J.W., Hansford, R.D., Reichle, A.D., and Sosnowski, J., "Unicracking—JHC Hydrocracking Catalysts and Processes—New Advances," API Div. of Ref'g., Philadelphia, May, 1973

Watkins, C.H., and Jacobs, W.L., *Oil & Gas Journal*, November 24, 1969, pp. 94–95

Watkins, C.H., "Providing Flexibility with Hydrocracking," UOP 1971 Technical Seminar, Arlington Heights, IL

Wood, F.C., Eubank, O.C., and Sosnowski, J., *Oil & Gas Journal*, June 17, 1968, pp. 83–89

CHAPTER 15

HYDROTREATING

Hydrotreating is a general term applied to processes where a feedstock is enhanced in some manner by passing it over a catalyst in the presence of hydrogen. In the process, there is essentially no reduction in molecular size of the feed. The objective has most often been to reduce the sulfur content of the feed. It is also practiced to reduce nitrogen content, saturate olefins, and to reduce aromatics. The latter is receiving much study due to the eminent requirement to lower the aromatic content of diesel fuel.

Sulfur is the easiest to remove, followed by nitrogen, and the aromatics. Saturation of olefins overlaps sulfur and nitrogen. Since hydrogen is the vital factor in this process, we find the hydrogen partial pressure required for desulfurization increasing from about 70psi to 100psi for virgin naphtha up to 350psi to 600psi for a vacuum gas oil. Cracked stocks require more hydrogen than do virgin stocks of the same boiling ranges. Temperatures employed are generally in the range of 625°F to 750°F. Hydrogen consumption in standard cubic feet per barrel (SCFB) ranges from 10 to 50 for virgin naphtha, up to 300 for heavy gas oil. Again, cracked stocks require greater amounts.

To reduce nitrogen to an acceptable level, as in cat feed hydrotreating, 600 SCFB to 1,100 SCFB may be required. It appears that the reduction of aromatics in diesel fuel to 10 to 12% could require a single–stage operation at about 1,800 psi or a two–stage operation at around 1,200 psi. Hydrogen consumption in either case would be about 1,100 SCFB.

Hydrotreating process description

A simplified process flow diagram of a typical hydrotreater is shown in Figure 15–1. The feed stream is combined with recycle hydrogen plus makeup hydrogen and heated in a fired heater on its way to a fixed bed reactor. The reactor effluent is separated in a high–pressure separator into a liquid phase and a recycle hydrogen stream. The high–pressure liquid is flashed into a low pressure separator producing a gas stream and a liquid feed to a fractionator. The gas is sent to a gas plant for recovery of C_3 and C_4 hydrocarbons. The liquid is fractionated into a hydrotreated product and lower boiling material produced in the reactor.

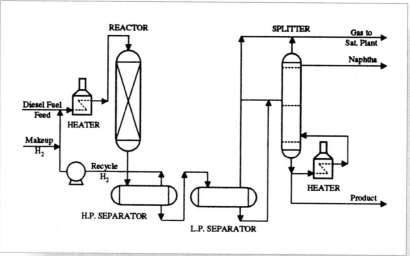

Fig. 15–1   Diesel Fuel Hydrotreating Unit

Hydrotreating data correlation

Data on hydrotreating gleaned from the literature have been separated and tabulated by type of feedstock as follows: Kerosene Data (Table 15–1), Furnace Oil Data (Table 15–2), Diesel Data (Table 15–3), Coker Gas Oil Data (Table 15–4), Cat Cycle Oil Data (Table 15–5), Thermal Cycle Oil Data (Table 15–6), Gas Oil Data (Table 15–7).

When considering addition of a hydrotreating unit to a refinery, it is important to determine very early the amount of hydrogen required by the

236

K	FEED API	%S	K	PRODUCT API	%S	SCFB H₂
11.9	48.2	0.0132	12.1	50.8		350
	47.7	0.0600		48.9	0.0004	280
11.9	47.5	0.2000	12.1	49.6	0.0001	400
	44.5	0.0260	12.0	46.0		1,000
	44.0	0.4000		44.8		40
	43.7	0.1400		44.6	0.0000	105
	43.7	0.1400		44.6	0.0000	321
	42.9	0.0240		43.6	0.0000	152
	42.9	0.0280		44.6	0.0003	379
11.9	42.9	0.0240		43.9	0.0002	155
	42.9	0.0240		43.8	0.0027	214
11.9	42.3	0.1000		44.8		400
11.7	41.8	0.3250	11.8	43.8	0.0127	220
11.5	41.0	0.0880	11.8	45.2	0.0003	340
	41.2	0.1300		43.9	0.0100	
	46.8	0.2700		47.0	0.0040	
	40.0	0.2200		44.6	0.0001	310
	44.2	0.0010		46.4	0.0001	350
	44.2	0.0010		48.4	0.0001	635
	37.4	1.2000		43.4		730
	44.9	0.2300		43.4	0.0200	
	38.5	0.0001		41.0	0.0000	434
	41.7	0.1000		42.0	0.0400	
	40.6	0.5500		41.5		
	47.7	0.0600		48.5	0.0004	280
	40.3	0.0600		41.3	0.0006	215
	40.3	0.0600		42.2	0.0002	300
	40.2	0.5300		41.3	0.0300	

Table 15–1 Kerosene Data

──FEED──		──PRODUCT──		SCFB
API	%S	API	%S	H$_2$
30.2	2.2000	33.2	0.35000	150
39.2	1.0200	40.6	0.10500	93
42.3	0.1800	42.8	0.01000	25
39.2	1.0200	40.6	0.10500	93
30.7	1.6000	33.0	1.00000	200
23.0	2.52	26.8	0.5	
38.9	1.13	40.4	0.33	
37.3	0.45	39.2	0.02	330
35.1	2.34	39.6	0.08	
32.9	0.36	34.8	0.04	
39.6	0.71	40.3	0.11	
33.6	1.49	35.7	0.18	
33.7	1.2	35.0	0.32	
33.7	1.2	35.1	0.25	
30.4	1.81	31.9	0.65	
30.4	1.81	31.8	0.49	
38.8	1.08	40.6	0.06	
34.4	1.64	37.2	0.17	
32.4	0.8	38.8	0.05	110
40.1	0.66	40.9	0.05	

Table 15–2 Furnace Oil Data

hydrotreater to reach the desired objective. It would be necessary to have a hydrogen balance for the refinery to know how much hydrogen may be available for the addition. The result may be that another source of hydrogen is required. This could represent a significant investment. The data tables listed above show actual chemical hydrogen consumption in SCFB. An allowance should be added to such data to account for the hydrogen left in the liquid phase from the separators. This is referred to as "solution loss." Actually it is not lost, but ends up in the refinery fuel gas where it is seldom practical to recover as a usable hydrogen stream.

Since most of the data are for deep desulfurization, usually in excess of 90%, it was thought that the hydrogen requirement could be correlated in terms of the API gravity and the sulfur content of the feed. Mixed results were obtained. Data for kerosene, thermal cycle oil, and virgin gas oil indicated an increase in hydrogen required as the API gravity of the feed increased. Diesel fuel and cat cycle oil indicated a decrease in hydrogen as

FEED		PRODUCT		SCFB
API	%S	API	%S	H₂
41.0	0.600	42.0	0.060	
41.0	1.000		0.100	178
40.3	0.660	41.4	0.030	
40.2	0.645	41.1	0.106	
39.2	1.020	40.6	0.105	93
38.7	0.900	39.8	0.040	
37.9	0.920	38.4	0.150	
37.9	0.920	38.4	0.500	
37.5	0.990	39.9	0.096	
35.9	1.320	37.9	0.210	115
34.8	1.400	37.6	0.084	35
33.2	1.750	35.9	0.000	200
33.1	1.400	36.2	0.100	
32.3	1.770	35.0	0.370	138
32.3	1.770	35.0	0.370	138
32.3	1.770	35.0	0.370	150
31.5	1.060	34.0	0.080	
31.3	0.720	38.0	0.002	1,060
31.3	2.100	34.8	0.140	140
31.1	1.040	31.7	0.450	
30.7	1.600	33.0	0.096	200
30.2	2.200	33.2	0.350	150
29.1	0.980	38.0	0.002	1,120
28.6	2.310	37.0	0.002	1,200
28.0	1.110	30.3	1.000	500
28.0	1.110	30.4	0.070	500
28.0	1.110	30.3	0.080	500
28.0	1.110	30.1	0.140	500
28.0	1.110	30.2	0.070	750
28.0	1.110	30.2	0.110	500
26.4	0.200	32.8	0.010	870
25.5	1.680	29.9	0.050	530
25.5	1.680	28.4	0.118	350
24.0	0.100	34.5	0.000	1,500
23.4	2.910	28.4	0.437	375
23.4	2.910	29.5	0.320	455
22.1	1.820	26.5	0.070	570
22.1	1.820	25.7	0.440	240
21.4	0.070	22.4	0.000	100

Table 15–3 Diesel Data

	FEED			PRODUCT		SCFB	
K	API	%S	K	API	%S	H₂	%DES
	40.0	0.2200	11.90	44.6		310	100.00
	39.1	0.5700	11.74	42.6		630	100.00
	38.6	0.6000	11.72	42.4		730	100.00
	38.2	0.6500	11.70	41.9		660	100.00
	38.1	0.1600	11.67	40.9		540	100.00
	37.4	1.2000	11.80	43.4		730	100.00
	37.4	0.7200	11.76	41.9		680	100.00
	35.0	0.9000			0.10000	500	89.00
11.25	29.4	0.5400	11.80	32.1	0.07000	1,500	87.04
	26.9	3.9000		39.0	0.05000	800	98.72
	26.5	2.3400		34.0	0.07000	625	97.01
11.70	25.1	2.4200	11.60	31.6		625	96.00
11.50	24.0	3.8000		30.5		700	100.00
11.50	18.7	2.7400		24.6	0.37000	475	86.50
11.50	18.7	2.7400		24.6	0.37000	475	86.50
11.50	15.4	1.7000		23.8	0.10000	1,070	94.12
11.50	11.3	2.4000		20.5	0.10000	1,180	95.83
11.50	10.5	4.1000		18.8		1,150	95.00
11.60	10.1	5.8000		20.1	0.10000	1,430	98.28
	-2.0	1.6100		2.2	0.66000	885	59.01
	19.3	2.54		26.3	0.06		
	26.9	3.9		39.0	0.05	200	
	34.9	0.53		36.9	0.08		
	40.9	0.11		41.1	0.006		

NOTE: %DES denotes the percent of feed sulfur removed

Table 15–4   Coker Gas Oil Data

FEED		PRODUCT		SCFB
API	%S	API	%S	H₂
28.6	2.3100	36.0	0.00200	1,200
29.1	0.9800	36.0	0.002	1,120
27.6	0.5300	30.1	0.01000	1,000
27.6	0.5300	30.6	0.01000	100
27.6	0.5300	30.3	0.01000	1,000
27.6	0.5300	30.6	0.01000	1,000
27.6	0.5300	30.0	0.01000	500
27.6	0.5300	29.9	0.01000	1,000
27.6	0.5300	30.1	0.01000	500
27.6	0.5300	30.0	0.01000	1,000
27.6	0.5300	30.6	0.01000	1,000
27.6	0.5300	29.9	0.01000	1,000

Table 15–6   Thermal Cycle Oil Data

SCFB H₂	FEED API	FEED S	PRODUCT API	PRODUCT S
450	26.2		32.2	0.020
371	29.7	0.19	33.2	0.050
408	19.3	0.20	26.2	0.160
165	19.3	0.20	29.9	0.009
190	31.2	0.21	38.9	0.010
350	31.9	0.22	35.1	0.005
150	27.0	0.38	29.3	0.030
1,150	15.8	0.39	17.8	0.220
500	15.8	0.39	19.0	0.220
800	27.6	0.40	30.6	0.070
3,380	37.3	0.45	39.2	0.020
1,430	26.5	0.47	32.1	0.070
408	27.2	0.78	35.0	0.060
850	27.2	0.78	31.0	0.210
1,500	26.2	0.83	31.7	0.015
1,000	26.8	0.84	28.7	0.080
270	30.3	0.85		0.050
160	30.3	0.85		0.300
1,500	18.7	0.88	25.1	0.030
1,500	21.0	0.89	26.6	0.020
170	29.0	0.93	32.3	0.038
190	29.6	1.04	33.0	0.110
70	31.0	1.10	33.0	0.090
110	26.0	1.15	30.0	0.180
	26.0	1.15	30.0	0.180
	26.3	1.17	28.3	0.130
	7.4	1.25	9.2	0.400
	22.7	1.37	25.8	0.260
	25.7	1.40	26.9	0.000
	17.7	1.42		0.300
	17.7	1.42		0.060
	21.5	1.52	25.7	0.090
	19.7	1.64	25.0	0.117
330	22.6	1.72	26.4	0.130
	25.9	1.80	29.2	0.080
180	22.6	2.10	26.0	0.360
	26.6	2.17	30.3	0.210
	19.3	2.25	23.4	0.400
750	19.1	2.49	25.0	0.040
1,500	23.4	2.91	28.4	0.440
	23.4	2.91	29.5	0.320
	22.6	3.06	35.1	0.240

Table 15–5 Cat Cycle Data

241

K	FEED API	%S	K	PRODUCT API	%S	SCFB H2
	31.9	0.2200		35.1	0.0050	450
	29.8	0.5100		31.9	0.0600	110
	29.4	0.5400		32.1	0.0700	1,500
	27.5	0.6200		30.3		445
	27.5	0.6200		29.8		380
	27.5	0.6200		31.0		500
	27.5	1.8000		32.0	0.1000	400
	26.9	3.9000		39.0	0.0500	800
	26.5	2.3400		34.0	0.0700	625
	25.5	1.6800		28.4	0.1200	350
	25.5	1.6800		29.9	0.0500	530
	25.5	1.6800		29.9	0.0500	530
	25.5	1.6800		28.4	0.1200	350
	25.1	2.4200		31.6	0.0900	625
	24.8	2.6600		30.2	0.2400	300
	24.7	1.9000			0.2000	470
11.95	24.6	2.3000	12.10	27.0	0.7900	183
11.95	24.6	2.3000	12.12	27.7	0.4300	240
	24.6	0.7200		27.2	0.1600	220
11.95	24.6	2.3000	12.18	29.3	0.0500	400
	24.6	0.3720		27.2	0.1600	220
11.95	24.6	2.3000	12.15	28.5	0.1600	277
	24.5	0.7800		25.5	0.4100	87
	24.1	2.2200		28.7	0.3400	520
	24.0	3.8000		30.5	0.0400	700
	24.0	2.2000		29.0		350
	23.6	1.9400		27.2	0.3900	170
	23.4	2.9100		29.6	0.3200	455
	23.4	2.9100		28.4	0.4400	375
	23.1	2.8000				375
11.60	23.1	1.2900	11.70	25.9		1,500
11.60	23.1	1.2900	11.83	27.7		1,500
	23.1	2.2800			0.2300	220
	23.0	2.6000				460
	23.0	3.0000				450
	22.7	1.3700		25.8	0.2600	150
	22.4	1.9100			0.1900	220
	22.2	3.0500			0.3000	280
	22.1	1.8200		25.7	0.4400	240
	22.1	1.8200		26.5	0.0700	570
11.45	21.8	1.1600	11.87	30.8	0.0500	675
11.45	21.8	1.1600	11.75	28.9	0.1200	400
	21.7	2.9700			0.3000	300
	21.5	1.5200		25.7	0.0900	500
	21.4	1.6800		24.5	0.2800	250
11.40	21.2	1.1800	11.75	27.9	0.0600	675
	21.0	0.8900		26.6	0.0200	800
	21.0	2.6000		24.5	0.5000	210
	18.8	2.7000		24.0	0.2000	425
	18.7	2.7400		24.6	0.3700	475
	15.4	1.7000		23.8	0.1000	1,070
	11.3	2.4000		20.5	0.1000	1,180
	10.1	5.8000		20.1	0.1000	1,430

Table 15–7 Virgin Gas Oil Data

242

the sulfur content of the feed increased. Only furnace oil and coker gas oil were well behaved.

The next attempt involved the use of sulfur content of feed and percent desulfurization as independent variables. Here again, mixed results were obtained. This time, only diesel fuel data gave a good correlation. The others indicated a decrease in hydrogen requirement as percent desulfurization increased.

Data for all the stocks were combined into one base for study. Unsatisfactory results were obtained when feed API and sulfur were used as independent variables. A good correlation resulted when feed sulfur and percent desulfurization were used. This correlation appears as Figure 15–2. It should be recognized that this correlation represents a compromise. Values read from this chart will be consistently high for some stocks, consistently low for others. For a given stock, increments read from the chart are probably good.

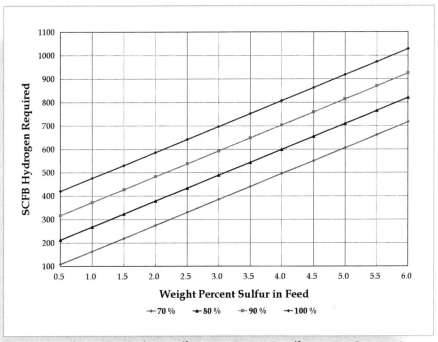

Fig. 15–2 Hydrogen Required in Desulfurization (Percent Desulfurization as Parameter)

243

Another item of interest in hydrotreating is the API gravity of the treated product. All of the stocks except thermal cycle oil gave good correlations using feed API and SCFB of hydrogen as independent variables. In effect, there were only four data sets for thermal cycle, not enough to resolve the problem. Rather than show the individual charts, all the data were combined and correlated in terms of feed API and SCFB of hydrogen vs. the increase in API as shown in Figure 15–3. Some of the correlation results appear in Table 15–8.

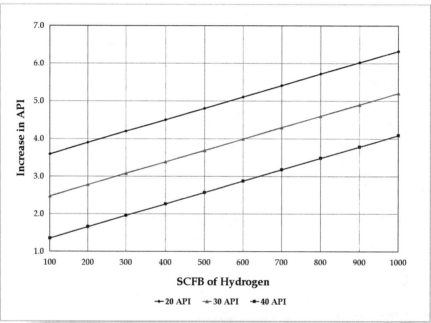

Fig. 15–3   API Increase in Hydrotreating (API Gravity of Feed as Parameter)

It should be borne in mind that when a large increase in API is reported for a stock (for a product with essentially the same boiling range as the feed), there is significant conversion of some of the feed to lower boiling material. Such instances are outside the scope of the present discussion.

Comparison with other correlations

No correlations of the sort being presented here were found in the literature on hydrodesulfurization. Several sources tabulated ranges of hydro-

DEPENDENT VARIABLE	KEROSENE	FURNACE OIL	DIESEL FUEL	COKER GAS OIL	CAT CYCLE OIL	THERMAL CYCLE	VIRGIN GAS OIL	ALL GAS OIL
HYDROGEN REQUIREMENTS—SCFB								
INDEPENDENT VARIABLE	FD API FD S	FD API FD S	FD API FD S	FD API FD S	FD API FD S	FD API FD S	FD API FD S	FD S %DES
R^2	0.884	0.822	0.962	0.801	0.918	1.00	0.902	0.91
SEE	50.3	32.19	47.13	137.2	45.79	0.00001	41.34	53.1
S1	4.845	4.535	2.744	3.509	2.82	0.00001	6.486	6.01
S2	54.38	33.9	19.84	29.94	24.91	0.00001	13.61	0.78
NO.	12	6	17	13	9	7	18	48
API GRAVITY OF PRODUCT								
INDEPENDENT VARIABLE	FD API	FD API	FD API	FD API	FD API	FD API	FD API	FD API SCFB H$_2$
R^2	0.9135	0.984	0.98	0.99	0.916	0.981	0.91	0.929
SEE	0.837	0.553	0.678	0.987	1.056	0.2915	0.857	2.28
S1	0.0648	0.0274	0.0226	0.0218	0.0455	0.2062	0.035	0.0005
S2								0.023
NO.	24	18	30	17	29	9	31	126

Table 15–8 Hydrotreating Correlation Results

gen required per pound or per weight percent of sulfur and other things such as nitrogen, olefins, aromatics, etc.[1,2,3]

Operating requirements in hydrotreating

The following tabulation of hydrotreating operating requirements is a fair consensus of the various sources consulted. Values are units per barrel of feed.

	Electric Water (kW)	*Fuel MBtu*	*Steam lb.*	*Cooling gal.*	*Hydrogen SCFB*
Kerosene	1.7	8.5	6.9	negl.	25–140
Diesel	1.7	8.3	6.8	negl.	50–200
Lt Gas Oil	1.5	34.5	7.2	negl.	100–200
Hy Gas Oil	1.3	16.3	7.8	negl.	200–300
Cat Cycle	1.4	24	7.2	400	100–900

In the case of cat feed hydrotreating, the objective is to reduce Conradson carbon residue and nitrogen. The quantity of hydrogen required can range from 600 SCFB to 1,100 SCFB.

Hydrotreating capital costs

Published data on capital cost of hydrotreating units were scaled and translated as with previous processes to 30,000 barrels of feed per day the first of 1991. A value of approximately $25 million appears reasonable for the lighter distillates such as kerosene and diesel. A value of $16 million is indicated for the gas oils. For a cat feed hydrotreater, a value of $37 million is indicated.

Notes

1. Edgar, M.D., Johnson, A.D., Pistorius, J.T., and Varadi, T., *Oil & Gas Journal*, June 4, 1984, pp. 67–70

2. Grote, H.W., Watkins, C.H., Poll, H.F., and Hendricks, G.W., *Oil & Gas Journal*, April 19, 1954, pp. 211–216

3. Kay, H., *Petroleum Refiner*, Vol. 35, No. 9, 1956, pp. 306–318

References

Abbott, M.D., Liedholm, G.E., and Sarno, D.H., *Oil & Gas Journal*, July 18, 1955, pp. 92–94

Abbott, M.D., Archibald, R.C., and Dorn, R.W., *Petroleum Refiner*, May, 1958, ff. 161

Aga, R.L., Debus, H.R., and Allen, E.R., "Jet Fuel Production by Arofining," 68th National Meeting of AIChE, Houston, March 1971

Anon., Brochure: "I.F.P. Fuel–Oil Desulfurization Process," Ref. 20 592, September, 1972

Anon., Lummus Co. Brochure: "Shell Hydrodesulfurization Process," May, 1973

Anon., *Oil & Gas Journal*, Jan. 7, 1980, pp. 75–77

Asim, M.Y., Keyworth, D.A., Zoller, J.R., Plantenga, F.L., and Lee, S.L., "Hydrotreating for Ultra–Low Aromatics in Distillates," NPRA Annual Meeting, San Antonio, 1990

Beaton, W.I., McDaniel, N.K., McWhirter, W.E., Petersen, R.D., and Van Driesen, R.P., *Oil & Gas Journal*, July 7, 1986, pp. 47–53

Bradley, W.E., Hendricks, G.W., Huffman, H.C., and Kelley, A.E., *Oil & Gas Journal*, June 8, 1959, pp. 194–198

Brown, C.L., Voorhies, A., Jr., and Smith, W.M., Industrial and Engineering Chemistry, February, 1946, pp. 136–140

Busch, R.A., Kociscin, J.J., Schroeder, H.F., and Shah, G.N., "Flexicoking + Hydrotreating Processes for Quality Products," AIChE National Meeting, Houston, April, 1979

Christensen, R.I., Frumkin, H.A., Spars, B.G., and Tolberg, R.S., "Low Sulfur Products from Middle East Crudes," NPRA Annual Meeting, San Antonio, 1973

Collins, J.M., and Unzelman, G.H., *Oil & Gas Journal*, May 30, 1983, pp. 71–78

Colvert, J.H., "Hydrogenation for Jet Fuel Manufacture," API Div. of Refining, Chicago, May 1969

Davidson, R.L., *Petroleum Processing*, November, 1956, pp. 116–138

Debus, H.R., Cahen, R.M., and Aga, R.L., *Hydrocarbon Processing*, September, 1969, pp.137–140

Eberline, C.R., Wilson, R.T., and Larson, L.G., *Industrial and Engineering Chemistry*, April, 1957, pp. 661–663

Eckhouse, J.G., Gerald, C.F., and de Rosset, A.J., *Oil & Gas Journal*, August 30, 1954, pp. 81–83

Fowler, C., *Hydrocarbon Processing*, September, 1973, pp. 131–133

Gilmartin, R.P., Horne, W.A., and Walsh, B.R., *Oil & Gas Journal*, February 6, 1956, pp. 85–88

Hallie, H., Yoe, J.R., Murff, S.R., Peterson, R.E., and Stanger, C.W., "Hydrotreating of Cracked Feedstocks," Ketjen Catalyst Seminar, Philadelphia, September, 1984

Hansford, R.C., Gaudio, D.A., Inwood, T.V., and Mavity, V.T., Jr., *Oil & Gas Journal*, May 5, 1969, pp. 134–136

Hansford, R.C., Inwood, T.V., Kouzel, B., and Mavity, V.T., Jr., "Unisar: Four Years of Proven Performance," NPRA Annual Meeting, San Antonio, April, 1973

Hendricks, G.W., Huffman, H.C., Kay, N.L., Stiles, V.E., Attane, E.C., and Inwood, T.V., *Petroleum Refiner*, February, 1957, pp. 135–139

Hoog, H., Klinkert, H.G., and Schaafsma, A., *Oil & Gas Journal*, June 8, 1953, pp. 92–94

Horne, W.A., McKinney, J.D., and Rice, T, *Petro/Chem Engineer*, May, 1960, pp. C–19 to C–23

Kellett, T.F., Sartor, A.F., and Trevino, C.A., *Hydrocarbon Processing*, May 1980, pp. 139–142

McAfee, J., and Horne, W.A., *Petroleum Processing*, April, 1956, pp. 47–52

McCulloch, D.C., Edgar, M.D., and Pistorius, J.T., *Oil & Gas Journal*, April 13, 1987, pp.33–38

Morbeck,R.C., *Oil & Gas Journal*, January 3, 1955, pp. 94–98

Mosby, J.F., McBride, W.L., and Moore, T.M., "Low Pressure Heavy Distillate Ultrafining," NPRA Annual Meeting, San Antonio, April, 1973

Moyse, B.M., Albjerg, A., and Cooper, B.H., *Oil & Gas Journal*, May 11, 1985, pp. 111–115

Murphy, H.C., Jr., Nejak, P.P., and Strom, J.R., "High Pressure Hydrogenation—Route to Specialty Products," API Div. of Refining, Chicago, May, 1969

Nash, R.M., *Oil & Gas Journal*, May 29, 1989, pp.47–56

NPRA Q and A Session, *Hydrocarbon Processing*, February, 1988, pp. 44–50

Ibid, pp. 113–116

Odasz, F.B., and Sheffield, J.V., *Oil & Gas Journal*, March 1955, pp. 203–204

Patterson, A.C., and Jones, M.C.K., *Oil & Gas Journal*, October 18, 1954, pp. 92–94

Poll, H.F., *Petroleum Refiner*, July 1956, pp. 193–198

Reynolds, G.P., *Hydrocarbon Processing*, May, 1965, pp. 160–164

Roeder, R.A., *Oil & Gas Journal*, October 15, 1973, pp. 123–126

Schuman, S.C., Chemical Engineering Progress, Vol. 57, No. 12, 1961, pp. 49–54

Siegmund, C.W., *Hydrocarbon Processing*, February 1970, pp. 89–95

Slyngstad, C.E., and Lempert, F.L., *Petro/Chem Engineer*, May, 1960, pp. C–13 to C–18

Stuckey, A.N., Fant, B.T., Duir, J.H., and Mickelson, G.A., "GO–FINING Goes Low Pressure," API Div. of Refining, New York City, May, 1972

Suchanek, A.J., *Oil & Gas Journal*, May 7, 1990, pp. 109–117

Sze, M.C., and Bauer, W.V., "Hydroprocessing of Pyrolysis Gas Oil," AIChE National Meeting, Houston, 1971

Tregilgas, E.T., and Crowley, D.M., "New Atlantic Richfield Hydrotreating Process Increases Jet Fuel Yields," API Div. of Refining, Chicago, May, 1969

Van der Giessen, J.A., *Hydrocarbon Processing*, August, 1970, pp. 113–114

Venuto, P.B., and Habib, E.T., Fluid Catalytic Cracking with Zeolite Catalysts, Marcel Dekker, Inc., New York, 1979

Watkins, D.H., and de Rossett, A.J., *Petroleum Refiner*, March, 1957, pp. 201–204

Watkins, C.H., and Czajkowski, C.J., *Chemical Engineering Progress*, August, 1971, pp. 75–80

Wolfson, M.L., Pelipetz, M.G., Damick, A.D., and Clark, E.L., *Industrial and Engineering Chemistry*, February, 1951, pp. 536–540

SECTION D:

LIGHT DISTILLATE PROCESSING

CHAPTER 16

NAPHTHA DESULFURIZATION

Naphtha hydrotreating is generally practiced to prepare feedstock for a catalytic reformer and many refiners treat the two processes as one unit.

All the stocks normally reformed except hydrocrackate contain sulfur, which is a poison to the noble metals present in reformer catalysts. The primary purpose of hydrotreating these naphthas is to reduce the sulfur content to a tolerable level.

An appreciation of the task facing the hydrotreater can be gained by reviewing data presented by J. T. Pistorius of American Cyanamid Co.:[1]

Maximum acceptable contaminant
levels for reformer feeds:

Sulfur	1 wt ppm
Nitrogen	0.5 wt ppm
Lead	10 wt ppb
Arsenic	2 wt ppb
Water	10 wt ppm
Chloride	1 wt ppm

Sulfur is readily removed from naphthas. Naphthas from some crudes will require more severe treatment to meet the nitrogen requirement. Where metals are expected to be a problem, one licensor advises the addition of an amount of sacrificial catalyst as a trap. Certain crudes contain sufficient levels of arsenic, for example, to require some such precaution.

Naphtha hydrotreater process description

The naphtha hydrotreater is essentially the same as the hydrotreater described in the previous chapter. It consists of a feed heater, a reactor, high– and low–pressure separators, a recycle compressor, and a treated naphtha splitter. In addition, when a highly unsaturated stock such as coker naphtha is being fed, a separate additional reactor may precede the main reactor (Fig. 16–1). This is to selectively saturate (under milder conditions than those in the main reactor) acetylenes and dienes in order to prevent runaway temperature increases due to the highly exothermic reactions when they are present.

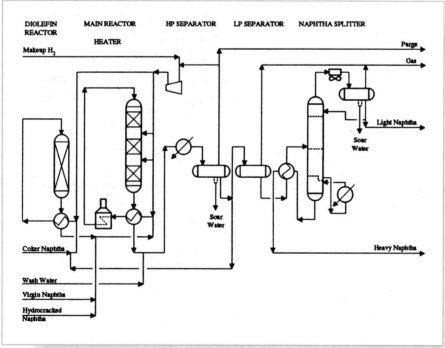

Fig. 16–1   Naphtha Hydrotreater

A hydrogen sulfide stripper is sometimes placed between the separators. If naphtha is being received from storage where there is no inert gas blanket, a reboiled stripper would be needed ahead of the heater to remove oxygen.

Naphtha hydrotreating correlations

Table 16–1 is a tabulation of data on naphtha hydrotreating assembled from the literature references listed. The data were segregated by type of stock for individual study. As in the case of hydrotreating the heavy distillates, some of the stocks did not correlate well (at least within the data populations at hand). In the case of hydrogen required, coker naphtha showed a decreasing hydrogen requirement as the sulfur content of the feed

FEEDSTOCK	–FEED – API	S	SCFB H₂	API	PRODUCT PROD S	LV %	FEED RON	PROD RON	FEED MON	PROD MON
HY CAT 278–402		0.34			0.17000		87.9	86.4		
SR		0.12			0.01000		63.5	64.0		
TML 300–450		0.28			0.12000		59.5	52.5		
CAT 152–297		0.28	500		0.00300	102.0				
CAT 250–442		0.24			0.07000		92.5	90.5		
COKR		1.03			0.15000		70.7	66.9		
TML 228–416		1.15	225		0.40000	101.0				
HY TML		0.28			0.12000		60.0	53.0		
SR 216–300		0.06			0.00600		40.0	45.0	43.0	46.0
CAT 152–297		0.28	500		0.00300	102.0				
SR 200–320		0.12			0.01000		63.5	64.0		
TML 228–416		1.15	225		0.40000	101.0				
CAR 286–427		0.50	138		0.20000	101.0				
HY CAT 250–442		0.24			0.07000		92.5	90.5		
HY TML 300–450		0.28			0.12000		59.5	52.5		
CAT 200–350		0.18			0.02000		88.9	85.9	77.5	77.5
SR 170–330		0.10			0.00400		48.0	50.0	51.0	51.0
HY CAT		0.20			0.14000		90.0	90.0	79.0	79.0
CAT 286–427		0.50	138		0.20000	101.0				
SR		0.15			0.00010		50.0	53.0		
TML/CAT	40.2	0.52	120	41.5	0.09000	100.0	87.0	81.0	76.0	73.0
TML/CAT 256–414	40.2	0.52	120	41.5	0.09000	100.0				
COKR 250–430	42.8	1.59		48.0	0.00700					
PYRO GASO	45.1	0.01		45.4	0.00020					
TML 229–400	45.6	2.47		47.4	0.03000		80.1	55.0		
SR/TML 221–440	46.1	1.90	435	51.1	0.00800	101.0				
PD 192–403	46.2	0.63		49.2	0.01800	100.0			73.0	59.0
PD 192–403	46.2	0.63		47.7	0.00000	96.0			73.0	62.0
VB 150–430	46.4	1.00		47.4	0.10000		71.2	63.6	66.2	62.4
VB/COKR 150–430	46.5	1.03		48.2	0.15000		70.7	66.9	66.0	64.9
SR 272–361	47.1	0.14	15	47.2	0.00220	99.9				

Table 16–1a Naphtha Hydrodesulfurization Database

FEEDSTOCK	–FEED – API	S	SCFB H$_2$	—PRODUCT— API	PROD S	LV %	FEED RON	PROD RON	FEED MON	PROD MON
COKER	47.8	1.90	400	55.0	0.00500					
CAT 200–350	48.1	0.18		49.0	0.02000		88.9	85.9	77.5	77.5
SR/CRKD 194–419	48.4	0.79		51.0	0.00500					
CRKD 215–408	49.1	0.12	420	51.0	0.00800	102.3				
VB 190–352	49.1	3.00		56.4	0.01000	97.0			70.0	60.5
SR 152–315	49.1	0.26		54.0	0.00200					
SOLVENT	49.2	0.01	375	52.0	0.00001	102.2				
SOLVENT	49.3	0.00	500	52.4	0.00005	101.4				
TML	49.7	0.08	260	51.8	0.00060					
TML	49.7	0.08	235	51.2	0.00480					
TML	49.7	0.08	275	51.6	0.00020					
TML	49.7	0.08	275	51.5	0.00100					
TML	49.7	0.08	270	51.2	0.00040					
TML	49.7	0.08	265	51.5	0.00020					
SR	49.8		470	52.6						
SR 302–408	49.9	0.14		49.9	0.00300					
SR/VB	50.0	0.31	170	51.0	0.00100					
SR/COKR	50.3	0.11		50.9	0.00100					
246–460	50.7	0.11	90	50.6	0.00110	100.0				
C3= TETRAMER	51.5	0.00	995	54.0	0.00020	102.6				
SR 200–331	51.9	0.14		52.0	0.00300					
SR 113–280	52.2	0.07		55.0	0.00130					
SR 119–420	52.9	0.64		55.3	0.02600		54.0	55.6		
SR	52.9	0.14	75	53.4						
TML 122–402	53.5	0.31		54.2	0.01500		65.8	61.0		
TML 122–402	53.5	0.31		56.7	0.00100		65.8	37.1		
COND	53.6	0.64	103	52.8	0.09000					
COKER	54.0	0.16	250	56.1	0.00120					
COKER	54.0	0.16	250	55.4	0.01800					
SR 260–378	54.3	0.08	15	54.4	0.00110	100.0				
	54.4	0.35		55.6	0.00060					
SR	55.0	0.06	10	55.3	0.00100					
	55.5	0.02	187	56.6	0.00010	101.1				
SR 104–402	55.5	0.28		58.1	0.00500		56.2	58.0		
	55.5	0.02	68	55.4	0.00020	100.0				
CAT 114–427	55.9	0.14		55.0	0.01500		91.0	83.0		
'SR 138–312	56.0	0.10		58.0	0.01000					
SR 163–398	56.4	0.14	30	56.5	0.00200	100.0				
COKER	56.7	0.48	800	64.4	0.00010	104.9				
COKER	56.7	1.17	890	64.9	0.00010	104.4				
COKER	57.4	0.34	780	65.0	0.00010	105.1				
SR 153–330	58.9	0.15	50	60.6	0.00160	100.4				
COKR	59.0	0.80	950	65.1						
CAT	60.0	0.16	145	61.3	0.01000					
SR/COKR C5–350	62.5	0.28		63.0	0.00700		69.4	64.2		
SR/VB 100–399	62.8	0.14	200	64.4	0.00800	100.0				
TML	65.6	0.01	80	66.1	0.00100					
COKR	66.4	0.20	600	66.8						
COKER	66.4	0.20		66.8	0.00090					
SR	67.5	0.02	10	67.5						

Table 16–1b   Naphtha Hydrodesulfurization Database cont'd

increased. In the case of the increase in API gravity of naphtha resulting from hydrotreating, it was the catalytic cracked naphtha that did not give a satisfactory correlation.

Good results were obtained when data for all the naphthas were combined. It was necessary to disregard some obvious "outliers" in the process. The results are shown in Table 16–2 and Figures 16–2 and 16–3. The API is needed only to permit weight balance(s) around the hydrotreater and/or the reformer. The hydrogen requirement is needed in making a hydrogen balance, with due allowance for solution losses, etc.

One matter that is of importance to the reformer is the composition of the feed with respect to hydrocarbon types. This does change in the course of hydrotreating, notably in the case of the cracked materials. Consequently, Table 16–3 was prepared to illustrate some of these changes and give the reader some guidance. As we shall see, the yield of reformate is strongly dependent on the PONA analysis of the reformer feed.

FEEDSTOCK	SR	TML	CAT	COKER	ALL
DEPENDENT VARIABLE	HYDROGEN REQUIREMENTS - SCFB				
INDEPENDENT VARIABLE	FD S % DES	FD S % DES	FD S % DES	FD S % DES	FD S % DES
R^2	0.929	0.908	0.996	0.775	0.901
SEE	9.015	21.5	15.55	174.2	41.12
S1	76.92	59.6	94.65	137.8	16.7
S2	5.548	1.96	0.879	19.4	0.762
NO.	6	9	5	6	18
C0	−1,963.4	−1,258.1	−1,946	−5,250.5	−30.64
C1	431.4	436.5	1,832.6	−230.6	191.43
C2	19.7	15.05	19.49	62.4	0.751
DEPENDENT VARIABLE	API GRAVITY OF PRODUCT				
INDEPENDENT VARIABLE	FD API SCFB	FD API SCFB	FD API SCFB	FD API SCFB	FD API SCFB
R^2	0.992	1.0		0.965	0.96
SEE	0.642	0.0402		1.18	1.331
S1	0.0423	0.00006		0.1259	0.0009
S2	0.0017	0.0026		0.00195	0.0398
NO.	8	8		6	32

Table 16–2 Some Results of Naphtha Hydrotreating Correlations

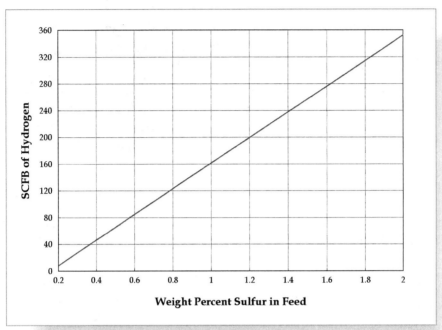

Fig. 16–2 Hydrogen Required in Naphtha Desulfurization

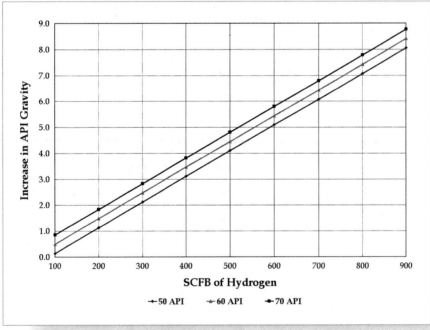

Fig. 16–3 API Increase in Naphtha Hydrotreating (API Gravity of Feed as Parameter)

	API	RON	P	O	N	A
STRAIGHT RUN NAPHTHAS						
BEFORE	52.9	23.5	64	1	20	15
AFTER	52.8	33.5	63	1	21	15
BEFORE	55.5	56.2	51	2	29	18
AFTER 98.2% DES	58.1	58.0	46	1	32	21
BEFORE	52.9	54.0	42	4	44	9
AFTER 95.9% DES	55.3	56.0	40	2	44	14
THERMAL NAPHTHAS						
BEFORE	53.5	66.0	44	27	17	11
AFTER 95.2% DES	54.2	61.0	42	24	21	12
AFTER 99.7% DES	56.7	37.0	51	2	34	13
BEFORE	45.6	80.0	39	——49——		12
AFTER 98.8% DES	47.4	55.0	45	1	40	13
BEFORE	47.9	70.0	25	43	17	15
AFTER 97% DES	50.4	46.8	45	2	38	15
BEFORE	55.3		40	34	14	12
BEFORE			27	46	14	13 AVE.
AFTER	50.4	46.8	45	2	38	15
VISBREAKER NAPHTHA						
BEFORE			23	45	11	11 AVE.
COKER NAPHTHAS						
BEFORE			13	58		29
BEFORE			42	38	10	10
BEFORE			37	33	15	15 AVE.
AFTER		49.0	49		39	12
AFTER			40	5	30	25
FLEXICOKER NAPHTHAS						
BEFORE	47.4		11	37	5	47
BEFORE		77.5	15	58	3	24
CAT CRACKER NAPHTHAS						
BEFORE		89.7	28.7	16	13.6	41.7
AFTER		85.0	31	——33——		36
BEFORE	36.6	90.2	17	29	8	46
BEFORE	55.9	91.0	33	44	2	21
AFTER 89.3% DES	55.0	83.0	34	38	6	22
HEAVY HYDROCRACKATE						
HVY HC		67.0	20	0	63	11
HVY HC		84.0	17	0	41	42
HVY HC		92.0	15	0	29	56
HVY HC	51.7	45.0	40	3	43	14
HVY HC	54.0	48.2	44	2	44	10

Table 16–3   Effect of Hydrotreating on Naphtha Properties

259

As we would expect, the properties shown in Table 16–3 for straight run naphthas are little affected by hydrodesulfurization. We see some slight increases in API gravity and octane, but little change in PONAs. A very different picture is seen for thermal naphthas. The API increases are still small, but the decrease in octane is very dramatic. This is due to saturation of olefins that generally have higher octane than the corresponding paraffin. The value shown as olefin includes cyclic olefins as well as alkenes. So we see an increase in naphthenes as well as paraffins. These same generalizations apply to the other cracked stocks.

Some of the PONAs are denoted as Ave. and can be used as representative where specific data are lacking. Data are included for heavy hydrocrackates. Even though they would not go through the hydrotreater, they are generally reformed to increase their octane.

Naphtha hydrotreating operating requirements

Operating requirements for naphtha hydrotreating are modest and consist of approximately the following:

Electric power, kW/b	2
Fuel, kBtu/b	30
Steam, #/b	15

Naphtha hydrotreating capital cost

An average cost of $16 million was estimated for a 30,000 BPD plant at the beginning of 1991.

Notes

1. Pistorius, J.T., *Oil & Gas Journal*, June 10, 1985, pp. 146–151

References

Abbott, M.D., Liedholm, G.E., and Sarno, D.H., *Petroleum Refiner,* June, 1955, pp. 118–122

Baeder, D.L., and Siegmund, C.W., *Oil & Gas Journal,* February 21, 1955, pp. 122–126

Bradley, W.E., Hendricks, G.W., Huffman, H.C., and Kelley, A.E., *Oil & Gas Journal,* June 8, 1959, pp. 194–198

Busch, R.A., Kocsin, J.J., Schroeder, H.F., and Shah, G.N., "Flexicoking + Hydrotreating Processes for Quality Products," AIChE National Meeting, Houston, April, 1979

Byrns, A.C., Bradley, W.E., and Lee, M.W., *Industrial and Engineering Chemistry,* November, 1943, pp. 1,160–1,167

Cole, R.M., and Davidson, D.D., *Industrial and Engineering Chemistry,* December, 1949, pp. 2,711–2,715

Davidson, R.L., *Petroleum Processing,* November, 1956, pp. 116–138

Edgar, M.D., Johnson, A.D., Pistorius, J.T., and Varadi, T., *Oil & Gas Journal,* June 4, 1984, pp. 67–70

Grote, H.W., Watkins, C.H., Poll, H.F., and Hendricks, G.W., *Oil & Gas Journal,* April 19, 1954, pp. 211–216

Harshaw Chemical Company, Hydrotreating Catalyst Brochure

Hoffman, E.J., Lewis, E.W., and Wadley, E.F., *Petroleum Refiner,* June, 1957, pp. 179–186

Kay, H., *Petroleum Refiner,* September, 1956, pp. 306–318

Kellett, T.F., Sartor, A.F., and Trevino, C.A., *Hydrocarbon Processing,* May, 1980, pp. 139–142

Kirsch, F.W., Heinemann, H., and Stevenson, D.H., *Industrial and Engineering Chemistry*, April, 1957, pp. 646–649

Komarewsky, V.I., Knaggs, E.A., and Bragg, C.J., *Industrial and Engineering Chemistry*, August, 1954, pp. 1689–1695

Murphy, H.C., Jr., Nejak, R.P., and Strom, J.R., "High–Pressure Hydrogenation—Route to Specialty Products," 35th Midyear Meeting of API Div. of Ref'g., Chicago, May, 1969

Patterson, A.C., and Jones, M.C.K., *Oil & Gas Journal*, October 18, 1954, pp. 92–94

Poll, H.F., *Petroleum Refiner*, July, 1956, pp. 193–198

Roeder, R.A., *Oil & Gas Journal*, October 15, 1973, pp. 123–126

Satchell, D.P., and Crynes, B.L., *Oil & Gas Journal*, December 1, 1975, pp. 123–124

Slyngstad, C.E., and Lempert, F.L., *Petro/Chem Engineer*, May, 1960, pp. C–13 to C–18

Stevenson, D.H., and Mills, G.A., *Petroleum Refiner*, August, 1955, pp. 117–121

Voorhies, A., Jr., and Smith, W.M., *Industrial and Engineering Chemistry*, September, 1947, pp. 1,104–1,107

CHAPTER 17

CATALYTIC REFORMING

Since its introduction, catalytic reforming has been the principal means of meeting increasing octane demands. In its early days, research octane numbers (RON) of 65 to 80 were commonly produced by the reformer. Today, the reformer is required to produce octanes of 100 and more. As a result primarily of catalyst developments, the reformer has been equal to the challenge. Like the FCC unit, the reformer is under pressure to further modify its product. Aromatics in general and benzene in particular must be reduced to meet the Clean Air Act stipulations. Since the primary function of the reformer has been, indirectly at least, to produce aromatics, this means a considerable adjustment. Raising the initial boiling point of reformer feed to exclude the main sources of benzene–namely cyclohexane and methylcyclopentane–can reduce benzene production. Some benzene would still arise from dealkylation of other rings. Production of other aromatics could be reduced by lowering the end point of either feed or reformate or by solvent extraction of the reformate. Any of these actions would result in a decrease in both octane and volume of reformate.

Another option would be to reduce the severity (octane of reformate produced), resulting in more reformate, but of lower octane.

Several graphs have been prepared to illustrate just what a typical reformate is like. Figure 17–1 is based on data published by Stine, *et. al.*[1] It shows the amount of each type of hydrocarbon species by carbon number contained in a typical reformate. Figure

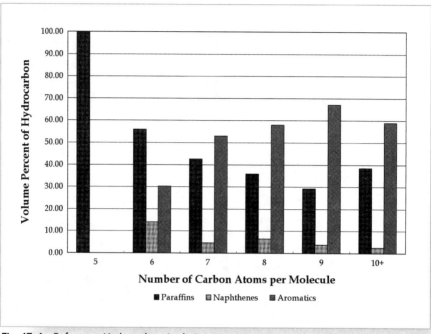

Fig. 17–1 Reformate Hydrocarbon Analysis

17–2 is a plot of aromatics content of reformate vs. clear research octane. Since very few naphthenes remain in reformate, it is evident that the increase in octane of reformate is due to the increase in aromatics content and the decrease in low octane paraffins.

The principal reactions occurring in catalytic reforming are:

- Dehydrogenation of naphthenes to aromatics

- Dehydrocyclization of paraffins to aromatics

- Isomerization of normal paraffins to isoparaffins and of naphthenes, such as methylcyclopentane to cyclohexane

- Hydrocracking of large paraffins into small paraffins

- Demethylation of paraffins

- Dealkylation of aromatics

264

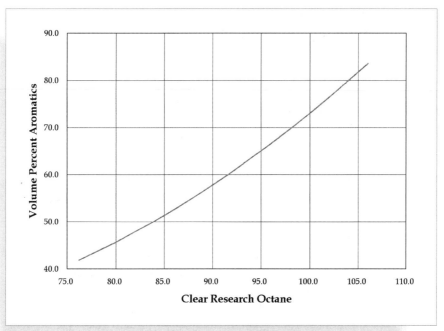

Fig. 17-2 Aromatics Content of Reformate

The first two reactions result in an increase in number of molecules in the forward direction of the reaction and are therefore sensitive to system pressure. The other reactions should be relatively unaffected by pressure. The first reformers operated at pressures as high as 900 psi. The operating pressure has declined over the years to the point that new units are designed for 100 psi or less. Hydrogen is still recycled to suppress the formation and deposition of coke on the catalyst.

In the early days of catalytic reforming it appeared that aromatics were being produced almost entirely by dehydrogenation of naphthenes and being concentrated in the C_{5+} reformate by hydrocracking of paraffins to butanes and lighter. With todays catalysts and lower pressure operation, it now appears that a substantial portion of net aromatic production comes from paraffins.[2,3]

Peer, *et. al.,* of UOP published the following typical operating conditions for the current design of UOP's continuous catalyst regeneration (CCR) process:[4]

Reactor pressure,	100 psig
Liquid hourly space velocity (LHSV)	1.6 hr^{-1}
H_2/HC, molal	2–3
Research octane number clear (RONC)	100–107

Catalytic reforming process description

Figure 17–3 is a simplified flow diagram of a conventional semi–regenerative reformer. Hydrotreated naphtha is combined with recycle hydrogen, heated to reaction temperature and introduced into the first reactor. The effluent from this reactor is reheated before entering the second reactor. This sequence is repeated for one or more additional reactors. Effluent from the last reactor is cooled and enters a separator where a hydrogen–rich stream is separated. A part of this stream is compressed and recycled back to combine with fresh naphtha feed. The remaining hydrogen is available to other hydrogen users in the refinery. Liquid from the separator is fractionated into a stabilized reformate as well as a stream of butanes and lighter materials. The catalyst is regenerated intermittently, whenever its activity falls to a predetermined level or sometimes when the unit is down for other reasons.

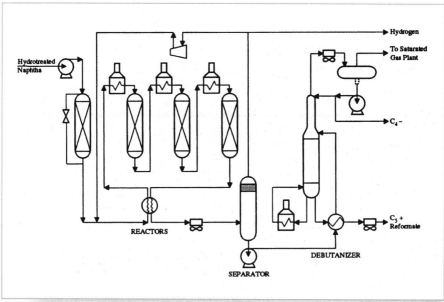

Fig. 17–3   Semiregenerative Catalytic Reformer

Many of the recently built reformers have employed continuous catalyst regeneration as licensed by IFP and UOP. Figure 17–4 is a simplified flow diagram of the UOP version in which the reactors are stacked vertically and the catalyst flows by gravity from one reactor to the other. The process stream is removed from one reactor, reheated and passed on to the next reactor. A portion of the catalyst is removed intermittently and lifted pneumatically to a regenerator. The catalyst flows by gravity through the regenerator and the regenerated catalyst is lifted pneumatically to the top reactor, completing the catalyst cycle. In this manner, the activity of the catalyst can be maintained at a high level.

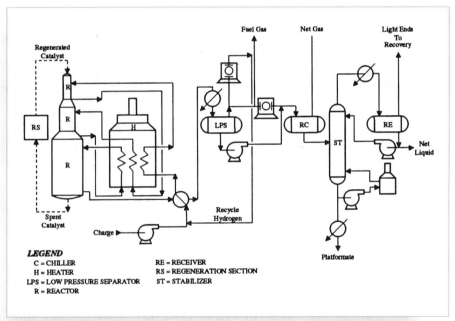

Fig. 17–4 UOP Continuous Platforming Process (© UOP; reprinted with permission)

Catalytic reforming data correlation

Because of its importance, a large amount of data on catalytic reforming is found in the literature. Table 17–1 is a compilation of that employed in this work.

		FEED PROPERTIES					REF PRESS	REF LV%	REF PROPERTIES			WEIGHT PERCENT					
API	IBP	EP	P	N	A	RON	PRESS	LV%	RON	API	%AROM	H2	C1	C2	C3	IC4	NC4
57.2	200.0	400.0	50.6	43.1	5.7	42.6	450	92.4	76.2	52.5	37.3						
54.5	218.0	370.0	51.5	33.8	14.7	43.5	200	80.7	100.0	41.0		2.3	1.2	2.2	3.1	2.3	3.2
49.5	202.0	388.0	32.0	48.0	20.0	69.5	200	85.0	102.0	38.0		2.4	0.7	1.5	2.0	1.5	2.3
40.7	255.0	420.0	34.5	23.2	42.3	77.3	250	93.6	100.0	36.7		1.2	0.3	0.5	1.0	0.7	1.0
45.9	180.0	350.0	21.8	39.6	38.6	85.0		90.5	103.0	37.0		2.13	0.41	0.5	0.58		
50.4	160.0	390.0	35.6	37.6	26.8	74.0		86.7	103.0	38.9		2.64	0.66	1.07	1.25		
54.6	160.0	370.0	44.6	42.3	13.1	54.2		81.2	103.0	38.8		3.31	1.13	1.73	2.15		
58.3	160.0	360.0	68.2	18.0	13.8	37.1		74.0	103.0	40.3		2.7	1.78	3.58	4.42		
56.6	172.0	358.0	52.0	35.0	13.0	50.9	150	82.0	100.0		72.0	1,370.0					
56.6	172.0	358.0	52.0	35.0	13.0	50.9	150	77.6	103.0		78.0	1,370.0					
45.1	224.0	380.0	25.0	36.0	39.0	81.7	150	91.8	100.0		72.0	940.0					
45.1	224.0	380.0	25.0	36.0	39.0	81.7	150	88.4	103.0		77.0	960.0					
45.1	224.0	380.0	25.0	36.0	39.0	81.7	150	83.6	106.0		84.0	950.0					
	202.0	394.0	26.0	55.5	18.5	59.4	300	91.8	90.0			900.0					
	202.0	394.0	26.0	55.5	18.5	59.4	300	83.8	100.0			1,100.0					
	206.0	361.0	50.0	41.5	8.5	44.5	200	90.5	85.0			1,080.0					
	206.0	361.0	50.0	41.5	8.5	44.5	200	84.1	95.0			1,260.0					
	206.0	361.0	50.0	41.5	8.5	44.5	300	89.9	85.0			970.0					
	206.0	361.0	50.0	41.5	8.5	44.5	300	82.8	95.0			1,150.0					
	149.0	377.0	77.0	15.0	8.0	45.3	300	84.5	85.0			540.0					
	149.0	377.0	77.0	15.0	8.0	45.3	300	80.0	90.0			590.0					
55.5	206.0	361.0	50.0	41.5	8.5	44.5		84.7	93.0	46.0							
48.1	202.0	394.0	26.0	55.5	18.5	59.4		86.0	98.0	40.0							
52.2	202.0	400.0	44.5	37.5	14.0	46.1		84.1	93.0	44.0							
	132.0	358.0	71.0	20.0	9.0			84.2	85.0			690.0					
50.8	200.0	400.0	33.0	53.0	14.0	51.0		78.7	97.0	44.3	66.0						
50.8	200.0	400.0	33.0	53.0	14.0	51.0		75.1	100.0	43.5	75.0						
68.1	136.0	362.0	77.6	14.0	8.4	73.9		87.5	73.9	68.4		0.5	0.15	0.3			
68.1	136.0	362.0	77.6	14.0	8.4	73.9		79.0	80.5	67.8		0.5	0.6	0.89			
67.4	114.0	351.0	62.4	30.3	7.3	64.4		81.2	88.5	62.9		1.2	0.34	0.68			
56.5	181.0	320.0	43.2	47.6	9.2	54.9		90.9	80.3	52.2		1.57	0.17	0.2			
56.5	181.0	320.0	43.2	47.6	9.2	54.9		86.7	86.2	50.5		1.78	0.27	0.37			
54.6	218.0	365.0	42.1	46.7	11.3	51.5		84.4	92.6	47.8		1.9	0.17	0.43			
54.6	218.0	365.0	42.1	46.7	11.3	51.5		86.7	89.2	47.8		1.91	0.12	0.3			
51.6	228.0	406.0	40.5	38.7	20.8	51.2		89.2	81.0	50.2		1.2	0.31	0.34			
51.5	242.0	348.0	42.6	36.5	20.9	56.9		91.0	83.4	48.1		1.18	0.08	0.15			
51.5	242.0	348.0	42.6	36.5	20.9	56.9		88.9	88.5	48.2		1.31	0.16	0.28			
47.9	317.0	401.0	48.2	36.1	15.7	36		89.5	83.0	46.1		1.11	0.21	0.45			
46.9	284.0	360.0	27.4	59.6	13	60.2		88.4	91.6	43.1		1.57	0.27	0.51			
50.3	232.0	374.0	39.5	32.5	28.0	60.3		93.1	84.8	47.2		1.14	0.09	0.14			
50.2	268.0	396.0	46.8	36.4	16.8	46.5		85.2	86.2	47.7		1.07	0.13	0.42			
53.1	196.0	382.0	49.2	34.9	16.8	47.4		87.4	82.2	49.7		1.34	0.23	0.52			
45.8	255.0	401.0	31.0	46.0	23.0	61.3		88.3	92.0	39.7		2.0	0.6	1.0			
45.8	255.0	401.0	31.0	46.0	23.0	61.3		80.6	101.0	36.5		1.8	1.8	2.9			
55.3	188.0	368.0	44.0	42.0	14.0	55.6		91.0	80.0	52.7		1.6	0.5	0.8			
55.3	188.0	368.0	44.0	42.0	14.0	55.6		83.5	92.0	45.8		1.3	1.0	2.4			
52.9	244.0	398.0	64.0	20.0	15.0	23.5		87.4	76.8	50.7		0.7	1.1	1.5			
52.9	244.0	398.0	64.0	20.0	15.0	23.5		80.4	90.0	47.7		0.7	2.2	2.8			
58.8	230.0	394.0	77.0	17.0	6.0	24.6		78.1	82.3	54.0		0.9	1.5	3.0			
58.8	230.0	394.0	77.0	17.0	6.0	24.6		80.8	80.1	53.2		1.1	1.1	2.3			
54.1	236.0	373.0	48.8	43.2	7.7	36.5	500	87.3	85.0	48.0		860.0	1.0	1.6			
54.1	236.0	373.0	48.8	43.2	7.7	36.5	200	88.7	85.0	44.8		910.0	0.6	1.1			
54.1	236.0	373.0	48.8	43.2	7.7	36.5	500	80.8	95.0	46.7		910.0	1.8	2.8			
54.1	236.0	373.0	48.8	43.2	7.7	36.5	200	83.4	95.0	42.3		1,115.0	1.2	2.0			
54.0	240.0	355.0	65.1	19.4	14.7	29.6	500	84.9	85.0	48.9		405.0	1.2	2.4			
54.0	240.0	355.0	65.1	19.4	14.7	29.6	200	86.7	85.0	50.9		555.0	0.9	1.8			
54.0	240.0	355.0	65.1	19.4	14.7	29.6	500	75.9	95.0	45.5		550.0	2.0	3.8			
54.0	240.0	355.0	65.1	19.4	14.7	29.6	200	79.7	95.0	42.3		760.0	1.5	2.9			
	202.0	394.0	26.0	55.5	18.5	59.4	300	91.8	90.0			900.0					

Table 17–1a   Catalytic Reforming Database

	FEED PROPERTIES						REF		REF PROPERTIES				WEIGHT PERCENT				
API	IBP	EP	P	N	A	RON	PRESS	LV%	RON	API	%AROM	H2	C1	C2	C3	IC4	NC4
	202.0	394.0	26.0	55.5	18.5	59.4	300	83.8	100.0			1,100.0					
	206.0	361.0	50.0	41.5	8.5	44.5	300	89.9	85.0			970.0					
	206.0	361.0	50.0	41.5	8.5	44.5	300	82.8	95.0			1,150.0					
	149.0	377.0	77.0	15.0	8.0	45.3	300	84.5	85.0			540.0					
	149.0	377.0	77.0	15.0	8.0	45.3	300	80.0	90.0			590.0					
58.1	182.0	365.0	53.8	40.4	5.8		125	77.2	99.9			1,470.0				1.9	3.5
55.3	212.0	345.0	55.1	31.7	13.2		190	82.6	96.4			1,048.0				1.8	3.1
53.5	217.0	368.0	36.2	58.7	5.1		200	78.3	103.0			1,550.0					
52.8	216.0	376.0	34.3	57.9	7.8		127	82.3	101.0			1,550.0					
51.8	238.0	383.0	40.0	49.5	10.5	50.0		87.4	91.7	45.5	53.4	980.0					
54.5	218.0	370.0	51.5	33.8	14.7	43.5	200	80.7	100.0	41.0		2.3	1.2	2.2	3.1	2.3	3.2
49.5	202.0	388.0	32.0	48.0	20.0	69.5	200	85.0	102.0	38.0		2.4	0.7	1.5	2.0	1.5	2.3
40.7	255.0	420.0	34.5	23.2	42.3	77.3	250	93.6	100.0	36.7		1.2	0.3	0.5	1.0	0.7	1.0
	218.0	370.0	51.5	33.8	14.7		350	77.7	100.0			2.0	1.6	2.9	4.4	3.1	4.3
	218.0	370.0	51.5	33.8	14.7		250	80.0	100.0			2.3	1.3	2.3	3.5	2.6	3.5
	218.0	370.0	51.5	33.8	14.7		150	82.4	100.0			2.6	1.0	1.8	2.7	1.9	2.7
	146.0	333.0	55.1	34.2	10.7		340	75.5	99.2			1.9					
	200.0	350.0	66.1	20.5	13.4		222	82.8	94.1			1.6					
53.3			49.2	42.6	8.2	38.0	200	71.9	95.0			580.0					
53.3			49.2	42.6	8.2	43.0	200	75.2	95.0			450.0					
55.0			47.5	44.0	8.5	48.8	200	75.2	95.0			640.0					
53.7			45.2	40.9	13.6	55.8	200	80.0	95.0			540.0					
55.5			48.7	41.0	10.3	52.5	200	77.4	95.0			320.0					
56.9	210.0	278.0	42.0	48.0	10.0	61.5		85.8	93.0	45.8	59.5	2.3	0.6	1.3	2.5		
56.9	210.0	278.0	42.0	48.0	10.0	61.5		81.4	97.0	44.4	65	2.5	1.0	2.1	4.0		
51.6	272.0	318.0	41.0	43.0	16.0	54.0		86.6	93.0	43.0	59.9	1.0	0.6	1.2	2.3		
51.6	272.0	318.0	41.0	43.0	16.0	54.0		82.2	97.0	40.7	70.6	2.0	0.9	2.0	3.7		
47.3	306.0	371.0	39.0	42.0	19.0	53.0		90.7	90.0	41.3	55.5	1.5	0.3	0.6	1.1		
47.3	306.0	371.0	39.0	42.0	19.0	53.0		88.4	93.0	40.5	62.0	1.7	0.4	0.9	1.7		
63.1	186.0	303.0	71.0	18.7	10.0	47.4		66.9	99.7	42.9		1.5	2.8	5.6	7.7		
63.1	186.0	303.0	71.0	18.7	10.0	47.4		74.0	94.5	46.5		1.3	2.0	4.4	6.4		
58.6	207.0	356.0	66.7	18.0	14.3	35.2		57.0	103.5	41.5		1.4	5.1	8.7	10.5		
58.6	207.0	356.0	66.7	18.0	14.3	35.2		61.0	100.4	45.0		1.3	3.4	6.9	9.4		
59.5	191.0	327.0	53.5	35.8	9.7	54.4		74.0	96.5	47.2		1.8	1.8	3.5	6.3		
59.5	191.0	327.0	53.5	35.8	9.7	54.4		76.6	93.8	48.5		1.5	1.5	3.3	5.5		
64.0	211.0	309.0	37.2	38.5	23.2	64.0		84.2	98.7	42.0		2.0	1.0	1.5	2.5		
64.0	211.0	309.0	37.2	38.5	23.2	64.0		87.5	95.7	43.4		1.9	0.8	1.0	1.6		
53.0	211.0	350.0	38.1	42.6	19.3	55.4	190	86.9	95.8			1,035.0					
53.1	205.0	348.0	39.0	43.0	18.0	55.8	192	83.8	98.2			1,570.0					
53.1	210.0	348.0	38.1	42.4	19.5	56.8	193	89.4	89.0			1,217.0					
59.6	185.0	318.0	53.0	36.9	10.1	58.7	196	79.8	96.5			1,170.0					
52.2	232.0	358.0	40.9	43.1	16.0		235	78.3	102.0			2.55					
52.2	232.0	358.0	40.9	43.1	16.0		85	83.2	102.0			3.05					
53.0	204.0	382.0	44.9	41.4	13.7		125	83.5	100.0		(1,425)	2.8	1.2	1.7	2.2		
45.9	210.0	330.0	21.8	6.0	38.6	85.0		90.5	103.0	37.0		2.13	0.41	0.5	0.58		
50.4	194.0	347.0	35.6	37.6	26.8	74.0		86.7	103.0	38.9		2.64	0.66	1.07	1.25		
54.6	202.0	345.0	44.6	42.3	13.1	54.2		81.2	103.0	38.8		3.31	1.13	1.73	2.15		
58.3	189.0	348.0	68.2	18.0	13.8	37.1		74.0	103.0	40.3		2.7	1.78	3.58	4.42		
53.6	205.0	355.0	39.0	46.5	14.5	59.8	180	84.0	98.0			1,440.0					
56.0	199.0	317.0	38.9	47.9	13.2	62.8	180	83.5	98.0			1,440.0					
47.9	303.0	361.0	40.5	41.1	18.4	48.1	180	86.0	98.0			980.0					
55.3	188.0	349.0	44.0	42.0	14.0	51.2		77.7	97.3								
55.3	188.0	349.0	44.0	42.0	14.0	51.2		88.5	91.0								
55.3	188.0	349.0	44.0	42.0	14.0	51.2		96.0	71.0								
55.3	188.0	368.0	44.0	42.0	14.0	55.6		85.8	89.1	48.2	55.0						
52.7	238.0	396.0	40.0	1.0	12.0	42.5		84.3	90.3	44.1	57.0						
51.6	220.0	396.0	34.0	52.0	14.0	51.5		86.6	90.2	47.1	55.0						
55.4	228.0	332.0	46.0	41.0	12.0	48.7	275	88.4	86.6	46.2	52.0	1,090.0	0.5	1.1	1.5		
55.4	228.0	332.0	46.0	41.0	12.0	48.7	275	85.0	93.1	44.3	60.0	1,210.0	0.9	1.5	2.1		

Table 17–1b   Catalytic Reforming Database cont'd

		FEED PROPERTIES						REF	REF PROPERTIES				WEIGHT PERCENT				
API	IBP	EP	P	N	A	RON	PRESS	LV%	RON	API	%AROM	H2	C1	C2	C3	IC4	NC4
55.4	228.0	332.0	46.0	41.0	12.0	48.7	275	80.0	98.3	41.5	68.0	1,380.0	1.5	2.5	3.2		
51.5	286.0	399.0	44.0	44.0	11.0	26.4		85.2	84.0	44.9	53.0						
51.5	286.0	399.0	44.0	44.0	11.0	26.4		74.3	96.9	41.6	68.0						
52.4	220.0	405.0	35.0	52.0	12.0	43.3		87.0	84.0	45.0							
52.4	220.0	405.0	35.0	52.0	12.0	43.3		76.4	96.7	41.2							
51.7	236.0	391.0	43.0	42.0	13.0	38.9		86.4	85.3	45.2	53.0						
51.7	236.0	391.0	43.0	42.0	13.0	38.9		78.0	95.3	42.0	65.0						
46.7	267.0	398.0	19.0	65.0	16.0	53.3		90.8	85.6	40.3	56.0						
46.7	267.0	398.0	19.0	65.0	16.0	53.3		83.6	95.2	38.0	67.0						
52.7	238.0	396.0	44.0	43.0	12.0	42.5		86.5	87.0	46.0	51.0						
52.7	238.0	396.0	44.0	43.0	12.0	42.5		81.0	95.0	43.0							
52.3	241.0	373.0	44.0	41.4	14.6		145	82.7	100.0	39.6		1,308.0	1.02	1.68	2.75		
52.8	172.0	376.0	45.0	42.4	12.6		145	76.2	105.0	35.4		1,508.0	1.53	2.51	4.09		
	274.0	392.0	21.5	57.0	21.5	57.0		91.0	90.0			875.0					
	274.0	392.0	21.5	57.0	21.5	57.0		84.0	100.0			1,050.0					
	206.0	361.0	50.0	41.5	8.5	44.5		89.9	85.0			970.0					
	206.0	361.0	50.0	41.5	8.5	44.5		82.8	95.0			1,150.0					
	119.0	378.0	55.0	38.0	7.0	49.4		88.7	85.0			870.0					
	119.0	378.0	55.0	38.0	7.0	49.4		78.3	95.0			1,060.0					
	111.0	375.0	77.0	15.0	8.0	47.8		82.8	85.0			530.0					
	111.0	375.0	77.0	15.0	8.0	47.8		77.5	90.0			600.0					
	202.0	394.0	26.0	55.5	18.5	59.4	300	91.8	90.0								
	202.0	394.0	26.0	55.5	18.5	59.4	300	83.8	100.0								
	206.0	361.0	50.0	41.5	8.5	44.5	200	90.5	85.0								
	206.0	361.0	50.0	41.5	8.5	44.5	200	84.1	95.0								
	206.0	361.0	50.0	41.5	8.5	44.5	300	89.9	85.0								
	206.0	361.0	50.0	41.5	8.5	44.5	300	82.8	95.0								
	149.0	377.0	77.0	15.0	8.0	45.3	300	84.5	85.0								
	149.0	377.0	77.0	15.0	8.0	45.3	300	80.0	90.0								
54.1	236.0	373.0	48.8	43.2	7.7	36.5	500	86.8	85.0			850.0	1.0	2.4	4.5		
54.1	236.0	373.0	48.8	43.2	7.7	36.5	500	79.0	95.0			740.0	1.5	3.0	6.5		
54.1	236.0	373.0	48.8	43.2	7.7	36.5	200	87.8	85.0			880.0	0.7	1.8	2.6		
54.1	236.0	373.0	48.8	43.2	7.7	36.5	200	83.3	95.0			1,150.0	0.9	2.1	3.6		
48.6	252.0	385.0	33.8	43.0	22.9	51.0	500	91.7	85.0			725.0	0.5	0.8	1.2		
48.6	252.0	385.0	33.8	43.0	22.9	51.0	500	85.8	95.0			683.0	1.2	1.9	3.1		
48.6	252.0	385.0	33.8	43.0	22.9	51.0	200	91.8	85.0			930.0	0.5	0.6	0.8		
48.6	252.0	385.0	33.8	43.0	22.9	51.0	200	86.8	95.0			830.0	0.8	1.3	2.0		
54.0	240.0	355.0	65.1	19.4	14.7	29.6	500	85.6	85.0			510.0	1.3	2.4	3.5		
54.0	240.0	355.0	65.1	19.4	14.7	29.6	500	75.6	95.0			470.0	2.2	4.5	7.2		
54.0	240.0	355.0	65.1	19.4	14.7	29.6	200	86.8	85.0			730.0	0.4	1.3	2.1		
54.0	240.0	355.0	65.1	19.4	14.7	29.6	200	81.2	95.0			875.0	1.1	2.4	3.4		
57.4	214.0	336.0	58.8	27.7	11.1	43.4	500	85.0	85.0			685.0	1.2	1.9	3.7		
57.4	214.0	336.0	58.8	27.7	11.1	43.4	500	74.8	95.0			590.0	1.9	3.0	6.7		
57.4	214.0	336.0	58.8	27.7	11.1	43.4	200	87.8	85.0			910.0	0.6	1.1	1.9		
57.4	214.0	336.0	58.8	27.7	11.1	43.4	200	82.2	95.0			1,150.0	1.0	2.1	3.6		
56.5	214.0	346.0	48.9	37.3	13.8	51.3	500	91.5	86.1			1.0	1.3	1.9	2.4	1.6	3.3
56.8	210.0	350.0	46.4	39.9	13.7	52.6	500	92.0	86.1			0.9	1.2	1.8	2.2	1.5	3.2
55.6	185.0	380.0	43.1	46.0	10.9	47.8	350	87.0	87.4	48.2	51.1	1.7	1.2	1.5	2.2	1.3	2.4
51.8	256.0	382.0	53.0	27.0	19.0	40.0		93.9	84.2	52.2	42.0	300.0					
51.8	256.0	382.0	53.0	27.0	19.0	40.0		92.1	90.1	52.4	49.0	391.0					
51.8	256.0	382.0	53.0	27.0	19.0	40.0		91.0	93.6	53.3	54.0	396.0					
51.8	256.0	382.0	53.0	27.0	19.0	40.0		88.4	96.1	52.8	64.0	423.0					
58.1	166.0	355.0	62.0	33.0	5.0	52.5	720	92.6	81.3	57.4	37.0	401.0				2.1	3.5
48.6	272.0	374.0	39.4	39.5	21.1	52.2		92.2	83.2	45.2	53.7	503.0					
48.6	272.0	374.0	39.4	39.5	21.1	52.2		90.0	87.6	45.0	58.1	518.0					
48.6	272.0	374.0	39.4	39.5	21.1	52.2		87.4	92.3	44.3	63.0	554.0					
48.6	272.0	374.0	39.4	39.5	21.1	52.2		83.9	96.1	43.4	69.0	559.0					
48.6	272.0	374.0	39.4	39.5	21.1	52.2		82.2	96.9	43.1	72.1	569.0					
48.6	272.0	374.0	39.4	39.5	21.1	52.2		79.9	98.7	42.9	74.5	508.0					

Table 17–1c Catalytic Reforming Database cont'd

	FEED PROPERTIES							REF	REF PROPERTIES				WEIGHT PERCENT				
API	IBP	EP	P	N	A	RON	PRESS	LV%	RON	API	%AROM	H2	C1	C2	C3	IC4	NC4
48.6	272.0	374.0	39.4	39.5	21.1	52.2		91.8	85.7	44.1	56.7	663.0					
48.6	272.0	374.0	39.4	39.5	21.1	52.2		90.2	89.1	43.8	61.0	684.0					
48.6	272.0	374.0	39.4	39.5	21.1	52.2		89.5	90.4	43.4	62.4	699.0					
48.6	272.0	374.0	39.4	39.5	21.1	52.2		57.2	93.3	42.9	66.4	772.0					
48.6	272.0	374.0	39.4	39.5	21.1	52.2		85.0	95.8	42.2	72.0	772.0					
48.6	272.0	374.0	39.4	39.5	21.1	52.2		81.6	98.4	41.8	75.8	702.0					
54.5	272.0	378.0	66.3	19.8	13.9	32.4		85.3	83.6	51.1	51.0	271.0					
54.5	272.0	378.0	66.3	19.8	13.9	32.4		84.1	86.2	50.7	51.1	401.0					
54.5	272.0	378.0	66.3	19.8	13.9	32.4		81.0	91.2	49.6	56.4	446.0					
54.5	272.0	378.0	66.3	19.8	13.9	32.4		76.6	95.2	48.1	59.6	456.0					
54.5	272.0	378.0	66.3	19.8	13.9	32.4		90.0	72.8	49.8	44.8	531.0					
54.5	272.0	378.0	66.3	19.8	13.9	32.4		86.7	80.0	49.2	50.6	566.0					
54.5	272.0	378.0	66.3	19.8	13.9	32.4		84.2	87.6	47.5	55.9	666.0					
54.5	272.0	378.0	66.3	19.8	13.9	32.4		80.7	92.8	46.0	62.5	726.0					
54.5	272.0	378.0	66.3	19.8	13.9	32.4		76.0	97.2	43.7	72.4	796.0					
55.7	194.0	392.0	54.0	36.0	10.0	45.7		86.4	83.0	49.5							
54.6	194.0	396.0	63.0	16.0	21.0	48.0		87.0	83.2	51.5							
55.0	200.0	384.0	40.0	52.0	8.0	48.0		83.5	93.0	49.0							
58.1	213.0	348.0	67.0	22.0	11.0			76.5	92.5	53.1							
58.1	213.0	348.0	67.0	22.0	11.0			77.0	93.5	50.5							
	198.0	376.0	41.0	43.0	16.0	52.0		90.6	94.0	48.1							
56.4	168.0	328.0	54.2	36.1	9.7	52.5			98.2	43.2	64.2						
	216.0	370.0	44.8	48.1	7.1	48.3		80.1	100.0	41.6		1,605.0	1.4	1.98	3.05		
								85.0	95.0	45.2		1,445.0	0.95	1.35	2.07		
								88.2	90.0	47.3		1,325.0	0.66	0.94	1.44		
63.2	175.0	340.0	70.0	21.0	9.0			77.8	98.0	48.0		1,282.0	1.07	2.8	3.91	2.47	3.56

Table 17-1d   Catalytic Reforming Database cont'd

It was found that the yields of the light hydrocarbons and hydrogen plotted well against the yield of reformate as shown in Figure 17–5. The composition of butanes may be assumed to be 40 volume percent or 44 weight percent isobutane.

The yield of hydrogen required introduction of another independent variable, octane of the reformate, to correlate satisfactorily. The result is shown in Figure 17–6, where hydrogen yield is expressed as weight percent of feed. Hydrogen yield is frequently expressed as standard cubic feet per barrel of feed (SCFB). Therefore, Figure 17–7 was prepared to facilitate conversion from weight percent to SCFB.

One of the earliest attempts by the author at correlating process yields was that of catalytic and thermal reforming.[5] This was performed during the period when the industry was converting from motor octane to research. Consequently, much of the data was based on motor octanes. In order to use more of the data, curves were developed for the sensitivity and lead susceptibility of both feed naphthas and reformates. These curves

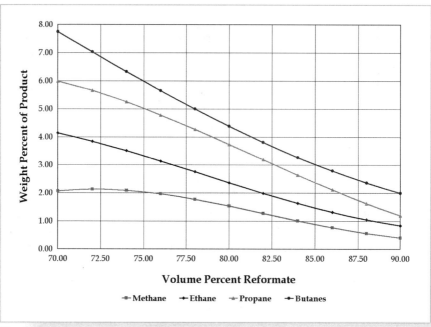

Fig. 17–5 Reforming Light End Yields

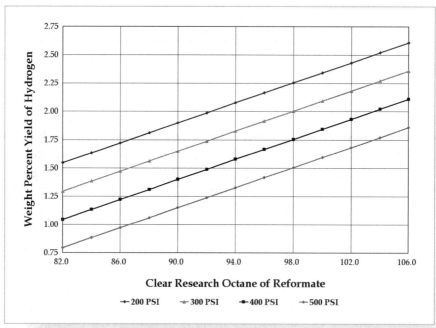

Fig. 17–6 Reformer Hydrogen Yield (Reactor Pressure as Parameter)

272

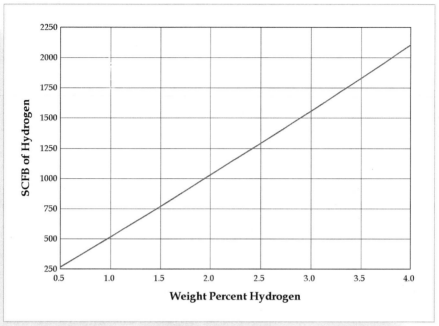

Fig. 17–7 SCFB of Reformer Hydrogen

proved to be of greater interest than the yield correlations and were very useful in other work as well.

The yield correlations were based on the assumption that the yield–octane relationship in reforming is independent of operating conditions, and that the incremental decrease in yield accompanying an incremental increase in octane of reformate is a function solely of the octane level at which the change occurs. These assumptions were fairly well substantiated by the correlations.

This earlier work was all performed manually, by plotting data points and eye–balling best curves. Using a larger database and multiple linear regression, Figure 17–8 was developed. This figure shows the relation between three variables—octane of feed, octane of reformate, and yield of reformate. Unfortunately, the octane of the feed is frequently omitted in literature reports on catalytic reforming.

Most correlations rely on the paraffins, olefins, naphthenes, aromatics (PONA) analysis of the feed. Some use N+2A as an independent variable, while others use N+A (or P). The author's correlation using N+2A as the

273

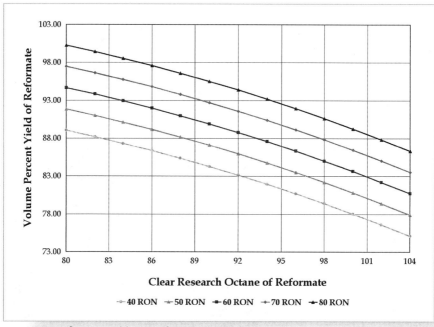

Fig. 17–8 Reformate Yield (RON of Feed as Parameter)

parameter is shown in Figure 17–9. With yield of reformate as the dependent variable, this situation was explored by introducing other variables successively to the octane of the reformate. The following matrix of correlation coefficients was derived, where LV% is the dependent variable:

	RON	P	N	A	Psi
LV%	.2351	.1321	.0505	.1020	.0236
RON		.5410	.3408	.4715	.2806
P			.1450	.1527	.1663
N				.1489	.0552
A					.1750

These results indicate the relative importance of the possible dependent variables in decreasing order of importance as RON > P > A > N > Psi. However, it should be noted that P, N, and A are not completely independent, since $P + N + A = 100$, because olefins are absent or essentially zero.

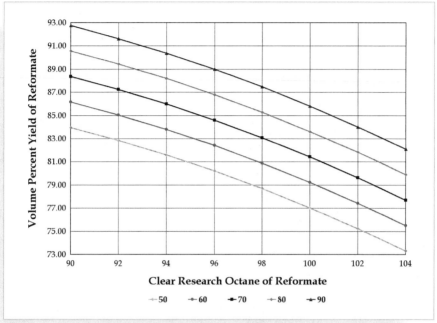

Fig. 17–9 Yield of Catalytic Reformate (N + 2A as Parameter)

Equations were derived for LV% yield of reformate in terms of all combinations of the five independent variables, taken from one to five at a time. It is thought that the previous correlation in terms of feed octane and reformate octane is better than any of these equations.

The following additional figures were developed: Motor Octane of Reformate from Research Octane (Fig. 17–10), API Gravity of Reformate from Research Octane (Fig. 17–11), Weight Percent of Reformate from Volume Percent (Fig. 17–12).

Comparison with other correlations

As mentioned previously, most correlations are in terms of N + 2A or N + A. However, these are proprietary and not in the public domain. Occasionally, a plot of volume percent reformate in terms of reformate octane does indicate the N + 2A of the feed. Little shows such a plot in his book *Catalytic Reforming*.[6] Table 17–2 compares values read from Little's small graph with values calculated from an equation in terms of N + 2A

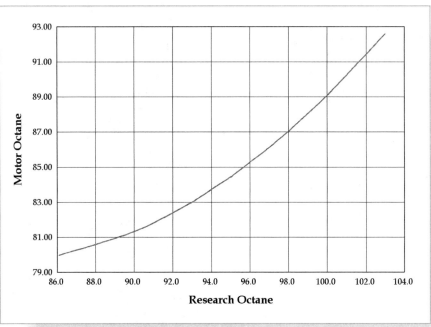

Fig. 17–10 Motor Octane of Reformate

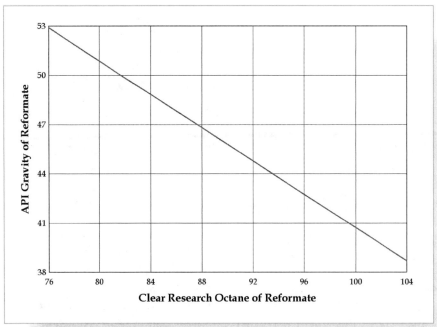

Fig. 17–11 API Gravity of Reformate

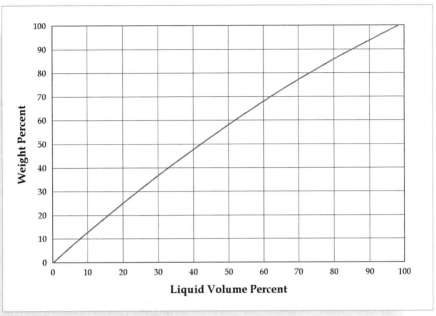

Fig. 17–12 Lv% vs. Wt% for Reformate

REF RON	N + 2A = 40		N + 2A = 60		N + 2A = 80	
	L	A	L	A	L	A
90	82.0	81.8	87.8	86.2	91.0	90.6
92	81.0	80.7	86.5	85.0	90.0	89.4
94	79.5	79.4	85.0	83.8	88.8	88.2
96	77.7	78.0	83.5	82.4	87.5	86.8
98	75.7	76.5	81.8	80.9	85.9	85.3
100	72.6	74.8	79.8	79.2	84.1	83.6

Note: *L* denote data from Little's graph; *A* denotes values calculated from the author's equation

Table 17–2 Comparison of Reformate Yield Correlations

developed by the author (Fig. 17–9). Overall, there appears to be a fair check. The results of the author's study indicate N + 2A to be marginally better than N + A as a parameter.

Several articles have appeared in the literature that discuss the effect of reaction pressure on yields in catalytic reforming. An article by Hughes, *et. al.,* of Chevron Research charts the effect of pressure on the yield of reformate, and of hydrogen for an Arabian naphtha.[7] The charts indicate increases in reformate of from two to seven volume percent over the range of octanes of 86 to 100 when reducing the pressure from 200psi to 100psi and increases in hydrogen yield of from 18 weight percent to 24 weight percent under the same circumstances. Note that these percentages are for a particular naphtha feed. Nelson published a chart providing a correction factor for reformate yield that is a function of a base yield of reformate of a given octane,[8] which in effect personalizes the relationship for the particular naphtha of interest. In the author's correlations, the coefficient for pressure was about -0.01, meaning a reduction of about 1% in reformate yield per 100psi increase in pressure. In the case of hydrogen, the coefficient was approximately -0.003, meaning a reduction in weight percent of hydrogen of 0.3 for each 100psi increase in pressure.

Zielinski published a paper describing pilot plant studies of catalytic reforming at Sun Oil Company and the correlation of the results.[9] He used a modified characterization factor (where the 50% point in the distillation was substituted for the average boiling point) as an independent variable (and to account for the properties of the feed). It is not known whether or not this correlation is still in use.

Catalytic reforming operating requirements

Operating requirements for catalytic reforming appear to be approximately as follows on a per–barrel–of feed basis:

Electric power	1.0 kWh
Fuel	300 kBtu
Cooling water	100 gal
Steam	<40>

Catalytic reformer capital cost

As with previous processes, published figures on catalytic reforming units were scaled and updated to yield an average value of $45 million for a 30,000 BPD unit at the beginning of 1991. The individual numbers varied from $37.3 to $54.5 million or -17% to +22% of $45 million.

278

Notes

1. Stine, L.O., DeVeirman, R.M., and Schuller, R.P., *Oil & Gas Journal,* May 31, 1971, pp. 49–53

2. McDonald, G.W.G., *Hydrocarbon Processing,* June, 1977, pp. 99, 147–150

3. Sterba, M.J., Weinert, P.C., Lickus, A.G., Pollitzer, E.L., and Hayes, J.C., *Oil & Gas Journal,* December 30, 1968, pp. 140–146

4. Peer, R.L., Bennett, R.W., and Bakas, S.T., *Oil & Gas Journal,* May 30, 1988, pp. 52–60

5. Maples, R.E., *Petroleum Refiner,* September, 1954, pp. 284–299

6. Little, D.M., *Catalytic Reforming,* PennWell Books, 1985, p. 25

7. Hughes, T.R., Jacobson, R.L., Schornack, L.G., and McCabe, J.R., *Oil & Gas Journal,* May 17, 1976, pp. 121–130

8. Nelson, W.L., *Oil & Gas Journal,* August 2, 1971, pp. 76–77

9. Zielinski, R.M., "Relationship of the Composition of Reformer Feed to Reforming Yield," Mtg. of Div. of Petroleum Chemistry of the American Chemical Society, New York, September, 1957

References

Aalund, L.R., *Oil & Gas Journal,* December 20, 1971, pp. 43–60

Anon., *Oil & Gas Journal,* July 20, 1953, pp. 64–65

Ibid., November 16, 1953, pp. 186–187

Ibid., November 19, 1956, pp. 238–243

Ibid., August 26, 1974, pp. 115–116

279

Berg, C., *Petroleum Refiner*, December, 1952, pp. 131–136

Berg, C., *Oil & Gas Journal*, March 23, 1953, pp. 286–293

Birmingham, W.J., *Petroleum Engineer*, April, 1954, pp. C–35 to C–37

Bland, R.E., *Petroleum Engineer*, April, 1954, pp. C–18 to C–21

Bozeman, H.C., *Oil & Gas Journal*, December 23, 1963, pp. 51–58

Burtis, T.A., and Noll, H.D., *Oil & Gas Journal*, April 28, 1952, pp. 75–100

Dart, J.C., Oblad, A.G, and Schall, J.W., *Oil & Gas Journal*, November 17, 1952, pp. 386–392

Decker, W.H., *Petroleum Engineer*, April, 1954, pp. C–30 to C–32

Decker, W.H., and Stewart, D., *Oil & Gas Journal*, July 4, 1955, pp. 80–84

Ibid., February 2, 1959, pp. 88–91

Engelhard Brochure on Magnaforming, 1970

Forrester, J.H., Conn, A.L., and Malloy, J.B., *Oil & Gas Journal*, April 12, 1954, pp. 139–142

Fowle, M.J., Bent, R.D., Milner, B.E., and Masologites, G.P., *Oil & Gas Journal*, May 26, 1952, pp. 181–185

Gumaer, R.R., and Raiford, L.L., *Oil & Gas Journal*, August 8, 1955, pp. 119–123

Haensel, V., and Donaldson, G.R., *Petroleum Processing*, February, 1953, pp. 236–239

Hatch, W.H., Cohen, S.J., and Diener, R., "Modern Catalytic Reformer Designs Help Reduce Cost of Low–Lead Gasoline," NPRA Annual Meeting, San Antonio, 1973

Heinemann, H., Schall, J.W., and Stevenson, D.H., *Oil & Gas Journal*, November 15, 1951, pp. 166–169

Jacobson, R.L., and McCoy, C.S., *Hydrocarbon Processing,* May, 1970, pp. 109–111

IFP Brochure on Catalytic Reforming

Kastens, M.L., and Sutherland, R., *Industrial and Engineering Chemistry,* April, 1950, pp. 582–593

Kirkbride, C.G., *Oil & Gas Journal,* July 5, 1951, pp. 60–79

Murphree, E.V., *Petroleum Refiner,* Dec., 1951, pp. 97–108

Nevison, J.A., Obaditch, C.J., and Dalson, M.H., *Hydrocarbon Processing,* June, 1974, pp. 111–114

Nix, H.C., *Oil & Gas Journal,* May 20, 1957, pp. 168–172

Pistorius, J.T., *Oil & Gas Journal,* June 10, 1985, pp. 146–151

Read, D., and Weinert, P.C., *Oil & Gas Journal,* April 23, 1956, pp. 105–108

Read, D., *Oil & Gas Journal,* June 23, 1952, pp. 82–89

Resen, E.L., *Oil & Gas Journal,* July 25, 1955, pp. 130–131

Ibid., February 16, 1959, pp. 105–120

Steel, R.A., Bosk, J.A., Hertwig, W.R., and Russum, L.W., *Petroleum Refiner,* May, 1954, pp. 167–171

Taylor, W.F., and Welty, A.B., Jr., *Oil & Gas Journal,* December 2, 1963, pp. 142–146

Teter, J.W., Borgerson, B.T., and Beckberger, L.H., *Oil & Gas Journal,* October 12, 1953, pp. 118–140

Thornton, D.P., Jr., *Petro/Chem Engineer,* May, 1969

Tripp, R.G., and Swart, G.S., *Oil & Gas Journal,* May 11, 1970, pp. 68–70

White, P.C., Johnston, W.F., and Montgomery, W.J., *Petroleum Refiner,* May, 1956, pp. 171–177

281

SECTION E:

LIGHT HYDROCARBON PROCESSING

CHAPTER 18

ISOMERIZATION

Isomerization is an intermediate, feed preparation-type of process. Two isomerization processes are of interest. For some time, the isomerization of normal butane to isobutane has been practiced as a means of supplementing that normally available to the refiner from refinery gases or field butanes. Isobutane is an essential ingredient in alkylation and is chronically short in most refineries, limiting alkylate output. Recently a demand has developed for isobutylene for the production of methyl tertiary butyl ether (MTBE). Isobutylene can be produced from isobutane by dehydrogenation. Hence, the butane isomerization unit can be a supplier directly to alkylation and/or indirectly to MTBE.

The other kind of isomerization is for the conversion of C_5 and C_6 normal paraffins into branched-chain hydrocarbons with much higher octanes.

Butane Isomerization

This is stuctural isomerization, wherein the carbon skeleton of the molecule is rearranged with no change in the molecular formula:

$$C-C-C-C \quad \rightarrow \quad \underset{\textit{isobutane}}{\overset{\displaystyle \overset{C}{|}}{C-C-C}}$$

n–butane   *isobutane*

This is an equilibrium reaction. The composition of the equilibrium mixture is a function of temperature, with lower temperature favoring isobutane as shown in Figure 18–1, which was calculated from data in Chemetron's *Physical and Thermodynamic Properties of Elements and Compounds.*[1]

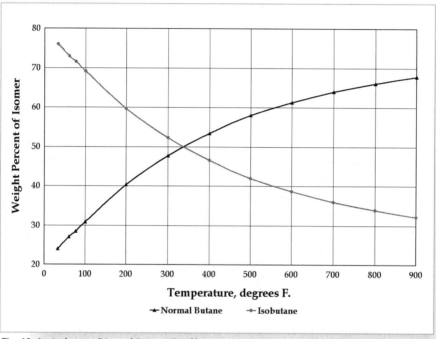

Fig. 18–1 Isobutane/Normal Butane Equilibrium

The rate at which equilibrium is attained is also determined by the temperature, the rate increasing with temperature. Therefore, the temperature at which the process operates is a compromise that gives a satisfactory conversion to isobutane at a reasonable rate.

Butane isomerization process description

While the actual isomerization unit itself is quite simple (Fig. 18–2), there are several ways it can be integrated into a refinery. A deisobutanizer tower (DIB) may be included in the scheme. The relative position of the DIB depends on the concentration of isobutane in the feed. If the concen-

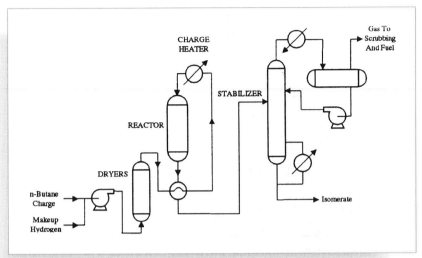

Fig. 18–2 Hot Butamer Flow Scheme (© UOP; reprinted with persmission)

tration is 30% or more,[2] it is advantageous to charge both fresh feed and recycle to the DIB. Then by taking a sidecut below the feed zone, a higher concentration of normal butane is available to feed the reactor.

If the DIB in the alkylation unit is not loaded, fresh feed could be sent to this tower, saving the expense of another DIB in case the isomerization unit is an add–on. A new refinery could be designed with a single, common DIB.

The usual catalyst is a platinum–bearing alumina–chloride type that must be protected from water, sulfur, and fluorides. Leaving the dryer, the feed is joined by a recycle gas stream to which hydrogen has been added to minimize deposition of coke on the catalyst. This combined stream is heated to reaction temperature and passed through a fixed catalyst bed to a separator. Separator gas is recycled. Separator liquid is stabilized to remove small quantities of light hydrocarbons resulting from side reactions. Since an organic chloride is introduced with the feed to replace the chloride removed from the catalyst as HCl, stabilizer light ends are scrubbed with caustic before going into the fuel system.

Butane isomerization yields

The volumetric yield of isobutane formed will be greater than the volume of butane converted due to the high selectivity of the catalyst and the specific gravity ratio of normal to iso of 1.038. With a reasonable DIB design, at least 95% of the isobutane can be recovered in an overhead stream containing at least 95% isobutane.

Butane isomerization operating requirements

For the isomerization unit alone, the requirements per barrel of feed are approximately 1 kWh of electric power, 12,000 Btu of fuel and 36 pounds of steam. With a DIB included, the total requirements become 3.5 kWh and 420,000 Btu with a fired reboiler on the DIB.

Butane isomerization capital cost

From the limited data found in the literature, it appears that a 10,000 BPD butane isomerization unit at the first of 1991 would have cost about $5 million. The addition of a DIB tower to the unit would have brought the cost to almost $20 million.

C_5/C_6 Isomerization

As stated previously, the purpose of C_5/C_6 isomerization is to produce a material with a significantly higher octane. Interest in C_5/C_6 isomerization languished until it became apparent that lead was going to be phased out of gasoline. This is because pentanes and hexanes have a high octane response to TEL (high lead susceptibility). An appreciation of the interest in converting normal paraffins to branched paraffins can be gained by examining the following tabulation:

Component	RON	MON	BP, °F
Isopentane	93.5	89.5	82.14
Normal pentane	61.7	61.3	96.93
Cyclopentane	101.3	85.0	120.68
2, 2 DMB	93.0	93.5	121.54
2, 3 DMB	104.0	94.3	136.38
2 MP	73.4	72.9	140.49

Component	RON	MON	BP, °F
3 MP	74.5	74.0	145.91
Normal hexane	30.0	25.0	155.74
MCP	95.0	80.0	161.27
Cyclohexane	83.0	77.2	177.33
Benzene	>100	>100	176.18

From the boiling points, it can be seen that most of the isopentane in a feed could be removed ahead of the reactor in a deisopentanizer (DIP) tower. Also, a split could be made between the two DMBs and 2 MP in a deisohexanizer (DIH) tower. This would allow the MPs and normal hexane to be recycled. A C_5/C_6 splitter could be inserted between the reactor and the DIH to permit recycling C_5's back to the DIP. Another option would be to send the reactor effluent to a molecular sieve to separate normal pentane and normal hexane for recycle.

The stepwise addition of these towers into a complex could result in increased production of isomerate with the same reactor.

C_5/C_6 isomerization process description

The basic process is essentially the same as the butane isomerization process. However, because of the large number of compounds involved, there are a number of arrangements that might be employed as described above. Figure 18–3 is an example of a once–through operation. Typical results for the various process schemes are:

Case	Arrangement	RON Relative	Capital Cost
I	Once–through	83	1.0
II	I plus DIP	84	1.2

Case	Arrangement	RON Relative	Capital Cost
III	II plus C_5/C_6 Splitter	86.4	1.54
IV	I plus Molecular Sieve	89	1.93
V	III plus DIH	92.5	2.34

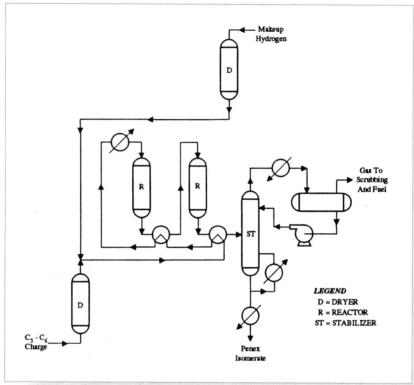

Fig. 18–3 UOP Hot Penex Process (© UOP; reprinted with persmission)

Figure 18–4 is a plot of the above data resulting from an earlier study (1978) by the author for these five cases. It is thought that the same general relationships hold today.

C_5/C_6 isomerization operating requirements

The operating requirements will, of course, vary with the scheme under consideration. For the once–through situation:

1 kWh/b electric power,   10,000 Btu/b fuel,   36 lb/b steam

C_5/C_6 isomerization capital cost

For a 10,000 BPD unit at the first of 1991, use $7 million for a once–through unit (Type I); $17 million for a Type IV unit.

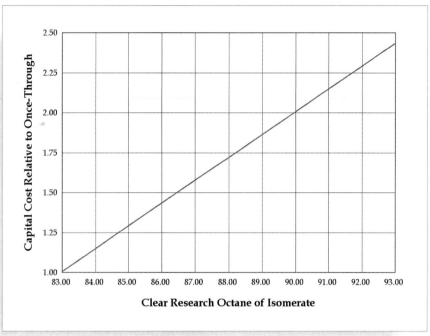

Fig. 18–4 Isomerization Cost vs. Octane

Notes

1. Anon., *Physical and Thermodynamic Properties of Elements and Compounds,* Chemetron Corporation, Louisville, 1969

2. Rosati, D., *Handbook of Petroleum Processes,* R.A. Meyers, Ed., McGraw–Hill Book Company, New York, 1986, pp. 5–40 to 5–46

References

Anon., *Hydrocarbon Processing,* April, 1984, p. 111

Ibid., April, 1990, p. 73

Bour, G., Schwoerer, C.P., and Asselin, G.F., *Oil & Gas Journal,* October 26, 1970, pp. 57–65

Chopey, N.P., *Chemical Engineering,* May 31, 1971, pp. 24–26

Cusher, N.A., in *Meyer's Handbook of Petroleum Refining Processes,* pp. 5–3 to 5–13, and 5–15 to 5–24

Ewing, R.C., *Oil & Gas Journal,* August 16, 1971, pp. 61–66

Greenough, P., and Rolfe, J.R.K., in *Meyer's Handbook of Petroleum Refining Processes,* pp. 5–25 to 5–37

Greenough, P., and Rolfe, J.R.K., *Handbook of Petroleum Processes,* Ed. R.A. Meyers, McGraw–Hill Book Company, New York, 1986, pp. 5–25 to 5–37

Raghuram, S., Haizmann, R.S., Lowry, D.R., and Schiferli, W.J., *Oil & Gas Journal,* December 3, 1990, pp. 66–69

Schmidt, R.J., Weiszmann, J.A., and Johnson, J.A., *Oil & Gas Journal,* May 27, 1985, pp. 80–88

Symoniak, M.F., and Holcombe, T.C., *Hydrocarbon Processing,* May, 1983, pp. 62–64

Ware, K.J., and Richardson, *Hydrocarbon Processing,* November, 1972, pp. 161–162

Weiszmann, J.A., in *Meyer's Handbook of Petroleum Refining Processes,* pp. 5–47 to 5–59

CHAPTER 19

ALKYLATION

The kind of alkylation considered here is the addition of isobutane to olefins, predominately butylenes, but also propylene and amylenes by some refiners. The process was commercialized during World War II as a means of supplying high-octane gasoline for military aircraft. With the conversion of military and other aircraft to jet engines, the demand for alkylate for that purpose plummeted. However, the demand for ever-greater quantities of ever-higher octane fuel for automobiles has maintained alkylate's prominence as a premium gasoline blending stock.

Alkylate promises to become even more important as environmental concerns are mandating the reduction of aromatics and olefins in gasoline. Alkylate contains neither of these.

Alkylation is catalyzed by a strong acid, either sulfuric (H_2SO_4) or hydrofluoric (HF). The sulfuric acid process operates at 40°F to 60°F under sufficient pressure to allow evaporation of hydrocarbons in the reactor to provide the refrigeration needed to maintain such temperatures. Hydrofluoric acid alkylation operates at temperatures attainable by cooling water (80°F to 110°F).

Since the olefins are reactive chemicals that readily polymerize given the chance, it is necessary to circulate a high ratio of isobutane to olefin (as high as 10:1 or more) through the reactor to minimize the odds of polymers forming. Since the reaction takes place in a liquid phase, it is necessary to have very intimate mixing because of unfavorable diffusion coefficients.

Maintenance of acid strength at a high level is important not only to obtain high alkylate quality, but to minimize corrosion of equipment. This is done by removing a slip stream from the circulating acid, replacing it with fresh makeup acid. Mixed butylenes produce better alkylate quality than does propylene, which produces better alkylate than do mixed amylenes.

In summary, we find that alkylate quality depends on:

- Isobutane/olefin ratio
- Temperature
- Degree of mixing
- Acid strength
- Feed composition

Though widely practiced, HF alkylation has fallen in favor recently due to some accidental releases of HF to the atmosphere. This problem and possible remedies have been discussed in the literature.[1,2,3]

Alkylation process description

The processes offered for license differ mechanically in the reactor/settler area. All however, provide intimate mixing of acid and hydrocarbon followed by separation of acid and hydrocarbon phases. The acid is recirculated while the hydrocarbon is fractionated to provide the high isobutane recycle required. Propane, normal butane, and alkylate are produced. See Figures 19–1 and 19–2. (Several companies are actively working to develop a viable solid–catalyst alkylation process.)

Preparation of the hydrocarbon feed is very important. Principal concerns are moisture and sulfur. More and more, selective hydrogenation units are found in alkylation units. The purpose is to convert diolefins to monoolefins. As a side benefit, there is some isomerization of butene–1 to butene–2 that produces a better alkylate.[4,5]

Where the refinery has a sulfuric acid alkylation unit and is considering adding a MTBE unit, there is an advantage to feeding the mixed butylenes to the MTBE unit before sending the remainder to the alkylation unit. This

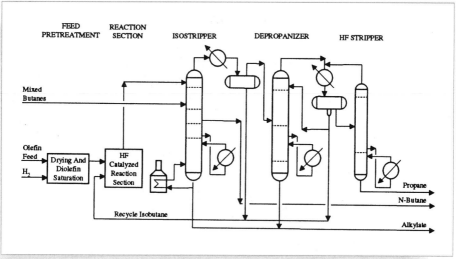

Fig. 19-1 Alkylation Unit

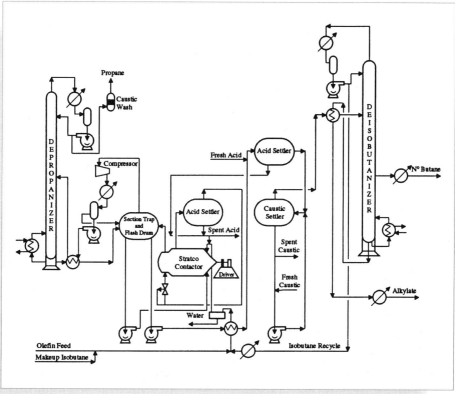

Fig. 19-2 Sulfuric Acid Alkylation Unit

is because the MTBE reaction is extremely selective for isobutylene that is the least desirable butylene in sulfuric acid alkylation.[6,7]

There is increased interest in alkylating amylenes, which would normally go directly into gasoline, because of the need to decrease olefins in, and RVP of, gasoline. The following tabulation illustrates this:[8]

	Mixed Amylenes	Amylene Alkylate
Mol. Wt.	70	110.9
(R+M)/2	87.4	89.7
RVP	19.4	1.0
Sp. Grav.	0.644	0.703
Relative Volume	1.0	1.51

Alkylation yield correlations

Most of the data in the literature is for single olefin species feeds with very little on actual refinery olefin mixes. No correlation attempts were made. Instead, Table 19–1 presents a fair consensus of the literature. Sulfuric acid does appear to have a slight advantage octane–wise due probably to its lower reaction temperature. Its acid consumption is higher, but it is a less expensive acid.

	H_2SO_4				HF				
FEED COMPOSTION, %									
PROPYLENE	45.0	100.0			100.0			44.0	18.5
BUTYLENES	55.0		100.0			100.0		56.0	72.2
AMYLENES				100.0			100.0		9.3
VOLUME RATIOS									
ISOBUTANE/OLEFIN	1.18	1.3	1.13	1.05	1.37	1.16	1.05	1.25	1.21
ALKLATE/OLEFIN	1.74	1.77	1.74	1.58	1.73	1.79	1.62	1.76	1.8
ALKYLATE OCTANE									
RON	93.3	90.5	96.0	92.5	92.0	95.5	91.5	93.5	94.0
MON	91.2	89.0	93.0	91.0	90.0	92.0	90.0		

Table 19–1 Sulfuric Acid Alkylation Unit

Alkylation operating requirements

Operating requirements are approximately as shown in the following tabulation:

	Electric (kWh/b)	Fuel (kBtu/b)	Steam (lb/b)	CW (gal/b)
Hydrofluoric	3.1	325	36	2300
Sulfuric	11.	—	180	1850

Alkylation capital cost

Unlike most process units, alkylation unit capacities are expressed in terms of alkylate production. From published data scaled to 10,000 BPD of alkylate and the first of 1991, we find the hydrofluoric unit costing $25 million and the sulfuric unit costing $29 million. However, both costs were within the range seen for the other.

Though finding in favor of hydrofluoric in this question, Anderson of UOP offers this caution, "As with many processes, however, establishment of basis and scope of design has more effect on capital cost than real process differences."[9]

Notes

1. Van Zele, R.L., and Diener, R., Hydrocarbon Processing, June, 1990, pp. 92–98.

2. Anon., "HF's Future is up in the Air," Chemical Engineering, May, 1990, pp. 39–41

3. Anon., "Studies Cover HF Spills and Mitigation," *Oil & Gas Journal,* October 17, 1988, pp. 58–62

4. Hammershaimb, H.U., and Shah, B.R., *Hydrocarbon Processing,* June, 1985, pp. 73–76

5. Heck, R.M., Patel, G.R., Breyer, W.S., and Merrill, D.D., *Oil & Gas Journal,* Jan. 17, 1983, pp. 103–113

6. Masters, K.R., and Prohaska, E.A., *Hydrocarbon Processing,* August, 1988, pp. 48–50

7. Anon., *Oil & Gas Journal,* November 18, 1991, pp. 99–100

8. Lew, L.E., Makovec, D.J., and Pfile, M.E., "Integrated Olefin Processing," 1991 NPRA National Meeting, San Antonio

9. Anderson, R.F., *Oil & Gas Journal,* February 11, 1974, pp. 78–82

References

Anon., *Hydrocarbon Processing,* September, 1978, p. 175

Ibid., p. 176

Anon., *Hydrocarbon Processing,* September, 1986, p. 101

Anon., "Phillips HF Alkylation Process" brochure

Ewing, R.C., *Oil & Gas Journal,* August 16, 1971, pp. 67–70

Hutson, T., Jr., and McCarthy, W.C., in *Meyer's Handbook of Petroleum Refining Processes,* pp. 1–23 to 1–28

Meyer, D.W., Chapin, L.E., and Muir, R.F., *Chemical Engineering Progress,* August, 1983, pp. 59–65

Shah, B.R., in *Meyer's Handbook of Petroleum Refining Processes,* pp. 1–3 to 1–22

CHAPTER 20

CATALYTIC POLYMERIZATION

The technology to be discussed here is the catalyzed reaction of propylene and/or butylenes to produce gasoline- and/or distillate-boiling-range materials. This became of interest to the refiner as a means of utilizing the olefins resulting from thermal cracking to produce more gasoline. Later, it was supplanted largely by the charging of these olefins to alkylation. The cat cracker is the primary source of olefins today. A second source is the byproduct C_4 stream from the thermal crackers producing ethylene (and propylene) as primary product for polymerization into much larger polymer molecules. The quantities available from this source are much more variable than those from the cat cracker.

The polymerization of butylenes for the above purposes is highly unlikely because of increasing demand for isobutylene for the production of MTBE and the increasing value of alkylate in future gasolines.

Though there has been much interest shown in dimerization of propylene to blend into gasoline, the future of this outlet is very uncertain. An examination of the following tabulation can be instructive in this regard:

Dimer	as produced	after saturation
Bromine No.	131	1.3
RON	94.5	63.7

Dimer	as produced	after saturation
MON	80.9	70.6
Average Octane	87.7	67.3

The dimer as produced is still an olefin as shown by the high bromine number and high sensitivity (RON–MON). On saturation, the bromine number drops dramatically, but so does the octane. With pressure on reduction of olefins in gasoline, the future for poly gas doesn't look bright.

Isobutylene can be selectively converted to dimer and trimer in the presence of other butanes.[1] These polymers can be hydrogenated to saturation without loss of RON, but with dramatic increase in MON making the product more or less equivalent to butylene alkylate. However, it is doubtful that units will be constructed for this purpose in the face of competition for isobutylene in existing alkylation units, and present and future MTBE units.

This technology can be, and has been, used to produce intermediates for the manufacture of synthetic detergents, alcohols, plasticizers, etc.

A lingering interest in this technology is for the production of distillate–boiling–range material. It has been demonstrated that propylene polymer (specifically the tetramer) has a better cetane number (20–21) than does the polymer from butylenes with similar boiling range. On hydrogenation, the cetane is 34–35. This material is still an inferior diesel, but could be blended into premium diesel fuel to some extent. In fact, one operator has 60 cetane number material into which he can blend 35% of the polymer and still produce a 45 minimum cetane diesel.[2]

Perhaps a more interesting use, is as a premium jet fuel blending stock as shown by the following tabulation:[3]

	Untreated Polymer	Hydrogenated	Specification Polymer
Bromine No.	—	77.9	0.019
Smoke Point, mm	42	—	25 minimum
Freeze Point, °C	-70	-68	-40 max

A study of these possibilities (jet fuel from LPG) was made for a client with very special circumstances. It involved the production of the olefin from paraffin by dehydrogenation. Even in the face of this very expensive requirement, the scheme proved attractive. Where the olefin (preferably propylene) is already a commodity (as in the U.S.), this technology should be considered.

Catalytic polymerization process description

Two processes will be described. The older process (still practiced today in modified form) uses a solid phosphoric acid catalyst on silica. The feed to the process needs to be treated to remove basic nitrogen and sulfur. The first is a poison to the catalyst and the second is not desired in the products. Moisture content needs to be controlled as well.

The treated feed is combined with recycle, exchanges heat with reactor effluent, and enters the reactor. Early reactors usually consisted of tubes filled with catalyst and surrounded by a boiling water bath to absorb the heat of reaction and maintain the desired temperature in the reactor. In a flash drum, recycle is separated from the polymer that is stabilized to remove remaining C_3 and C_4 hydrocarbons. The polymer is then separated by fractionation into the different products with dimer and trimer going to gasoline if desired or recycled to increase production of heavier materials, such as the tetramer and pentamer.

Reactors today are more likely to have the catalyst disposed in horizontal beds in a vertical vessel with the temperature controlled by the introduction of propane and/or butane between the beds as a quench. Figure 20–1 illustrates this scheme.

The other process uses a homogeneous catalyst system that is more or less miscible with the hydrocarbon phase. The catalyst complex includes an aluminum–alkyl and a nickel coordination complex. According to Andrews and Bonnifay,[4] "The reaction rate for propylene dimerization is high enough to obtain conversion rates above 90% in a single reaction." The catalyst is injected continuously in very low concentrations into the reactor. The catalyst is separated from the reaction product(s) by the addition of ammonia into the stream followed by a water wash. From the separator, the aqueous phase goes to waste water treatment and the hydrocarbon phase is stabilized and fractionated into LPG and dimer as shown in Figure 20–2.

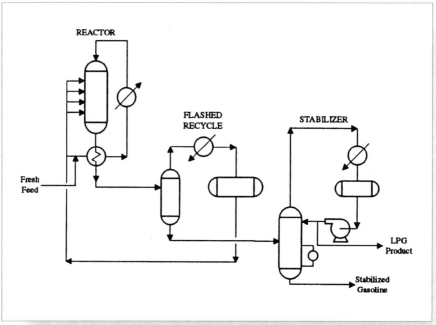

Fig. 20–1   Typical Catalytic Condensation Unit Flow   (© UOP; reprinted with permission)

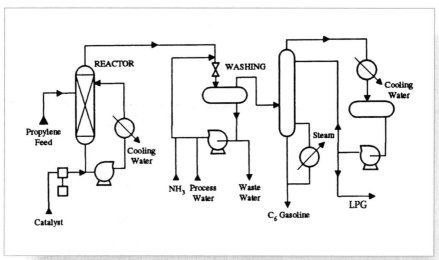

Fig. 20–2   IFP Dimersol for Hexenes

Catalytic polymerization yield correlations

No correlations were made in this area, since the results are very straightforward for the several processes and product objectives as illustrated in the following tabulations:

Reference	5	6	7
Olefin in Feed, wt %	78	59	71.6
Products, Wt%			
LPG	22	41.3	32
Gasoline	10	10.5	68
Distillate	67.5	47.0	—
Heavy Polymer	0.5	1.2	—

Product Properties	Gaso	Dist	Gaso	Dist	Gaso
API	58	52.2	65	49.9	73.5
RVP	—	—	3	—	6.5
RON	—	94.5	94	97	—
MON	—	80.9	82	82	—
Cetane No.	—	—	—	—	18
Freeze Point, °C	—	-70	—	—	—

Catalytic polymerization operating requirements

The requirements tabulated below are representative of these operations:

Operation	Phosphoric Acid Catalyst Gasoline	Homogeneous Catalyst Distillate	Gasoline
Electric, kWh/b	1.5	8	2
Fuel, kBtu/b	—	250	—
Steam, lb/b	250	350	60
CW, gal/b	—	—	20

Catalytic polymerization capital cost

Based on the meager data in the literature, the capital cost of a unit producing 3,000 BPD of product at the start of 1991 would be about as follows:

Homogeneous catalyst process	$3.6 million
Phosphoric acid process	
Gasoline production	$14 million
Distillate production	$33.5 million

However, it is doubtful that units will be constructed for this purpose in the face of competition for olefin in existing alkylation units and present and future MTBE units.

Notes

1. Scharfe, G., *Hydrocarbon Processing,* April, 1973, pp. 171–173

2. Maples R.E., and Jones, J.R., *Chemical Engineering Progress,* February, 1983, pp. 55–59

3. Anon., "Chevron Bulk Acid Polymerization Process," Chevron Research Company, San Francisco, 1972

4. Andrews, J.W., and Bonnifay, P.L., "The IFP Dimersol Process for Dimerization and Codimerization," 1973 NPRA Annual Meeting, San Antonio

5. Swart, J.S., Czajkowski, G.J., and Conser, R.E., *Oil & Gas Journal,* August 31, 1981, p. 86

6. Tajbl, D.G., in *Meyer's Handbook of Petroleum Refining Processes,* McGraw–Hill Book Company, New York, 1986, pp. 1–43 to 1–53

7. Benedek, W.J., and Mauleon, J.L., *Hydrocarbon Processing,* May, 1990, pp. 143–149

References

Andrews, J.W., and Bonnifay, P.L., *Hydrocarbon Processing,* April, 1977

Chauvin, Y., Gaillard, J., Leonard, J, Bonnifay, P., and Andrews, J.W., "New Processing Techniques as an Outgrowth of IFP Dimersol Technology," 1982 NPRA Annual Meeting, San Antonio

Kohn, P.M., *Chemical Engineering,* May 23, 1977, pp. 114–116

Weismantel, G.E., *Chemical Engineering,* June 16, 1980, pp. 77–80

CHAPTER 21

CATALYTIC DEHYDROGENATION

In this section, the interest is primarily in the production of isobutylene from isobutane, and secondarily, in the production of propylene from propane. Isobutylene is (or will be) needed to supplement the existing sources of isobutylene (mostly catalytic cracking) for the production of MTBE and probably ETBE, since alkylation will continue to be a prime source of gasoline blending stock and a major consumer of isobutylene.

Propylene has many uses. It can be polymerized to polypropylene. It can be alkylated if isobutane is available. It can be polymerized to the tetramer and hydrogenated to produce a premium jet fuel blend stock.

The following discussion, while addressing isobutane dehydrogenation specifically, applies generally to propane dehydrogenation as well.

In isobutane dehydrogenation, the objective is the production of isobutylene. The desired reaction involves the removal of one molecule of hydrogen from each molecule of isobutane:

$$H_3C-\underset{\underset{CH_3}{|}}{C}H-CH_3 \;\leftrightharpoons\; H_3C-\underset{\underset{CH_3}{|}}{C}=CH_2 + H_2$$

A number of side reactions can and do occur. The extent of these reactions depends on the selectivity of the catalyst at the

307

operating conditions. Products resulting from these side reactions include methane, ethane, ethylene, propane, propylene, other butylenes, heavier hydrocarbons, and coke.

Conversion and selectivity are the most important factors in determining the yield in a catalyzed reaction. Conversion is a measure of the degree to which the feed is reacted per pass through the reactor. This in turn determines the amount of recycling of unconverted material required. Selectivity is a measure of the amount of desired product produced per unit amount of feed reacted.

In isobutane dehydrogenation, conversion increases with temperature—selectivity decreases. Since there is an increase in number of molecules in the forward reaction, dehydrogenation is affected by reaction pressure, conversion increasing with decreasing pressure. The price of reduced pressure operation is increased equipment size and compression cost. The effect of pressure on conversion at a given temperature is illustrated in Figure 21–1. Thus, for the designer it becomes a matter of selecting temperature and pressure levels to optimize capital and operating costs.

The dehydrogenation of isobutane is highly endothermic (-52,732 Btu/lb mol at 1000°F).

Dehydrogenation process description

There are five different dehydrogenation technologies available for license: UOP's *oleflex,* Houdry's *catofin,* Phillips *star,* Coastal's *isobutane cracking* and Snamprogetti/Yarsintez's *isobutane dehydrogenation.*

Each of these processes is unique with respect to its reactor and regenerator technology.[1,2,3,4,5] Table 21–1 summarizes many of the features of the catalytic processes, excluding the Coastal process that is a non–selective, thermal process. All four catalytic processes include compression of the reactor effluent to an appropriate level for the separation of light ends–recycle and product.

Dehydrogenation operating requirements

The utility requirements found in the literature vary widely. Fortunately, they represent a small fraction of the cost of producing olefins. The varia-

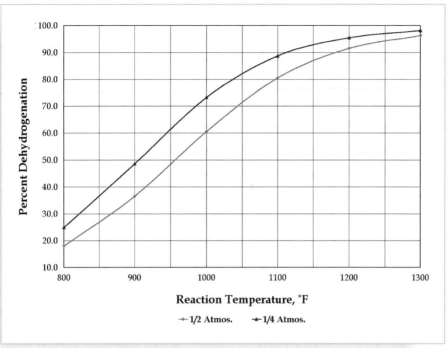

Fig. 21-1 Isobutylene/Isobutane Equilibrium

	HOUDRY	**UOP**	**SNAMPROGETTI**	**PHILLIPS**
Pressure, psia	4.6–14.2	36.5–19.8	17.1–21.3	Steam Dilution
Temperature, °F	1,100–1,175	1,031–1,209	1,020–1,110	900–1,150
Conversion, %	60	49	48	55
Selectivity, %	89	88	88	91
Reactor Type	Fixed Bed	Moving Bed	Fluidized Bed	Tubular
No. Reactors	2 minimum, usually 4 or 5	3 for IC4, 4 for C3	One	8
No. Regenerators	Regen. in situ	One	One	Regen. in situ
Catalyst Type	Chromic oxide on Alumina	Platinum on Alumina	Chromia-Alumina	Promoted Noble Metal
Catalyst Life	1.5 years	2–2.5 years	3 years	1 year
Cycle Time	7–15 minutes	Continuous	Continuous	7 hours process, 1 hour regeneration
Reaction Heat Source	Coke on Catalyst plus Fuel	Reheat Furnaces	Coke on Catalyst plus Fuel	Fire outsie Tubes

Table 21-1 Characteristics of Catalytic Dehydrogenation Processes

309

tion results from the fact there is a choice to be made for the drivers for two big machines, namely the main product compressor and the air (for catalyst regeneration) blower. These drivers may be electric motor, steam turbine, or gas turbine. The heat of reaction must be supplied, but much of this comes from the combustion of carbon on catalyst during regeneration.

The following data were taken from Meyer's Handbook and converted to a per–Metric–ton basis:[1,3]

	Electric kWh	Fuel kBtu	Steam k#	Cooling Water kgal	Licensor
Propylene	68	1,800	—	56	Houdry
Isobutylene	71.7	1,100	4.6	22.7	Phillips

Table 21–2 is based on data from Meyer's *Handbook* and shows the breakdown of the production cost of each of these olefins in percent. It is readily apparent that the principal cost factor is feedstock consumption. Therefore, anyone considering installing a dehydrogenation facility should investigate carefully the licensors' claims regarding selectivity and conversion and the experience of operators of the processes. The figures in this table are generally consistent with results obtained by the author in a recent confidential isobutylene technology study. In this study, feedstock cost ranged from 72% to 86% of the total; catalyst, from 0.6% to 3.0%.

	PROPYLENE	ISOBUTYLENE
Feedstock	63.49	76.74
Utilities	2.82	3.79
Catalyst	3.00	0.74
Labor	1.36	0.72
Capital Related	10.99	7.57
20% ROI	18.35	10.44
Total Cost	100.00	100.00
Licensor	Houdry	Phillips

Table 21–2 Dehydrogenation Production Cost Breakdown

Dehydrogenation capital cost

There is even less information in the literature on capital costs of dehydrogenation units than there is on operating requirements. It appears that a unit producing 300,000 tons per year of isobutylene would have cost $50 to $65 million the first of 1991. In the aforementioned study by the author, the range for a plant this size was from $46 to $64 million. Published data on propane dehydrogenation capital costs are confusing. They appear to vary considerably in what is included and what is not.

Notes

1. Craig, R.C., and Spence, D.C., *Meyer's Handbook of Petroleum Refining Processes,* McGraw–Hill Book Company, New York, 1986, pp. 4–3 to 4–22

2. Friedlander, R.H., *Meyer's Handbook of Petroleum Refining Processes,* McGraw–Hill Book Company, New York, 1986, pp. 4–23 to 4–28

3. Hutson, T., Jr., and McCarthy, W.C., *Meyer's Handbook of Petroleum Refining Processes,* McGraw–Hill, Book Company, New York, 1986, pp. 4–29 to 4–34

4. Buonomo, F., Fusco, G., Sanfilippo, D., Kotelnikow, G.R., and Michailov, R.A., *DeWitt 1990 Petrochemical Review,* Houston, March, 1990

5. Soudek, M., and Lacatena, J.J., *Hydrocarbon Processing,* May, 1990, pp. 73–76

References

Anon., *Hydrocarbon Processing,* September, 1980, p. 210

Ibid., April, 1984, p. 112

Ibid., November, 1985, p. 165

Bakas, S.T., Pujado, P.R., and Vora, B.V., AIChE Summer Meeting, San Diego, August, 1990

Berg, R.C., Vora, B.V., and Mowry, J.R., *Oil & Gas Journal,* November 10, 1980, pp. 191–197

Brinkmeyer, F.M., Rohr, D.F., Olbrich, M.E., and Drehman, L.E., *Oil & Gas Journal,* March 28, 1983, pp. 75–78

Clark, R.G., Gussow, S., and Schwartz, "Propylene and Butyleneby Selective Production," Sixth Congreso Argentino de Petroquimica, Bahia Blanca, November, 1982

Craig, R.C., Delaney, T.J., and Dufallo, J.M., *DeWitt 1990 Petrochemical Review,* Houston, March, 1990

Craig, R.C., Penny, S.J., and Schwartz, W.A., *Oil & Gas Journal,* July 25, 1983, pp. 161–163

Craig, R.C., and Spence, D.C., *Houdry Technology Reports No. 100.2*

Craig, R.C., and White, E.A., *Hydrocarbon Processing,* December, 1980, pp. 111–114

Dunn, R.O., and Anderson, R.L., *DeWitt 1990 Petrochemical Review,* Houston, March, 1990; 1990 Summer National Meeting, AIChe, San Diego, August, 1990; Washington National ACS Meeting, August, 29,1990

Gussow, S., Spence, D.C., and White, E.A., *Oil & Gas Journal,* December 8, 1980, pp. 96–101

Vora, B.V., and Berg, R.C., "Catalytic Dehydrogenation of Propane and Butanes," ACHEMA '82, Frankfurt, June, 1982

Whitehead, R.T., Dufallo, J.M., Spence, D.C., and Tucci, E.L., "The Catofin Process: The Catalytic Source of Propylene and Isobutylene," 1990 Summer National AIChE Meeting, San Diego, August, 1990

Wilcher, F.P., Vora, B.V., and Pujado, P.R., *DeWitt 1990 Petrochemical Review,* Houston, March, 1990

Section F:

OXYGENATES

CHAPTER 22

OXYGENATES

The two classes of compounds to be considered here are alcohols and ethers. The alcohols were first considered as a means of increasing gasoline octane quality to offset the phasing out of lead antiknock compounds. Later, the U.S. government considered ethanol by fermentation from agricultural products as a means of reducing our dependence on imported crude oil for energy. As an incentive, because it was not otherwise economic to add ethanol to gasoline, a subsidy program was established. This resulted in about 50,000 BPD of ethanol being blended into gasoline today.[1] Methanol, along with a necessary co-solvent, has been tried as a gasoline blending agent. Several problems connected with the use of these alcohols have been identified and will be discussed later.

The amended CAA required that by November 1, 1992, oxygen content of gasoline be a minimum of 2.7 weight percent in the wintertime in carbon monoxide (CO) non-attainment areas. Initially, it was expected that most of this oxygen would be supplied, by methyl tertiary butyl ether (MTBE).

The following tabulation shows some of the properties of the compounds most likely to find their way into gasoline, together with typical values for their blending behavior:[2]

Blending Values

	Sp. Gr.	BP, F	RVP	Wt% O2	RVP	(R+M)/2
Alcohols						
Methanol	0.796	149	4.6	49.9	50–60	116
Ethanol	0.794	172	2.3	34.7	17–22	113
TBA	0.791	181	1.8	21.6	10–15	101
Ethers						
MTBE	0.744	131	7.8	18.2	8–10	109
ETBE*	0.747	161	1.5	15.7	3–5	110
TAME**	0.770	187	1.5	15.7	3–5	104.5
DIPE***	0.725	155	4	15.7	4–5	105

* Ethyl tertiary butyl ether
** Tertiary amyl methyl ether
*** Diisopropyl ether

(In a later discussion, the octane blending values of these materials vary with the base stock into which they are blended.)

From these data it is readily apparent that addition of methanol or ethanol to gasoline is counter to the trend toward lower RVP to reduce evaporative emissions. In fact, the amended CAA provides a waiver of one pound greater pressure (10 RVP vs. 9 RVP) for alcohol-gasoline blends.

Alcohols

Problems with utilization of ethanol and methanol as motor fuels, stem primarily from their miscibility with water.

When the concentration of ethanol in gasoline exceeds about 9%, an aqueous phase forms. This phase is heavier than gasoline and remains on the bottom of the gasoline container.

Brazil has used ethanol extensively as a motor fuel. When the author was residing in Brazil in the 1950s, the government ministry in charge of sugar and alcohol determined the percent of alcohol put in gasoline. It was not unusual for a motorist to fill-up and receive mostly aqueous phase that shortly killed his engine. It was soon realized that this problem could be remedied by minimizing moisture throughout the gasoline system, from refinery to terminals to stations to automobiles.

More recently, Brazil has utilized alcohol alone as motor fuel in specially equipped cars with more satisfactory results. Figure 22–1 illustrates the phase behavior of a typical ethanol–gasoline–water system.[7]

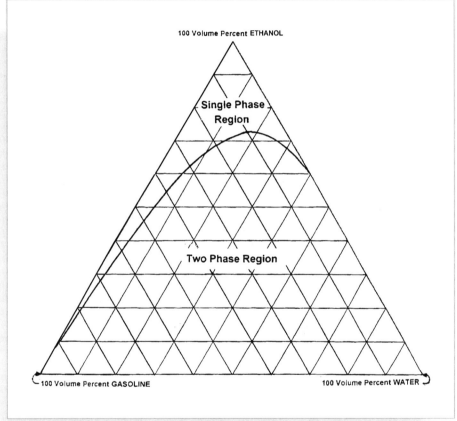

Fig. 22–1 Equilibrium Phase Diagram for Ethanol, Water, and Gasoline at 76°F

The use of methanol together with a co–solvent in gasoline has been tried, but with unsatisfactory results, due to corrosion, volatility, and water sensitivity. It appears that methanol will find its way into gasoline only in the form of one of the ethers—MTBE or TAME. The future of ethanol in gasoline hinges on the continuance of tax subsidies.

Ethanol

The two general classes of ethanol are fermentation and synthetic. Fermentation of grain has not been considered a petroleum refinery process, so it will not be discussed in detail here. Synthetic ethanol is produced either by the direct or indirect hydration of ethylene. Again, neither method has been considered a petroleum refinery process, though the principal producers have been chemical branches of major oil companies.

"All of the world's potable alcohol and 75% of its industrial alcohol are produced via the fermentation route."[3] This statement was made in 1973, but is probably reasonably representative of the situation today. At that time 75% of the world's synthetic capacity was in the U.S. In the face of the decreasing production of synthetic since then, this percentage may be significantly higher today.

Indirect hydration of ethylene was the earlier method employed to produce synthetic ethanol. It involved the reaction of ethylene with sulfuric acid producing an ester that could be reacted with water to produce ethanol and sulfuric acid.

Ethylene is hydrated directly in the vapor phase by use of a catalyst (phosphoric acid) and elevated pressure (1,000psig).[4] This method has largely replaced indirect hydration, although all production of synthetic alcohol has decreased more or less steadily as the cost of ethylene has made synthetic alcohol less competitive with fermentation alcohol.

"Ethanol is the only alcohol which is used in significant quantities in gasoline/alcohol blends in the U.S. 1988 consumption of ethanol in gasoline blends (10% ethanol) was about 800 million gallons."[3] This is alcohol produced by fermentation of grain.

Since the production of synthetic ethanol in the U.S. peaked in the early 1970s, there has been excess capacity. As a result, the author found no capital cost data in recent literature.

On the other hand, there is an abundance of data in the literature of the 1980s on fermentation plant costs. From these data, it appears that a 50 million gallon (95% alcohol) per year ethanol plant would have cost about $94 million the first of 1991.

The largest element of cost of producing ethanol by fermentation is that of the raw material. In a comprehensive study by Raphael Katzen Associates for the DOE in 1978,[5] the cost of grain ranged from 65.5% of production cost for a 10 million gallon per year plant to 85.7% for a 100 million gallon plant. This difference reflects the decrease in effect of fixed costs (labor and capital related) as the plant size is increased.

Paul J. Johnston of Union Carbide was quoted as saying a ton of ethanol requires 2 tons of sugar, 3.3 tons of corn or 4 tons of molasses. For synthetic ethanol, 0.6 tons of ethylene is required.[6] Since raw material represents such a large percentage of production cost, it is evident that these processes are very sensitive to the vagaries of the market.

Methanol

Production of methanol traditionally has not been considered a petroleum refinery process. This situation may change as ethers for blending into gasoline become important refinery products. Most methanol plants in the U.S. today are based on natural gas as feed and fuel, but in the past plants in Europe have been based primarily on naphtha. Raffinate from aromatic extraction is a good feed. With the reformulation of gasoline, some light naphthas may be seeking such an outlet.

The early methanol synthesis plants were small and employed reciprocating compressors in order to operate methanol synthesis loops at very high pressures (in excess of 5,000psi). As plant size increased above 500 tons per day (of methanol) and a more active catalyst was developed, it became possible to operate the synthesis loop at approximately 750psi and achieve satisfactory conversion. This permitted the use of the more efficient centrifugal compressor and the consumption of much less power.

The first commercial plant employing this low–pressure process started operation in 1966. Since then, many low pressure plants have been erected. In addition, there have been a number of medium pressure (1,500psi–2,500psi) plants constructed. The choice of design is a complex matter involving among other things fabrication and transport of large reactor vessels and matching compression requirements to the capacities of standard compressor designs.

Methanol is synthesized by reacting hydrogen and oxides of carbon in the presence of a catalyst:

$$2H_2 + CO \rightarrow CH_3OH$$
$$3H_2 + CO_2 \rightarrow CH_3OH$$

Since the forward reaction involves a molecule shrinkage, the importance of pressure in the process is apparent.

The synthesis gas is made by the same conventional steam–methane reforming as is used in ammonia synthesis or hydrogen manufacture.

Methanol synthesis is another process that is very sensitive to raw material cost. The operating requirements for methanol synthesis from natural gas average per ton of methanol are:

Electric power	35 kWh
Cooling water	15 kgal
Natural gas	29–30 million Btu

Variable cost (primarily natural gas) represents about 68% of production cost. The capital cost of a 1,000 ton per day methanol–from–natural gas plant would have cost about $85 to $90 million the first of 1991.

Ethers

MTBE

The use of MTBE in gasoline has been embraced by the oil companies and a large number of them are building or planning to build MTBE units within their refineries or chemical plants. The synthesis of MTBE involves the reaction of methanol with isobutylene:

$$CH_2 = \overset{\overset{\displaystyle CH_3}{|}}{C} - CH_3 + CH_3OH \rightarrow CH_3 - \overset{\overset{\displaystyle CH_3}{|}}{\underset{\underset{\displaystyle CH_3}{|}}{C}} - O - CH_3$$

Because of the reactivity of the double bond on the tertiary carbon in isobutylene, the reaction is highly selective and at one time was considered a way of producing polymer grade isobutylene (by decomposition of MTBE into methanol and isobutylene) from mixed butenes. The reaction is conducted in the liquid phase over an acidic ion–exchange resin. Since the reaction is exothermic, (-17,250 Btu/lb–mol), temperature control must be addressed in some manner in the design.

The feed to the MTBE unit can vary widely depending on its source. FCC butane/butylene streams contain 10% to 20% isobutylene. Depending on severity of the operation, the C4 fraction (by–product) from a steam cracker charging liquid feedstock can contain from 10% to 32% before butadiene removal or 34% to 44% after removal. The effluent from an isobutane dehydrogenation unit can exceed 50% isobutylene. In view of the variability in production of by–product C_4's from steam crackers and the fact that there will not be sufficient isobutylene from FCCs to meet MTBE requirements, it is not surprising that some oil companies are considering the installation of a complex to convert abundant normal butane to isobutane and dehydrogenate isobutane to isobutylene to produce MTBE. Figure 22–3 is a simplified process flow diagram of a typical MTBE unit. The integration of such a unit into a refinery is illustrated in Figure 22–4. The addition of a methanol plant to this scheme would render the refiner completely independent of other processors.

Not surprisingly, the cost of facilities for MTBE production reflects the isobutylene content of the feed to the unit. In a paper by Andre and Clark,[8] there appeared to be a nearly linear relation between capital cost and percent isobutylene with a slope of -$33,000 per percent isobutylene for a 2,500 BPD plant in 1986.

Operating requirements for MTBE manufacture average:

Electric power	1.8 kWh/b
Steam	225 lb/b
Cooling water	1200 gal/b

This is another process that is very sensitive to raw material price that represents 85% to 95% of production cost. The remainder of production cost is divided more or less equally between variable (ex raw material) and fixed costs. To produce a barrel of MTBE requires 0.66 barrels of isobutylene and 0.37 barrels of methanol. A plant to produce 12,500 BPD of MTBE from isobutylene would have cost $23 million at the first of 1991.

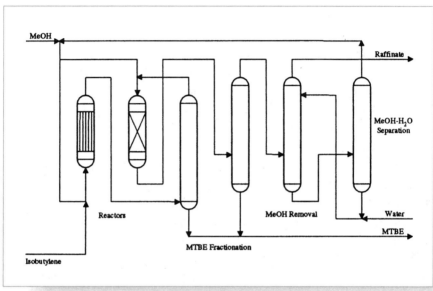

Fig. 22–3   Simplified MTBE Process Flow Diagram

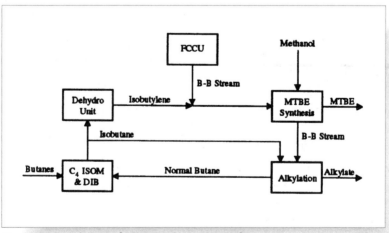

Fig. 22–4   Introduction of MTBE Production into Refinery

ETBE

"The rate of reaction of isobutylene with alcohols decreases with increasing molecular weight of the alcohol."[9] One would expect a greater capital cost for the same isobutylene consumption. Operating costs should be slightly higher as well. In addition, raw material cost will reflect the higher cost of ethanol over methanol. All this is borne out in SRI's PEP Review No. 88–3–2.

TAME

TAME can be produced in the same equipment as MTBE, but the conversion over the same volume of catalyst will be significantly lower, indicating the rate of reaction also varies with the molecular weight of the olefin. Isoamylene is present in the C_5 fraction of FCC light naphtha to the extent of about 25%. This light naphtha currently is blended directly into gasoline. Conversion of the isoamylene to a saturated compound with a very high blending octane number could aid the refiner significantly in complying with anticipated requirements for future gasolines.

Notes

1. Nierlich, F., Vora, B.V., Luebke, C.P., and Pujado, P.R., "Fuels/UOP Technology for ETBE/MTBE Production," *1989 Petrochemical Review,* DeWitt & Company, Houston

2. Unzelman, G.H., "Oxygenates in Gasoline—the '90 Decade", 1991 NPRA Annual Meeting, San Antonio

3. Anon., *Chemical & Engineering News,* March 5, 1973, pp. 7–8

4. Hatch, L.F., *Ethyl Alcohol,* Enjay Chemical Company, New York, 1962

5. DOE Contract No. EJ–78–C–01–6639, "Grain Motor Fuel Alcohol Technical and Economic Assessment Study," by Raphael Katzen Associates, December 31, 1978

6. Anon., *Chemical Week,* January 12, 1977, pp. 26–28

7. Maples, R.E., "Phase Behavior of Ethanol–Gasoline–Water Blends," AIChE 1997 Spring Meeting, Houston, March, 12, 1997, paper no. 57g

8. Andre, R.S., and Clark, R.G., "Butane–Derived MTBE Can Fill the Octane Gap," Air Products and Chemicals, Inc., 1986

9. Florez, M.P., and Greenaway, D., PEP Review No. 88–3–2, "ETBE Versus MTBE," SRI International

References

Anderson, E.V., *Chemical & Engineering News,* January 10, 1977, pp. 12–13

Anon., *Chemical & Engineering News,* December 16, 1985, p. 11

Anon., *Chemical Engineering,* September 17, 1984, p. 42

Ibid., October 5, 1984, p. 121

Anon., *Chemical Marketing Reporter,* September 13, 1976

Ibid., July 16, 1979

Ibid., December 21, 1981, p. 5

Ibid., May 24, 1982

Anon., *Chemical Week,* August 10, 1983, pp. 10–11

Ibid., March 9, 1968, pp. 49–51

Ibid., February 18, 1987, p. 32

Ibid., August 6, 1980, p. 13

Ibid., September 3, 1980, p. 17

Ibid., October 8, 1980, p. 11

Ibid., October 15, 1980, p. 15

Ibid., October 29, 1980, p. 9

Ibid., September 5, 1984, p. 46

Anon., *Hydrocarbon Processing,* November, 1985, p. 49

Ibid., December, 1984, p. 34

Anon., *Oil & Gas Journal,* September 29, 1980, p. 67

Ibid., August 4, 1980, p. 32

Ibid., February 15, 1988, p. 22

Rock, K.L., Dunn, R.O., and Makovec, D.J., "Automotive Fuels for an Improved Environment—How Does MTBE Contribute?," 1991 NPRA Annual Meeting, San Antonio

Methanol references

Anderson, E.V., *Chemical and Engineering News,* October 9, 1972, pp. 8–9

Ibid., June 20, 1983, pp. 20–21

Anon., *Chemical and Engineering News,* November 23, 1970, p. 21

Ibid., February 17, 1986, p. 19

Ibid., February 4, 1985, p. 13

Ibid., January 30, 1984, p. 13

Ibid., April 4, 1983, pp. 16–17

Ibid., April 7, 1980, P. 16

Ibid., May 8, 1967, pp. 24–25

Anon., *Chemical Engineering,* June 29, 1970, p. 19

Ibid., November 4, 1968, p.76

Ibid., February 11, 1980, p. 49

Ibid., December 11, 1972, p. 44

Anon., *Chemical Week,* January 6, 1968, pp. 34–36

Ibid., July 12, 1972, p. 35

Ibid., August 17, 1983, p. 28

Ibid., May 23, 1984, p. 18

Ibid., April 8, 1981, p. 16.

Anon., *European Chemical News,* August 28, 1970, p. 26

Ibid., July 18, 1975, p. 23

Ibid., September 27, 1974, p. 38

Ibid., November 29, 1974, p. 18

Ibid., February 16, 1973, p. 25–26

Anon., *Hydrocarbon Processing,* November, 1981, pp. 182–184

Ibid., November, 1983, pp. 111–113

Ibid., November, 1985, pp. 144–146

Ibid., March, 1991, p.164

Anon., *Oil & Gas Journal,* February 26, 1968, pp. 51–52

Ibid., March 17, 1975, pp. 112–118

Ibid., June 26, 1978, pp. 168–170

Ibid., April 21, 1980, p. 34

Ibid., July 17, 1967, p. 52

Ibid., *Oil & Gas Journal,* September 17, 1990, p. 82

Ibid., May 16, 1983, pp. 37–39

Anon., *Sources and Production Economics of Chemical Processes,* McGraw–Hill Publications, New York, 1979, pp. 204–205

Bare, B.M., and Lambe, H.W., *Chemical Engineering Progress,* May, 1968, pp. 23–30

Bolton, D.H., and Hanson, D., *Chemical Engineering,* September 22, 1969, pp. 154–156

Burke, D.P., *Chemical Week,* September 24, 1975, pp. 33–42

Cohen, L.H., and Muller, H.L., *Oil & Gas Journal,* January 28, 1985, pp. 119–124

Davis, J.C., *Chemical Engineering,* June 25, 1973, pp. 48–50

Duhl, R.W., *Chemical Engineering Progress,* July, 1976, pp. 75–76

Dutkiewicz, B., *Oil & Gas Journal,* April 30, 1973, pp. 166–178

Ganeshan, R., *Oil & Gas Journal,* July 24, 1972, pp. 61–62

Harris, W.D., and Davison, R.R., *Oil & Gas Journal,* December 17, 1973, pp. 70–71

Hedley, B., Powers, W., and Stobaugh, R.B., *Hydrocarbon Processing,* September, 1970, pp. 275–280

Ibid., August, 1970, pp. 117–119

Hiller, H., and Marschner, F., *Hydrocarbon Processing,* September, 1970, pp. 281–285

Minet, R.G., "A Technological Solution to Air Pollution Problems—Synthetic Fuel," Third Joint Meeting of the American Institute of Chemical Engineers—Institute Mexicano de Ingenieros Quimicos, Denver, 1970

Morrison, J., *Oil & Gas Journal,* July 3, 1972, pp. 60–61

Petzet, G.A., *Oil & Gas Journal,* September 6, 1982, pp. 35–38

327

Riegel, E.R., *Industrial Chemistry*, Reinhold Publishing Corporation, New York, 3rd ed., 1937

Rogerson, P.L., "ICI's Low Pressure Methanol Plant," 64th AIChE National Meeting, New Orleans, March, 1969

Royal, M.J., and Nimmo, N.M., *Oil & Gas Journal*, February 5, 1973, pp. 52–55

Soedjanto, P., and Schaffert, F.W., *Oil & Gas Journal*, June 11, 1973, pp. 88–92

Strelzoff, S., "Methanol—Its Technology and Economics," Sixty–Fourth National AIChE Meeting, New Orleans, 1969

Winter, C., and Kohl, A., *Chemical Engineering*, November 12, 1973, pp. 233–237

Zech, W.A., "Design, Development, and Operation of High Pressure Syn Gas Compressors in Methanol Production," 64th AIChE National Meeting, New Orleans, March, 1969

MTBE references

Ancillotti, F., Pescarollo, E., Szatmari, E., and Lazar, L., *Hydrocarbon Processing*, December 1987, pp. 50–53

Andre, R.S., Gussow, S., and Schwartz, W.A., "MTBE—the Feedstock Makes the Difference," 1981 NPRA Annual Meeting, San Antonio

Anon., C_4 *Monitor*, published by CTC International, Montclair, New Jersey, May, 1990

Anon., *Chemical and Engineering News*, February 12, 1990, p. 14

Ibid., July 27, 1987, p. 8

Anon., *Chemical Engineering*, May 25, 1987, p. 9

Anon., *Chemical Week*, June 4, 1986, pp. 36–37

Ibid., December 18, 1985, p. 7

Ibid., July 29, 1987, p. 26

Ibid., Feb. 25, 1987, p. 27

Anon., *Hydrocarbon Processing,* November, 1990, pp. 126 and 128

Ibid., p. 37

Ibid., August, 1990

Ibid., September, 1989, p. 39

Ibid., September, 1982, p. 177

Ibid., November, 1979, p. 197

Anon., MTBE Brochure, Atlantic Richfield Company, 1985

Anon., *Oil & Gas Journal,* March 25, 1991, pp. 26–27

Ibid., May 27, 1991, pp. 32–33

Ibid., June 25, 1990, p. 32

Ibid., February 12, 1990, p. 31

Ibid., August 7, 1989, p. 31

Ibid., June 19, 1989, p. 31

Ibid., March 21, 1988, p..31

Ibid., April 20, 1987, p. 30

Ibid., August 3, 1987, p. 31

Ibid., January 19, 1987, p 21

Ibid., May 26, 1986, p. 44

Ibid., March 12, 1990, p. 30

Ibid., May 27, 1985, pp. 92–93

Ibid., December 22, 1980, p. 61

Ibid., June 26, 1978, p. 62

Ibid., March 12, 1990, p. 30

Anon., *Sources and Production Economics of Chemical Products,* McGraw–Hill Publications Co., New York, 2nd ed., 1979, pp. 209–211

Bakas, S.T., Gregor, J.H., and Cottrell, P.R., "Integration of Technologies for the Conversion of Butanes into MTBE," 1991 NPRA Annual Meeting, San Antonio

Bitar, L.S., Hazbun, E.A., and Piel, W.J., *Hydrocarbon Processing,* October, 1984, pp. 63–66

Chase, J.D., and Galvez, B.B., *Hydrocarbon Processing,* March, 1981, pp. 89–94

Chase, J.D., and Woods, H.J., *Oil & Gas Journal,* April 9, 1979, pp. 149–152

Clementi, A., Oriani, G., Ancillotti, F., and Pecci, G., *Hydrocarbon Processing,* December, 1979, pp. 109–113

Glazer, J.L., Penny, S.J., and Gussow, S. "Convert Refinery Butanes to Oxygenates," 1984 NPRA Annual Meeting, San Antonio

Jones, J.R., Ludlow, W.I., Miller, K.D., and Acosta, T.A., "MTBE—A Practical Private Sector Route to Clean Fuels," 1989 Annual AIChE Meeting, San Francisco

Muddarris, G.R., and Pettman, M.J., *Hydrocarbon Processing,* October, 1980, pp. 91–95

Nierlich, F., Vora, B.V., Luebke, C.P., and Pujado, P.R., "Fuels/UOP Technology for ETBE/MTBE Production," *1989 Petrochemical Review,* DeWitt & Company, Houston, March, 1989

Verdol, J.A., and Hunt, M.W., "TBME Process," 85th National AIChE Meeting, Philadelphia, June 1978

Other ethers

Anon., *Chemical and Engineering News,* May 26, 1986, p. 26

Anon., *Oil & Gas Journal,* November 21, 1988, pp. 41–42

Ibid., October 19, 1987, p. 64

Short, H., *Chemical Engineering,* June 23, 1986, pp. 34–35

SECTION G:

TREATING AND OTHER AUXILIARY PROCESSES

CHAPTER 23

AROMATICS EXTRACTION

Aromatics extraction has long served as a means of obtaining aromatics for premium gasoline blending and for chemical manufacture. It now appears that decreasing the aromatics content of gasoline may become an additional, if not primary, function.

The most widely used technology for separating aromatics from non–aromatics combines liquid–liquid extraction and extractive distillation. This technology is most frequently used in producing benzene, toluene, and xylene (BTX), but is also used in reducing aromatic content of kerosenes and jet fuels, and for several other purposes. Figure 23–1 is a simplified process flow diagram of the Carom process, typical of this technique.

There are a number of solvents available today for the separation of aromatics from non–aromatics. Some of these, together with the references discussing them, are: Dimethylformamide (DMF),[1] N–Formylmorpholine (FM),[2] Dimethylsulfoxide (DMSO),[3] Sulfolane,[4] and Ethylene Glycols (UDEX, TETRA, CAROM).[5]

The operating requirements of an aromatic extraction unit are a function of many variables including properties of the solvent, composition of the feed, product purity requirements, and extent of recovery of aromatics desired. The principal items of consumption are electric power, steam, and solvent. Average values for electric power and steam per barrel of feed for the sulfolane process appear to be 0.83 kWh and 150 lb. respectively. Unfortunately, solvent consumption is generally stated as dollars

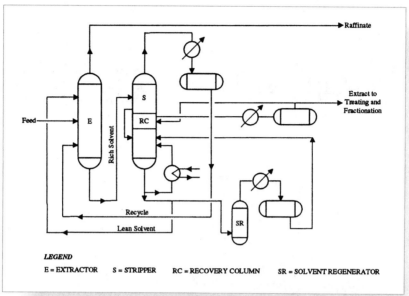

Fig. 23–1 Extraction Section Hydrocarbon Circuit of the Carom Process
(© UOP; reprinted with permission)

per day (by licensors) rather than as some volume or weight quantity of solvent on which a value could be placed at some other point in time.

The reader need not be disturbed by the occasional absence of some of the operating requirements for this and other of the following processes. These values generally represent minor fractions of the total cost of production. The feedstock usually represents the major element of cost. Where the process is a part of a complex, the requirements of the other processes will generally overshadow those of the process in question.

A different situation exists when comparing two or more processes that perform the same or similar function. In such a situation, one may be dealing with small differences between small (or large) numbers. This is the time to consult licensors of the technologies involved for definitive numbers to use in the comparison.

A sulfolane unit feeding 10,000 BPD would have cost about $6 million the first of 1991.

Notes

1. Durandet, J., Mikitenko, P., Cohen, G., Graco, F., Bonnifay, P., and Andrews, J.W., *Oil & Gas Journal,* August 18, 1975, pp. 112–114

2. Cinelli, E., Noe, S., and Paret, G., *Hydrocarbon Processing,* April, 1972, pp. 141–144

   Stein, M., *Hydrocarbon Processing,* April, 1973, pp. 139–141

   Preusser, G., Stein, M., and Franzen, J., *Oil & Gas Journal,* July 16, 1973, pp. 114–118

3. Choffe, B., Raimbault, C., Navarre, F.P., and Lucas, M., *Hydrocarbon Processing,* May, 1966, pp. 188–192

4. Deal, G.H., Jr., Evans, H.D., Oliver, E.D., and Papadopoulos, M.N., *Petroleum Refiner,* September, 1959, pp. 185–192

   Asselin, G.F., and Persak, R.A., "BTX Aromatics Extraction," 1975 UOP Technology Conference

   Broughton, D.B., "Extraction: Molex for Normal Paraffins— Sulfolane and/or Udex for Aromatics," UOP 1971 Technical Seminar, Arlington Heights, Illinois

   Wheeler, T., *Meyers' Handbook of Petroleum Refining Processes,* McGraw–Hill Book Company, New York, 1986, pp. 8–54 to 8–60

   Gary, J.H., and Handwerk, G.E., *Petroleum Refining Technology and Economics,* Marcel Dekker, Inc., New York, 2nd ed., 1984, pp. 247–255

   Anon., *Aromatics Extraction Process,* Brochure by Atlantic Richfield Company

5. Anon., *Chemical Week,* May 18, 1968, pp. 77–85

References

Anon., *Hydrocarbon Processing,* March, 1991, p. 141

Anon., *Oil & Gas Journal,* September 23, 1963, pp. 262–264

Lackner, K., and Emmrich, G., *Hydrocarbon Processing,* October, 1988, pp. 67–68

CHAPTER 24

HYDROGEN MANUFACTURE

Hydrogen is commonly made today by steam reforming of natural gas, LPG, or naphtha. Some refiners faced with adding to their hydrogen supply are giving some consideration to partial oxidation. Our consideration here will be limited to steam reforming of natural gas. Figure 24–1 is a simplified process flow diagram of a typical steam–methane reforming hydrogen plant.

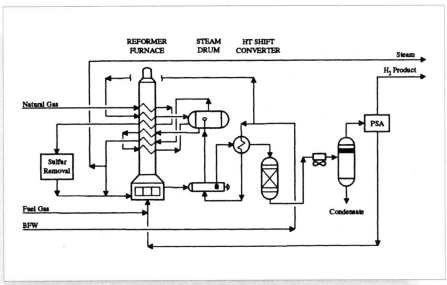

Fig. 24–1   Steam-Methane Reforming Hydrogen Unit

It appears that the consumption of natural gas for feed will average about 315,000 Btu per mscf of hydrogen—for fuel, about 182,000. Steam and electric requirements vary with the type of compressor drive employed.

Capital cost for a plant producing 100 mmscfd of hydrogen would have cost about $60 million at the first of 1991, including CO_2 removal and methanation.

References

Anon., *Hydrocarbon Processing,* April, 1982, pp. 161–163

Ibid., April, 1984, p. 108

Anon., *Hydrogen,* The Girdler Corporation Brochure, 1946

Anon., "Topsoe Technology in Hydrogen Production," Haldor Topsoe, Inc. Brochure, 1983

Buividas, L.J., Schmidt, H.R., and Viens, C.H., *Chemical Engineering Progress*, May, 1965, pp. 88–92

Gary, J.H., and Handwerk, G.E., *Petroleum Refining Technology and Economics,* Marcel Dekker, Inc., New York, 2nd ed., 1984, pp. 203–208

James, G.R., *Chemical Engineering,* December 12, 1960, pp. 161–166

Lee, G.T., Leslie, J.D., and Rodekohr, H.M., *Oil & Gas Journal,* May 11, 1964, pp. 154–156

CHAPTER 25

SOUR WATER STRIPPING

Water containing H_2S and/or NH_3 and sometimes phenols is produced by many of the units in the refinery including crude distillation, naphtha stabilizer, naphtha hydrotreater, hydrocracker, FCC, etc. A typical simplified process flow diagram appears in Figure 25-1.

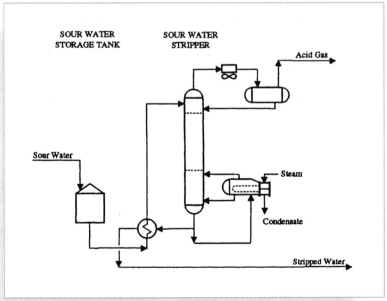

Fig. 25–1   Sour Water Stripping

Capital cost and operating requirements of a sour water stripper vary with the volume of water being treated, the concentration of H_2S and/or NH_3 in the water and whether each gas is recovered separately:

Electric power	0.6–1.8 kW per gpm
Steam	100–200 lb/h per gpm
Capital cost	$10–20 million for 1000 gpm

References

Annessen, R.J., and Gould, G.D., *Chemical Engineering,* March 22, 1971, pp. 67-69

Anon., *Oil & Gas Journal,* June 24, 1968, pp. 49-50

Anon., "Optimization of Sour Water Stripper Designs," Exxon Research and Engineering Company, June, 1979

Armstrong, T.A., *Oil & Gas Journal,* June 17, 1968, pp. 96-98

Bucklin, R.W., and Mackey, J.D., *Chemical Engineering Progress,* June, 1984, pp. 63-67

E. M. Blue of Chevron Research Company, personal communication re: Chevron Waste Water Treating Process, October 29, 1975

Klett, R.J., *Hydrocarbon Processing,* October, 1972, pp. 97-99

CHAPTER 26

SWEETENING

Sweetening here refers to the conversion of mercaptans (RSH) to disulfides (RSSR). In the case of LPG (C_3's and C_4's), the mercaptan in extracted in one column and the solution regenerated in a second column where the disulfide forms a separate layer from the regenerated solution. The sulfur content of the LPG is reduced by the amount of mercaptan extracted. A typical example of this process is illustrated in Figure 26–1.

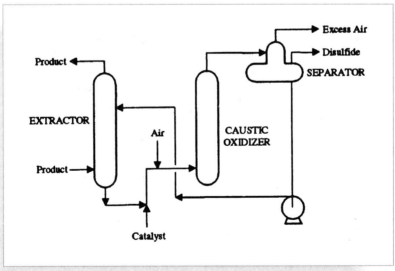

Fig. 26–1   Merox Treating—LPG unit

In the case of naphthas, jet fuels, etc., the mercaptans are converted in a fixed-bed reactor. The disulfides remain in the sweetened hydrocarbon liquid, so there is no decrease in sulfur content. Figure 26–2 is representative of this process.

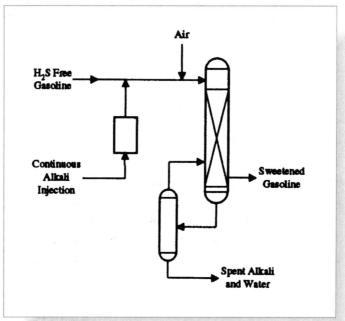

Fig. 26–2 Merox Liquid Treater

Operating requirements for sweetening consist of electric power and chemicals. Electric power is less than 0.1 kWh per barrel of feed and chemical cost is in the cents per barrel range. Capital cost is about $2 million for a 10,000 BPD LPG treater the first of 1991 and approximately $3 million to sweeten 10,000 barrels of naphtha or jet fuel.

References

Anon., *Hydrocarbon Processing,* April, 1982, p. 124

Ibid., April, 1988, p. 67

Anon., *Processing Guide,* UOP Brochure, 1975

Asselin, G.F., and Stormont, D.H., *Oil & Gas Journal,* January 4, 1965, pp. 90-93

Brown, K.M., "Commercial Results with the UOP Merox Process," presented before Francaise des Techniciens du Petrole, Deauville, June, 1960

Brown, K.M., Verachtert, T.A., Asselin, G.F., and Salazar, J.R., "Applications and Developments in the UOP Merox Process," 1977 Technology Conference

Cromwell, C.A., *Hydrocarbon Processing & Petroleum Refiner,* April, 1962, pp. 154-156

Embry, C.A., Tindle, A.W., and Wood, J.F., *Hydrocarbon Processing,* February, 1971, pp. 125-126

Salazar, J.R., *Meyers' Handbook of Petroleum Refining Processes,* McGraw-Hill Book Company, New York, 1986, pp. 9-4 to 9-13

Staehle, B.H., Verachtert, T.A., and Salazar, J.R., *Merox 1984,* UOP Process Division publication, October 29, 1984

CHAPTER 27

ACID GAS REMOVAL

There are three general categories of processes available for removal of acid gases from refinery gas streams:

- Chemical solvent (amines or potassium carbonate)
- Physical solvent (propylene carbonate, methanol, glycol ethers, etc.)
- Solid adsorbents

There are several amines used singly or in combination to meet the requirements of the individual processor for the selective or non-selective removal of H_2S and CO_2. The chemical solvents in general have a higher heat requirement (to decompose the chemical compounds formed) than do the physical solvents, but have lower power requirements due to lower solution circulation rates. The reader is referred to chapter endnotes for guidance on appropriate selection.[1,2,3,4] The simplified process flow diagram in Figure 27–1 is typical of an amine type sweetening unit.

Some typical values in terms of mmscfd of acid gas removed are as follows:

	Electric, kWh	Steam, klb
Physical	380	60
Chemical	4–6	8,000–10,000

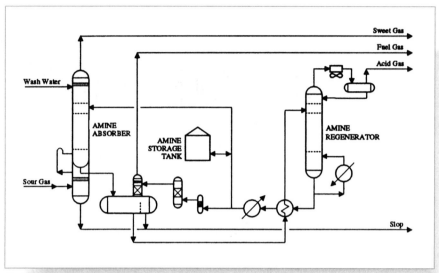

Fig. 27–1   Gas Sweetening Plant

Capital cost of a unit to remove 100 mmscfd of acid gas would have been about $15 million the first of 1991.

Notes

1. Samdani, G., *Chemical Engineering,* September, 1991, pp. 41-47

2. King, J.C., Stanbridge, D.W., Ide, Y., Trinker, T.A., and Gupta, S.R., *Oil & Gas Journal,* September 8, 1986, pp. 101-110

3. Mohr, V.H., and Ranke, G., *Chemical Engineering Progress,* October, 1984, pp. 27-34

4. Anon., *Engineering Data Book,* Gas Processors Suppliers Association, Tulsa, 9th ed., 1981

References

Anon., *Oil & Gas Journal,* November 3, 1986, pp. 72-73

Bartoo, R.K., *Chemical Engineering Progress,* October, 1984, pp. 35-39

Fitzgerald, K.J., and Richardson, J.A., *Oil & Gas Journal,* October 24,1966, pp. 110-118

Gary, J.H., and Handwerk, G.E., *Petroleum Refining Technology and Economics,* pp. 210-215

Maddox, R.N., and Burns, M.D., *Oil & Gas Journal,* September 18, 1967, pp. 113-121

Stephen M. Turner of Norton Company, personal communication re: Selexol, April 3, 1984

CHAPTER 28

SULFUR RECOVERY

This refers to the conversion of H_2S in a gas stream to elemental sulfur. The Claus process is considered here. It involves the combustion of one-third of the H_2S to SO_2 (by limiting the air input), which is then combined with the remaining two-thirds and passed over a catalyst where molten sulfur forms and is separated from the gas stream. The gas stream is cooled (by steam generation) and passed over another catalyst bed. This cycle is repeated for as many as four catalyst beds in some instances. The gas stream leaving the Claus unit still contains H_2S and/or SO_2, requiring further treatment to meet federal and/or state environmental regulations. Figure 28–1 pictures a typical sulfur recovery unit.

Operating requirements for a Claus unit in terms of long ton of sulfur produced are approximately as follows:

- 40 kWh of electric power
- 5,000 gallons of cooling water
- 750 gallons of boiler feed water

Capital cost of a unit producing 100 long tons of sulfur per day would have been about $5 million the first of 1991.

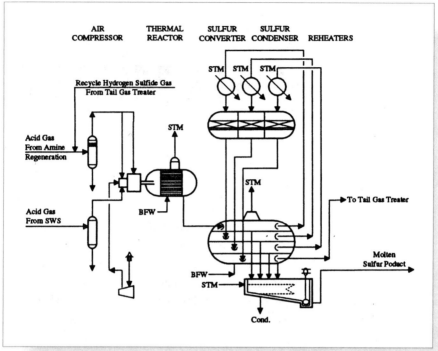

Fig. 28–1   Sulfur Recovery Unit

References

Anon., *Oil & Gas Journal,* October 28, 1968, pp. 88-101

Anon., *Oil & Gas Journal,* August 7, 1978, pp. 92-99

H. J. Gearin of The Fluor Corporation Ltd., personal communication re: Mathieson-Sasco Process, July 25, 1951

Kohl, A.L., and Fox, E.D., *Oil & Gas Journal,* February 25, 1952, pp. 154–177

CHAPTER 29

TAIL GAS CLEANUP

This refers to a unit designed to further reduce the sulfur content of the tail gas from a Claus unit. It is impractical to attempt to recover more than 98% of the contained sulfur in the Claus unit itself—94% to 96% is typical.[1,2] With states requiring 99% to 99.8% removal, some additional processing of the off gas is required. As a result, a family of 12 or so commercial processes has been developed for this purpose. Figure 29–1 is an example of a tail gas cleanup unit.

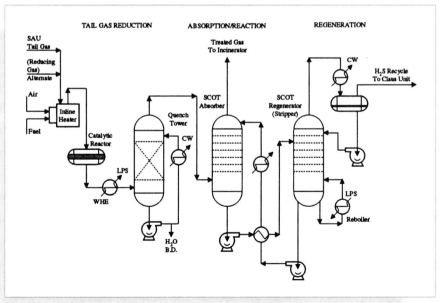

Fig. 29–1 Tail Gas Treating Unit

Since operation requirements vary so greatly with the specific situation and process, no generalization is presented. Again, these requirements are minor when studying a complex and can be ignored.

The capital cost of a tail gas cleanup unit appears to be about equal to the cost of the Claus unit when it is an add-on; about 75% of that, when installed along with the Claus unit.[1,2]

Notes

1. Doerges, A., Bratzler, K., and Schlauer, J., *Hydrocarbon Processing,* October, 1976, pp. 110-111

2. Naber, J.E., Wesselingh, J.A., and Groenendaal, W., *Chemical Engineering Progress,* December, 1973, pp. 29-34

References

Anon., *Oil & Gas Journal,* November 22, 1976, pp. 142-144

Ibid., August 28, 1978, pp. 160-166

Ibid., September 11, 1978, pp. 88-91

Anon., *Hydrocarbon Processing,* April, 1990, p. 97

CHAPTER 30

WASTE TREATMENT AND WASTE DISPOSAL

The principal process wastewater is from the sour water stripper. The stripped sour water will be recycled to the desalter for make-up to the extent possible. The excess will go to the wastewater treatment system as shown in Figure 30–1.

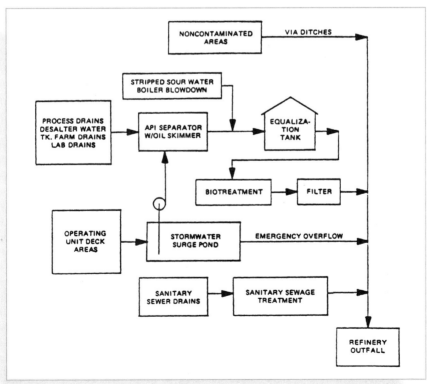

Fig. 30–1 Wastewater Treatment

355

Noncontaminated surface areas drain directly to the refinery outfall through ditches. Oily water drains are routed through API separators. Separator effluent water will be equalized with stripped sour water and boiler blowdown in an equalization tank before biotreatment and final filtration.

Storm water from operating unit deck areas will be accumulated in a storm water surge pond and gradually pumped into the contaminated water treatment system as treating capacity is available. During periods of heavy rainfall, when the capacity of the surge pond is exceeded, overflow storm water will pass through an overflow-underflow weir arrangement to trap surface oil before discharge; sanitary sewerage will be drained to a lift station and then routed through a package treating unit.

SECTION H:

BLENDING

CHAPTER 31

BLENDING

Blending is the combining of two or more materials to produce a new material. In the case of refinery products, the materials to be blended are complex mixtures of hydrocarbons (mostly) with the components frequently varying widely in properties. Since the final product of blending must meet certain specifications, it becomes necessary to be able to estimate a priori certain properties of a proposed blend. Otherwise, a trial-and-error procedure could prove costly in time and materials. Also, because of the complexity of the problem, there may be an infinite (or at least very large) number of blends that will meet a particular required specification. Usually there are several specifications to be met. Thus the problem becomes even more complicated. As a result, many refiners resort to a linear program to optimize their blends, particularly in the case of gasoline.

Estimating a property becomes a problem when the particular property is not additive, which is usually the case. A property is considered additive if the property of a blend is the average of that same property of each of the components in the blend (averaged on a weight-, volume- or mol-fraction basis). In other words, the property of a 50–50 blend would be the average of that property for the two components of the blend.

Specific gravity is an example of an additive property. The specific gravity of a blend can be estimated very accurately from the specific gravities and volume fractions of the components. (The change in volume that sometimes occurs in mixing hydrocarbons is too slight to be of concern in most cases.)

The API gravity on the other hand is not an additive property. In effect, API gravities of components are converted to specific gravities or densities in determining the specific gravity of a blend that may then be converted to API.

In addition to specific gravity, properties that are additive include:[1]

- Boiling point based on values from a TBP distillation
- Vapor pressure on a mol percent basis
- Aniline point
- Sulfur content

A number of properties of interest to the refiner, but that are not additive on a volume basis include:

- Octane number
- Viscosity
- Flash temperature
- Pour point
- Reid vapor pressure
- Smoke point

In discussing average boiling points (ABP), Maxwell lists properties that are additive when a particular ABP of each component is used:[2]

ABP	Property
Volumetric	Viscosity, liquid specific heat
Weight	Critical temperature
Molal	Pseudo critical temperature, liquid thermal expansion
Mean	Molecular weight, characterization
Factor (K), specific gravity, pseudo	Critical pressure, heat of combustion

360

When a property does not blend linearly (is not additive), one technique used is to substitute a blending number or blending index that does blend linearly. This practice was mentioned in chapter 5 in connection with octane numbers. It was pointed out that blending octane values are used to represent the apparent octane of a component in the usual gasoline blends. Many refiners develop their own sets of blending values through experience. A table of some octane blending values was included in chapter 5, Table 5–1.

Though some refiners still blend products batch–wise, most now have facilities for continuous or in–line blending.[3,4] With much improved continuous analyzers for octane and volatility coupled with computers, refiners can confidently blend directly to tankers and pipelines at considerable savings over batch blending due to reduced material in inventory, and closer approach to specifications (less quality give–away).

Gasoline Blending

Octane blending numbers

The blending value of a component in a base material is obtained by determining the octane rating of the base gasoline with and without the component in question. The value is then calculated by the following equation:

$$BOV = \frac{O_{blend} - O_{base}\, x(100-x)}{x}$$

where:

BOV =   Blending octane value of component

O_{blend} =   Octane of blend

O_{base} =   Octane of base gasoline

X =   Volume fraction of component

Any significant change in the composition of the base gasoline could result in a significant change in the BOV of the component in question.

Many methods have been proposed for predicting the octane number of a gasoline blend. Two of the most successful were developed by the Ethyl Corporation and the DuPont Company and will be reviewed here.

The answer to this dilemma developed by Ethyl Corporation,[5] was a set of equations employing the research octane, the motor octane, and the olefin and aromatic contents of each component in a blend to arrive at research and motor octanes of the blend. A simple blend was chosen by the author to illustrate this method and is depicted in Tables 31–1 and 31–2. The following equations are employed:

$$R = a_1[\overline{rj} - (\overline{r})(\overline{j})] + a_2[(\overline{O^2}) - (\overline{O})^2] + a_3[(\overline{A^2}) - (\overline{A})^2]$$

$$M = \overline{m} + b_1[\overline{mj} - (\overline{m})(\overline{j})] + b_2[(\overline{O^2}) - (\overline{O})^2] + b_3[\frac{(\overline{A^2}) - (\overline{A})^2}{100}]^2$$

Where the terms represent volumetric average values of given properties of components as follows:

r   Research octane
m   Motor Octane
j   Sensitivity (RON – MON)
rj   Research octane × Sensitivity
mj   Motor octane × Sensitivity
O   Volume percent olefins
A   Volume percent aromatics
R, M   Research, motor octane respectively of blend

Ethyl published two sets of coefficients for the above equations based on two sets of blend data: one for 135 blends, the other for 75. The coefficients for the 135 set without TEL are:

$a_1 = 0.03324$ $a_2 = 0.00085$ $a_3 = $ zero
$b_1 = 0.04285$ $b_2 = 0.00066$ $b_3 = -0.00632$

COMPONENT	LSR	ALKY	FCC	REF	N-C4	BLEND RON	BLEND MON
RON	66.4	94.9	90.4	91.0	94.0		
MON	66.4	91.5	78.7	82.8	89.1		
RVP	9.1	8.9	4.8	6.2	52.0		
OLEF	0.0	0.0	30.3	0.0	0.0		
AROM	5.0	0.0	32.8	41.0	0.0		
VOL FRAC	0.157	0.129	0.397	0.285	0.032		
AV RON	10.425	12.24	35.89	25.94	3.01	87.50	
AV MON	10.425	11.80	31.24	23.60	2.85		79.92
AV SENS	0.000	0.44	4.64	2.34	0.16	7.58	7.58
R*S	0.000	41.62	419.90	212.67	14.74	688.93	
M*S	0.000	40.13	365.55	193.50	13.97		613.16
AV O	0.000	0.00	12.03	0.00	0.00	12.03	
AV O^2	0.000	0.00	364.48	0.00	0.00	364.48	364.48
AV A	0.785	0.00	13.02	11.69	0.00	25.49	25.49
AV A^2	3.925	0.00	427.11	479.08	0.00	910.12	910.12
		RON COEFF		MON COEFF			
1		0.03324		0.04285		0.86	0.32
2		0.00085		0.00066		0.19	0.15
3		ZERO		−0.00632		0.00	0.00
ETHYL METHOD						88.55	80.40

Table 31–1 Ethyl Blending Method (135 Blends)

COMPONENT	LSR	ALKY	FCC	REF	N-C4	BLEND RON	BLEND MON
RON	66.4	94.9	90.4	91.0	94.0		
MON	66.4	91.5	78.7	82.8	89.1		
RVP	9.1	8.9	4.8	6.2	52.0		
OLEF	0.0	0.0	30.3	0.0	0.0		
AROM	5.0	0.0	32.8	41.0	0.0		
VOL FRAC	0.157	0.129	0.397	0.285	0.032		
AV RON	10.42	12.24	35.89	25.94	3.01	87.50	
AV MON	10.42	11.80	31.24	23.60	2.85		79.92
AV SENS	0.00	0.44	4.64	2.34	0.16	7.58	7.58
R*S	0.00	41.62	419.90	212.67	14.74	688.93	
M*S	0.00	40.13	365.55	193.50	13.97		613.16
AV O	0.00	0.00	12.03	0.00	0.00	12.03	12.03
AV O^2	0.00	0.00	364.48	0.00	0.00	364.48	364.48
AV A	0.79	0.00	13.02	11.69	0.00	25.49	25.49
AV A^2	3.93	0.00	427.11	479.08	0.00	910.12	910.12
		RON COEFF		MON COEFF			
1		0.03224		0.0445		0.84	0.34
2		0.00101		0.00081		0.22	0.18
3		ZERO		−0.00645		0.00	−0.04
ETHYL METHOD						88.56	80.39

Fig. 31–2 Ethyl Blending Method (75 Blends)

Researchers at DuPont developed "The Interactive Approach to Gasoline Blending."[6,7] This method requires the determination of the property being studied for each component and for 50–50 blends of all possible pairs of components to be included in the blend. It appears to give very good results for RVP, ASTM distillation, and V/L ratio in addition to octanes. The same simple gasoline was used to illustrate this method as shown in Table 31–3. This method requires interaction values for 50–50 blends of each possible pair of the blend components. These are listed in Table 31–3 as I–RON and I–MON. These values are multiplied by the volume fractions in the blend of the corresponding components. The sum of these products is added algebraically to the volume average of the corresponding octane (RON or MON). In the example: RON = 87.5–0.02 = 87.5 and MON = 79.9+0.34 = 80.2.

The results of these calculations are summarized in Table 31–4. It should be appreciated that this is just one example and does not necessarily represent the relative results to be expected by these two methods in every case.

COMPONENT	LSR	ALKY	FCC	REF	N-C4	BLEND			
RON	66.4	94.9	90.4	91.0	94.0				
MON	66.4	91.5	78.7	82.8	89.1				
RVP	9.1	8.9	4.8	6.2	52.0				
OLEF	0.0	0.0	30.3	0.0	0.0				
AROM	5.0	0.0	32.8	41.0	0.0				
VOL FRAC	0.157	0.129	0.397	0.285	0.032				
VOL AV RON	10.424	12.24	35.888	25.935	3.008	87.5			
VOL AV MON	10.424	11.80	31.243	23.598	2.8512	79.9			
						I-RON	INTER ACTIVE	I-MON	INTER ACTIVE
	X			X		−2.1	−0.0940	3.3	0.1476
	X		X			4.6	0.2867	5.8	0.3615
	X	X				−3.4	−0.0689	0.4	0.0081
	X				X	0.0			
			X	X		−1.7	−0.1923	2.4	0.2715
		X		X		−1.1	−0.0404	−4.3	−0.158
				X	X	0.0			
		X	X			1.7	0.0871	−5.6	−0.286
			X		X	0.0			
		X			X	0.0			
INTERACTIVE TERM							−0.0218		0.3439
BLEND							87.5		80.3

Table 31–3 DuPont Octane Blending Method

METHOD	RON	MON
VOLUMETRIC AVERAGE	87.5	79.9
DUPONT INTERACTIVE	87.5	80.3
ETHYL CORPORATION (75)	88.6	80.4
ETHYL CORPORATION (135)	88.5	80.4

Table 31–4 Gasoline Blend Octane Comparison

Lacking the detailed information required by either the Ethyl or the DuPont method or other specific data, the reader is left with general data such as that in chapter 5, Table 5–1. This is not a serious problem so long as the gasoline blends being studied do not differ significantly from the conventional blends on which the tabulated data are based.

The principal change taking place in gasoline blends currently is the addition of oxygenates in general and MTBE in particular. Some data have been found in the literature showing the octanes of base gasolines before and after the addition of certain percents of MTBE. These data have been correlated in terms of the sensitivity (RON–MON) of the base gasoline and the percent of MTBE added. It was found that the variation of blending octane value with amount of MTBE was very small compared to the effect of sensitivity. The results have been plotted and are shown in Figure 31–1. It is suggested that these curves be used in studies involving MTBE since the wide variation in blending octane value (BOV) of MTBE with base gasoline sensitivity is evident from the curves.

Reid vapor pressure (RVP)

Gasoline RVP is under close scrutiny in connection with emissions reduction. A simple method used by Chevron for calculating the RVP of a blend involves taking the sum of the products of the RVP of each component raised to the 1.25 power times its volume fraction.[8] This apparently assumes all components behave in a similar manner regardless of composition. The result for our simple gasoline blend is shown in Table 31–5.

Probably more accurate is the interactive method of DuPont. This is illustrated in Table 31–6. As in the case of octanes, the interactive factors (I–RVP)

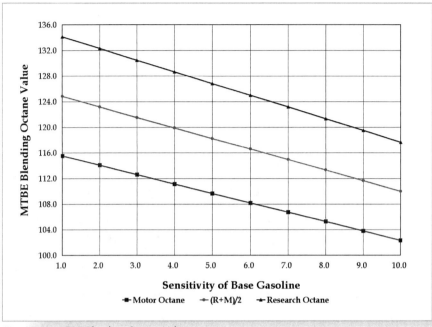

Fig. 31–1 MTBE Blending Octane Value

COMPONENT	LSR	ALKY	FCC	REF	N-C4	SUM
RON	66.4	94.9	90.4	91.0	94.0	
MON	66.4	91.5	78.7	82.8	89.1	
RVP	9.1	8.9	4.8	6.2	52.0	
OLEF	0.0	0.0	30.3	0.0	0.0	
AROM	5.0	0.0	32.8	41.0	0.0	
VOL FRAC	0.157	0.129	0.397	0.285	0.032	
FRAC RVP	1.4	1.1	1.9	1.8	1.7	7.9
BRVP	15.8	15.4	7.1	9.8	139.6	
FRAC BRVP	2.5	2.0	2.8	2.8	4.5	14.5
BLEND RVP						8.5

Table 31–5 Chevron RVP Blending Example

COMPONENT	LSR	ALKY	FCC	REF	N-C4	BLEND
RON	66.4	94.9	90.4	91.0	94.0	
MON	66.4	91.5	78.7	82.8	89.1	
RVP	9.1	8.9	4.8	6.2	52.0	
OLEF	0.0	0.0	30.3	0.0	0.0	
AROM	5.0	0.0	32.8	41.0	0.0	
VOL FRAC	0.157	0.129	0.397	0.285	0.032	
VOL AV RON	10.424	12.24	35.888	25.935	3.008	87.5
VOL AV MON	10.424	11.80	31.243	23.598	2.8512	79.9
VOL AV RVP	1.43	1.15	1.91	1.77	1.66	7.9

						I-RVP	INTER ACTIVE TERM
X				X		1.0	0.0447
X			X			0.0	0.0000
X	X					0.6	0.0122
X					X	−1.2	−0.0050
			X	X		0.2	0.0226
	X			X		0.0	0.0000
				X	X	14.4	0.1313
	X	X				0.2	0.0102
			X		X	7.0	0.0889
	X				X	6.2	0.0256

INTERACTIVE TERM	0.3307
DU PONT INTERACTIVE RVP BLEND VALUE	8.2

Table 31–6 DuPont RVP Blending Example

are multiplied by the volume fractions of the two components represented by the factor. The sum of these interactive terms is added to the straight volumetric average RVP to obtain the RVP of the blend (7.9+0.33 = 8.23).

Volatility

This refers to the distillation of the product. ASTM[9] has defined five volatility classes with temperature limits stipulated for certain percent points in the distillation of a gasoline. The class that applies in a particular instance varies geographically and seasonally in accordance with Table 2 in ASTM D 439. The specifications for the different classes are listed in Table 31–7.

DISTILLATION TEMPERATURES, °F.	A	B	C	D	E
10 VOL % MAX	158	149	140	131	122
50 VOL % MIN	170	170	170	170	170
50 VOL % MAX	250	245	240	235	230
90 VOL % MAX	374	374	365	365	365
END PT MAX	437	437	437	437	437
FOR 20 MAX V/L	140	133	124	116	105
RVP MAX, PSI	9	10	11.5	13.5	15

Table 31–7   ASTM Volatility Specifications for Gasoline

There is a nomograph in D 439 (Fig. X2.10) that enables the calculation of the temperature at which the vapor/liquid (V/L) ratio for a gasoline is 20. The equation for this nomograph is:

$$T_{V/L} = 114.6 - 4.1 RVP + 0.20 T_{10} + 0.17 T_{50}$$

One of the simpler methods for obtaining the partial effect of a component on the distillation of a blend has been presented by Decker, *et al.*[10] They present curves showing the relation between the distillation temperature in question minus the volumetric average boiling point of the component and a *percent evaporated blending value* for both narrow–boiling components and full–boiling components. Unfortunately, they do not give a numerical definition of narrow–boiling and full–boiling.

Values were read from their curves and were correlated by linear regression. The calculated values are plotted in Figure 31–2 and tabulated in Table 31–8. Assuming our simple gasoline blend is to meet specifications for Class C, percents evaporated at four temperatures were tabulated for each component. Applying the respective volume fractions to these values, the volume average percents were calculated. The temperature differences between the four specification temperatures and the VABP of each component were calculated to give blending values. The volume average percent evaporated was calculated for each of the four temperatures. All these results together with the RVP previously calculated and the specifications for Class C are displayed in Table 31–9. It can be seen that the blend meets

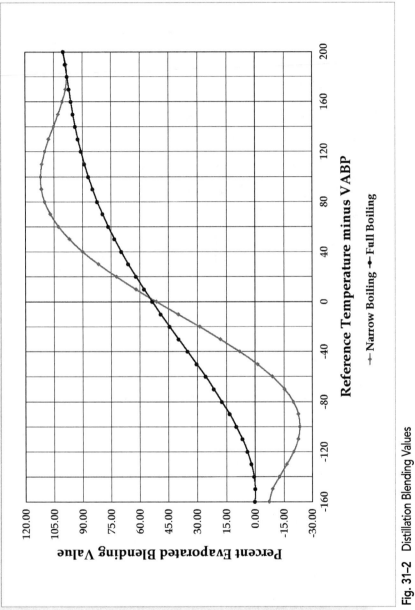

Fig. 31-2 Distillation Blending Values

| TEMP-VABP | PERCENT EVAPORATED BLENDING VALUES | |
	FULL BOILING	NARROW BOILING
−160	0.21	−7.13
−150	−0.02	−9.13
−140	0.61	−12.66
−130	2.00	−16.59
−120	4.07	−20.07
−110	6.72	−22.48
−100	9.87	−23.38
−90	13.44	−22.54
−80	17.36	−19.82
−70	21.54	−15.26
−60	25.93	−8.96
−50	30.46	−1.12
−40	35.08	8.01
−30	39.73	18.12
−20	44.37	28.91
−10	48.94	40.04
0	53.41	51.17
10	57.75	62.00
20	61.92	72.23
30	65.89	81.62

Table 31–8   Distillation Blending Values

all the specifications listed. The calculated percents are not greatly different from the volume average values.

Twu and Coon calculate gasoline blends using their universal interactive coefficients.[11]

Sweat and Maman describe a program for maximizing margin and minimizing quality giveaway when calculating blends for multiple products and multiple time periods using linear and non–linear programming.[12]

Vermeer, et al., use multivariable, model predictive technology to optimize blends.[13]

COMPONENT	LSR	ALKY	FCC	REF	N-C4	BLEND	
VOL. FRACT.	0.157	0.129	0.397	0.285	0.032	1.000	
VABP	130	217	233	254	31		
TEMP., °F							
140	62	22	10	10	117		
170	88	37	21	23	109		
240	99	67	52	55	100		
365	100	96	92	93	100		
VOL AV %							
140	9.73	2.84	3.97	2.85	3.74	23.1	
170	13.82	4.77	8.34	6.55	3.49	37.0	
240	15.54	8.64	20.64	15.68	3.20	63.7	
365	15.70	12.38	36.52	26.51	3.20	94.3	
TEMP-VABP							
140	10	−77	−93	−114	109		
170	40	−47	−63	−84	139		
240	110	23	7	−14	209		
365	235	148	132	111	334		
BLENDING VALUES							
140	58	18	13	6	94		
170	70	31	24	16	96		
240	89	63	57	47	100		
365	100	94	93	89	100		
							CLASS C
PERCENT EVAP							**SPECS**
140	9.1	2.3	5.2	1.7	3.0	21.3	10 % min
170	11.0	4.0	9.5	4.6	3.1	32.1	50 % max
240	14.0	8.1	22.6	13.4	3.2	61.3	50 % min
365	15.7	12.1	36.9	25.4	3.2	93.3	90 % min
DU PONT RVP - TABLE						8.2	11.5 max

Table 31-9 Distillation Blending

Distillate Blending

The properties of particular interest in blending distillates include distillation, cetane number, flash, pour, cloud, smoke, and viscosity.

Distillation

The volatility of a distillate blend as indicated by its distillation has been studied by DuPont's interactive method.[14] This has resulted in a series of

371

multipliers to be applied to the percents evaporated at various temperatures. For a blend of two components, the equation is:

$$PE = a_1 x_1 + a_2 x_2 + b_{12} x_1 x_2$$

where:

PE = Percent evaporated of blend

a_i = Percent evaporated of component i

x_i = Volume fraction of component i

b_{12} = Interaction coefficient (multiplier) for components 1 and 2

The equation for blending more than two components would include a first order (a_x) term for each component and an interaction term ($bx_i x_j$) for each pair of components. Table 31–10 shows the set of multipliers based on

TEMP.,°F	MULTIPLIER	TEMP.,°F	MULTIPLIER
400	–0.30	520	0.16
410	–0.35	530	0.19
420	–0.40	540	0.21
430	–0.34	550	0.27
440	–0.28	560	0.28
450	–0.22	570	0.28
460	–0.16	580	0.26
470	–0.09	590	0.24
480	–0.03	600	0.22
490	0.04	610	0.23
500	0.11	620	0.24
510	0.14	630	0.24
		640	0.24

Regression Output:

Constant	158.5738
Std Err of Y Est	0.020075 mult.
R^2	0.994081
No. of Observations	25
Degrees of Freedom	20

Variable	t	t^2	t^3	t^4
X Coefficient(s)	–125.004	36.28293	–4.60648	0.216438
Std Err of Coef.	12.18211	3.564843	0.460177	0.022114

Table 31–10   Diesel Fuel Interaction Coefficients

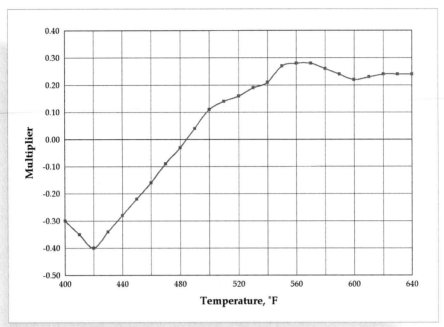

Fig. 31–3 Multipliers for Interaction Coefficients (Diesel Fuel Blends)

89 diesel fuel blends. Figure 31–3 is a plot of these multipliers plus the trace of the fourth order fit to the data.

Cetane number

In lieu of an actual engine test, the cetane number may be estimated from the cetane index or from the aniline point. There are two ASTM methods for calculating cetane index—D 976 and D 4737. The latter method requires the 10%, 50%, and 90% distillation temperatures plus the density of the blend. It appears more complicated than is justified in preliminary studies. D–976 requires only the mid–boiling (50%) temperature and the density and can be obtained readily from a nomograph. This method is on its way out as an ASTM method unfortunately.

There are several equations for calculating cetane number from cetane index. The one developed by Ethyl Corporation is:[15]

$$CN = 5.28 + 0.371(CI) + 0.0112(CI)^2$$

373

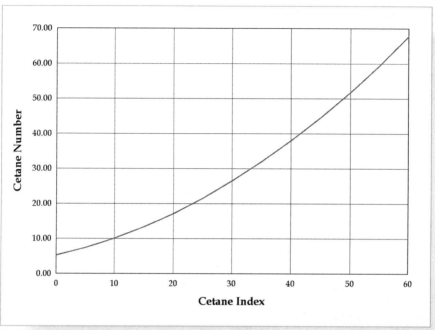

Fig 31–4 Cetane Number from Cetane Index

It "has been termed the best of those reviewed by the ASTM Cetane Prediction Task Force..."[16] Figure 31–4 is a plot of this equation.

The cetane number may be estimated from the aniline point by means of the following equation:

$$CN = 16.419 + 1.1332\left(\frac{AP}{100}\right) + 12.9676\left(\frac{AP}{100}\right)^2 + 0.205\left(\frac{AP}{100}\right)^3$$

This equation was derived by the Ethyl Corporation and was found to be a definite improvement over previous equations. Figure 31–5 is a plot of this equation. "Aniline point is a powerful predictor that is not currently in favor because it is not normally used as a quality control test. Results can vary widely if the aniline is not absolutely fresh as recommended by ASTM D 611."[16]

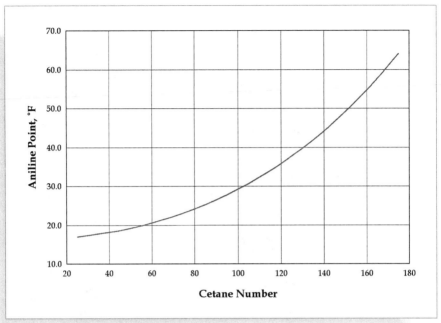

Fig 31–5 Cetane Number from Aniline Point

Flash, pour, and cloud points

Hu and Burns[17] presented equations and charts for estimating these properties of blends. They employed the concept of substituting blending index values for given properties, where the blending values are additive even though the property values are not. Tables 31–11, 31–12, and 31–13 are tabulations of index values calculated using their recommended general exponents. For a specific set of blend components, different exponents may give better results. For purposes of preliminary studies, however, the tabulated values are recommended.

Smoke point

This is an important specification for kerosenes and kerosene type jet fuels. It is determined by ASTM Method D 1322. In this test the smoke point is the height to which the flame in a lamp may be increased without smoking of the chimney. The Appendix to D 1322 has equations for calculating smoke point from luminometer number or the reverse. Jenkins and

375

FLASH POINT	FLASH PT INDEX	FLASH POINT	FLASH PT INDEX
0	10,000.00	145	103.55
5	8,349.87	150	90.27
10	6,985.51	155	78.78
15	5,855.13	160	68.82
20	4,916.75	165	60.20
25	4,136.25	170	52.70
30	3,485.82	175	46.19
35	2,942.79	180	40.53
40	2,488.59	185	35.60
45	2,108.00	190	31.30
50	1,788.54	195	27.54
55	1,519.94	200	24.26
60	1,293.70	205	21.39
65	1,102.84	210	18.88
70	941.56	215	16.68
75	805.07	220	14.75
80	689.36	225	13.05
85	591.13	230	11.56
90	507.61	235	10.25
95	436.49	240	9.09
100	375.84	245	8.08
105	324.05	250	7.18
110	279.77	255	6.38
115	241.84	260	5.68
120	209.32	265	5.06
125	181.40	270	4.52
130	157.40	275	4.03
135	136.73	280	3.60
140	118.92	285	3.22
		290	2.88

Table 31–11   Flash Point Index

POUR POINT	POUR POINT INDEX
–80	33.14
–75	39.03
–70	45.86
–65	53.77
–60	62.93
–55	73.50
–50	85.69
–45	99.70
–40	115.80
–35	134.27
–30	155.41
–25	179.57
–20	207.14
–15	238.57
–10	274.33
–5	314.96
0	361.06
5	413.31
10	472.42
15	539.24
20	614.65
25	699.65
30	795.35
35	902.97
40	1,023.85
45	1,159.45
50	1,311.40
55	1,481.49
60	1,671.67
65	1,884.09
70	2,121.08
75	2,385.24
80	2,679.36
85	3,006.53
90	3,370.10
95	3,773.74
100	4,221.43
105	4,717.53
110	5,266.78
115	5,874.32
120	6,545.75
125	7,287.15
130	8,105.12
135	9,006.82
140	10,000.00

Table 31–12 Pour Point Index

CLOUD POINT	CLOUD POINT INDEX
−80	1.07
−75	1.39
−70	1.80
−65	2.33
−60	2.99
−55	3.84
−50	4.90
−45	6.25
−40	7.94
−35	10.07
−30	12.72
−25	16.03
−20	20.16
−15	25.27
−10	31.60
−5	39.43
0	49.06
5	60.92
10	75.46
15	93.25
20	114.99
25	141.50
30	173.74
35	212.88
40	260.30
45	317.66
50	386.89
55	470.30
60	570.63
65	691.07
70	835.41
75	1,008.10
80	1,214.37
85	1,460.33
90	1,753.16
95	2,101.22
100	2,514.29
105	3,003.76
110	3,582.91
115	4,267.14
120	5,074.35
125	6,025.28
130	7,143.96
135	8,458.17
140	10,000.00

Table 31–13   Cloud Point Index

Walsh[18] presented four useful charts for estimating smoke point, luminometer number, aromatics content, or hydrogen content of jet fuels from the gravity and aniline point (Fig. 5–15). Since aniline point is additive, if one has aniline point data—and gravity, of course, for the blend components, one can easily estimate these other properties.

Viscosity

Viscosity blending has been investigated extensively. The ASTM temperature–viscosity charts have long been used to determine the viscosity of a blend manually. Blending indices have proven easier to use and more practical for computer simulation and linear programming. The Refutas method[19] employs the following relations:

$$VBI = 10.975 + 14.535 \ln \ln (Cs + 0.8)$$

$$Cs = exp(exp((VBI - 10.975)/14.535)) - 0.8$$

where:

VBI = Viscosity blending index

Cs = Viscosity in centistokes

This method is said to be for weight blending, but the author knows of one purveyor of a system for process simulation and LP optimization that uses it on a volume basis.

Chevron at one time used the Refutas method. They now have a system for volumetric blending.[19] The corresponding equations are:

$$VBI = \ln (Cs)/\ln (1000 \, Cs)$$

$$Cs = exp(VBI \ln (1000)/(1 - VBI))$$

where the variables have the same significance as before.

Al–Beshara, et al., described a method for determining the viscosity of blends of oils.[20] Abdel–Waly developed a method for estimating the viscosity of paraffinic oils.[21]

Semwal and Varshney describe a method for LP solution for minimum cost pour–point blending of diesel fuels.[22] Kahn presents a correlation for calculating cloud point, pour point, and cold flow point of diesel fuel blends.[23]

379

Crude Oil Blending

It is not unusual for a refiner to be faced with the possibility or the actuality of charging a new blend of crudes to the refinery. The assays of these crudes probably vary in the cut temperatures of the fractions collected and characterized during the assays. A series of equations can be derived to characterize the yields and properties of each of the crudes in the new blend.

The technique for doing this is described in an article by the author that was cited in chapter 7.[24] Having the necessary equations, the refiner is in a position to estimate the yields and properties of the streams distilled from the new blend having designated cut temperatures.

Notes

1. Nelson, W.L., *Petroleum Refinery Engineering*, New York, 4th ed., 1958, ff. 106

2. Maxwell, J.B., *Data Book on Hydrocarbons*, D. Van Nostrand Company, Inc., New York, 1950, pp. 10–12

3. Morris, W.E., *Oil & Gas Journal*, September 8, 1986, pp. 112–114

4. Wenzel, F.W., Serpemen, Y., and Hubel, A., "On–Line Gasoil Blending: An Important Tool to Improve Refining Profitability," 1991 NPRA Annual Meeting, San Antonio, March, 1991 and *Oil & Gas Journal*, March, 18, p. 62 and *Oil & Gas Journal*, April 1, p. 54

5. Healy, W.C., Jr., Maassen, C.W., and Peterson, R.T., "Predicting Octane Numbers of Multi–Component Blends," Report Number RT–70, Ethyl Corporation, Detroit, April 1, 1959

6. Morris, W.E., "The Interaction Approach to Gasoline Blending," NPRA 73rd Annual Meeting, San Antonio, March, 1975

7. Morris, W.E., *Oil & Gas Journal*, March 18, 1985, pp. 99–106

   Ibid., January 20, 1986, pp. 63–66

8. Anon., "31.0° API Iranian Heavy Crude Oil," Chevron Oil Trading Company, 1971

9. Anon., "Annual Book of ASTM Standards," American Society for Testing and Materials, Philadelphia, 1990

10. Anon., "Curves Predict Distillation Blending Behavior," *Oil & Gas Journal,* June 1, 1970, pp. 66–69, based on paper by R.R. Decker, J.R. Deckman, and L.W. Schneider presented at 1970 NPRA Annual Meeting, San Antonio

11. Twu, C.H., and Coon, J.E., *Hydrocarbon Processing,* March, 1997, ff 65

12. Sweat, B.K., and Naman, B.T., *Hydrocarbon Engineering,* March 1999, ff 16

13. Vermeer, P.J., Pedersen, C.C., Canney, W.M., and Ayala, J.S., *Oil & Gas Journal,* July 28, 1997, ff 74

14. Morris, W.E., *Oil & Gas Journal,* April 25, 1983, pp. 71–74

    Ibid., *Oil & Gas Journal,* September 23, 1985, pp. 119–122

15. Unzelman, G.H., *Oil & Gas Journal,* November 14, 1983, pp. 178–201

16. Collins, J.M., and Unzelman, G.H., *Oil & Gas Journal,* June 7, 1983, pp. 148–160 and June 13, 1983, pp. 128–131

17. Hu, J., and Burns, A.M., *Hydrocarbon Processing,* November, 1970, pp. 213–216

18. Jenkins, G.I., and Walsh, R.P., *Hydrocarbon Processing,* May, 1968, pp. 161–164

19. Baird, C.T., IV, *Guide to Petroleum Product Blending,* HPI Consultants, Inc., Austin, 1989

20. Al–Beshara, J.M., Akashah, S.A., and Mumford, C.J., *Oil & Gas Journal,* March 6, 1989, ff 50

21. Abdel–Waly, A.A., *Oil & Gas Journal,* June 16, 1997, ff 61

22. Semwal, P.B., and Varshney, R.G., *Oil & Gas Journal,* June 6, 1994, ff 89

23. Kahn, H.U., *Oil & Gas Journal,* June 29, 1994, ff 51

24. Maples, R.E., *Oil & Gas Journal,* November 2, 1997, ff 72

References

Anon., *MTBE Octane Enhancer,* Atlantic Richfield Company Brochure, 1985

Bott, D.J., and Piel, W.J., "Oxygenates for Future Fuels," 1991 AIChE Spring National Meeting, Houston

Chase, J.D., and Galvez, B.B., *Hydrocarbon Processing,* March, 1981, pp. 89–94

Chase, J.D., and Woods, H.J., *Oil & Gas Journal,* April 9, 1979, pp. 149–152

Ring, T.A., Bowers, K.E., and McGovern, L.J., *Oil & Gas Journal,* April 30, 1984, pp. 47–52

Unzelman, G.H., *Oil & Gas Journal,* April 10, 1989, pp. 33–37

Unzelman, G.H., and Michalski, G.W., "Processes for Blending Ethers—TAME and MTBE," 1984 NPRA Annual Meeting, New Orleans

SECTION I:

PROCESS ECONOMICS

CHAPTER 32

ECONOMICS

This chapter draws from previous chapters the information needed to perform process comparisons, technology evaluation, conceptual process design, and feasibility studies. One of the first tasks to be completed at the outset of any of these studies is to establish the bases to be employed, including the yield correlations, prices of raw materials and products, product properties, financing terms and conditions, process operating requirements, etc. This will help minimize later introductions of bias into a study and reduce interruptions due to missing information.

Refinery Economic Factors

At the beginning of the study of a proposed project, one of the first questions asked is: "What will it cost?"

With essentially no engineering other than selection of the process or process scheme and the throughput rate, the only estimate possible is that known as a *curve estimate*. This name derives from the fact that when cost data (after adjusting for time) for various capacities are plotted on log–log paper, the *best* line through the data is a straight line. Thus, costs can be read from the curve for various capacities.

Cost data for the refinery processes are presented in Table 32–1. Shown are a base capacity, the cost of that capacity for January, 1991, the Lang exponent, and a complexity factor. The use

Process	Base Capacity (BPSD)	Jan '91 Cost (MM$)	Scale Exponent	Stream Factor	Unit Complexity
Atmospheric Distillation	100,000	38	0.7	0.95	1.00
Vacuum Distallation	60,000	30	0.7	0.95	0.85
Solvent Deasphalt	30,000	34	0.6	0.95	2.03
Visbreaker	25,000	24	0.6	0.95	2.03
Delayed Coker	20,000	46	0.6	0.9	1.52
Fluid Coker	20,000	46	0.7	0.9	2.74
Fluid Catalytic Cracker	50,000	86	0.6	0.93	2.79
Heavy Oil Cracker	30,000	93	0.7	0.9	0.00
Hydrocracker	30,000	95	0.65	0.9	0.00
Hydrotreater	–	–	–	0.95	–
Kerosene/Jet	30,000	25	0.6	–	2.19
Diesel	30,000	25	0.6	–	2.19
Gas Oil	30,000	16	0.6	–	1.40
Naphtha Hydrotreater	30,000	16	0.6	0.95	1.40
Catalytic Reformer	–	–	0.6	0.95	–
Semi-regenerative	–	–	–	–	–
Cont. Catalytic Regen.	30,000	45	–	–	3.95
Isomerizer	–	–	0.6	0.95	–
Butane	10,000	20	–	–	5.26
Naphtha – Once-through	10,000	7	–	–	1.84
Naphtha – Recycle	10,000	17	–	–	4.47
Alkylation	–	–	0.6	0.95	–
Hydrofluoric acid	10,000	25	–	–	6.58
Sulfuric acid	10,000	29	–	–	7.63
Catalytic Polymerization	–	–	0.6	0.9	–
Gasoline	3000	14	–	–	12.28
Distillate	3000	33.5	–	–	29.39
Dehydrogenation	(300,000 T/Y)	50–65	0.7	–	–
Ethanol	(50 MM Gal/YR)	94	0.7	–	–
Methanol	(1,000 T/D)	85–90	0.7	–	–
MTBE	12,500	23	0.7	0.95	4.84
Amine Treater	(100 MMSCFD)	15	0.6	0.95	–
Sour Water Strupper	(1,000 gpm)	10–20	0.6	0.95	–
Sulfur Plant	(100 LT/D)	5	0.6	0.95	–
Scot Unit	(100 LT/D)	5	0.6	0.95	–
Hydrogen Plant	(100 MMSCFD)	60	0.6	0.95	–
Aromatics Extraction	10,000	6	0.6	0.95	1.58
Merox	–	–	0.6	0.95	–
LPG	10,000	2	–	–	0.53
Jet Fuel	10,000	3	–	–	0.79
Catalytic Naphtha	10,000	3	–	–	0.79

Table 32–1   Capital Cost Summary, Scaling Exponents, and Complexity

386

of the Lang exponent for scaling capacities is explained in chapter 1. These cost data are *curve* numbers and have a nominal accuracy of plus or minus 30%. This means a project estimated to cost $100 million could cost between $70 million and $130 million. The manner in which the accuracy of the estimate improves with completion of engineering is depicted in Figure 32–1.

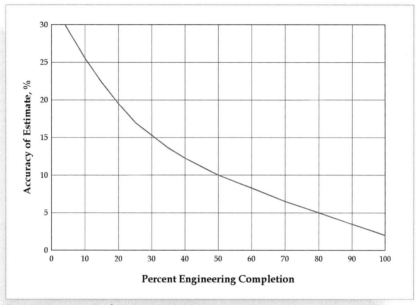

Fig,. 32–1 Accuracy of Engineering Estimate

An estimate can be adjusted to another time of completion other than January, 1991 by means of the Nelson–Farrar Refinery Construction Index (NFRCI). The base costs in Table 32–1 are for a NFRCI of 1,241.7. The adjusted cost is obtained by multiplying the base cost by the ratio of an appropriate NFRCI selected from Table 32–2 divided by 1,241.7. The most recent Nelson–Farrar Index is published in the *Oil & Gas Journal* in the first issue each month. Figure 32–2 is a plot of Table 32–2.

The complexity figures appearing in Table 32–1 are obtained by dividing the cost of a process unit by the cost of a crude distillation unit of the same capacity. The partial complexities included in the tabulation are obtained by multiplying the unit complexities by the corresponding ratio of a unit's capacity to the capacity of the crude distillation unit. This is equal to the ratio of the

X	194X	195X	196X	197X	198X	199X
0	77.6	146.2	228.3	364.9	822.8	1225.7
1	80.0	157.2	232.7	406.0	903.8	1252.9
2	83.7	163.6	237.6	438.5	976.9	1277.3
3	86.6	173.5	243.6	468.0	1026.0	1310.8
4	88.1	179.8	252.1	522.7	1061.0	1349.7
5	89.9	184.2	261.4	575.5	1074.4	1392.1
6	100.0	195.3	273.0	615.7	1089.9	1418.9
7	117.0	205.9	286.7	653.0	1121.5	1449.2
8	132.5	213.9	304.1	701.1	1164.5	1477.6
9	139.7	222.1	329.3	756.6	1195.9	1497.2

NOTE: *These data were compiled by the author from the Indexes published in the Oil & Gas Journal. Summaries of early years can be found in these issues: Nov. 29, 1976, p.70 and Jan. 30, 1978, P. 193. Data for more recent years can be found in the first issue for each month. The index was developed by W. L. Nelson. It has been continued by G.L.Farrar since the death of Mr. Nelson in 1978.*

Table 32–2   Nelson-Farrar Refinery Construction Index

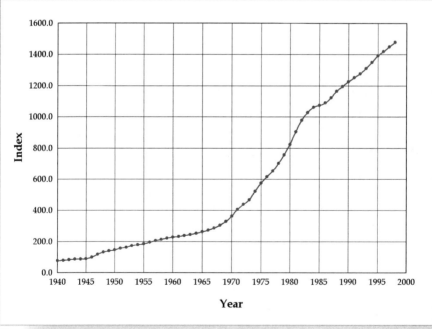

Fig. 32–2   Plot of Nelson-Farrar Refinery Index

cost of the given unit to the cost of the crude distillation unit. These partial complexities can be totaled for a process scheme to yield a process complexity, that in turn is used to estimate a plant complexity.[1] This relationship is plotted in Figure 32–3. In preliminary studies with essentially no engineering, the plant complexity is used to estimate the cost of plant and equipment for the entire refinery (including offsites). This is done by first calculating the total battery limits (BL) costs of process units and the process complexity, then multiplying the BL cost by the ratio of plant complexity to process complexity for the cost of total plant and equipment.

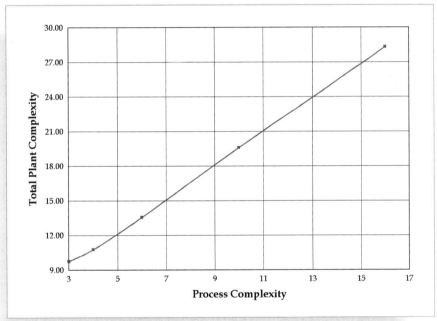

Fig. 32–3   Total Plant Complexity

The data presented in the following tables are average values obtained from numerous projects and are suitable for preliminary estimates. The data in Table 32–3 permit the calculation of the operating requirements of the refinery. The data in Table 32–4 provide some of the necessary additional costs defined as percents of BL cost. Table 32–5 provides additional data for individual processes.

As stated earlier, the economic bases to be employed in a study should be established at the outset. Table 32–6 is an example of this. Note that the

PROCESS	ELECTRIC (kWh/bbl)	FUEL (kBtu/bbl)	STEAM (lb/bbl)	CW (gal/bbl)
ATM DIST	0.5	100	25	
VAC DIST	0.5	100	50	
SOLV DEASPHALTER	2.0	80	60	
VISBREAKER	0.5	80	(50)	
DELAYED COKER	3.6	120	(40)	1
FLEXICOKER	13.0		(200)	30
FCCU	1.0	80	(20)	400
HEAVY OIL CRACKER	0.3		(80)	
HYDROCRACKER				
1,000 SCFB	8.4	93		
2,000 SCFB	13.1	214		
3,000 SCFB	17.9	335		
HYDROTREATING, SCFB				
KEROSENE, 25–140	1.7	9	7	
DIESEL, 50–200	1.7	8	7	
LT GAS OIL, 100-200	1.5	35	7	
HVY GAS OIL, 200-300	1.3	16	8	
CAT CYCLE, 100-900	1.4	24	7	400
NAPHTHA HYDROTREAT	2.0	30	15	
CAT REFORMER	1.0	300	(40)	100
ISOMERIZER	1.0	10	36	
ALKYLATION				
HYDROFLUORIC	3.1	325	36	2,300
SULFURIC	11.0		180	1,850
CAT POLYMERIZATION				
GASOLINE	1.5		250	
DISTILLATE	8.0	250	350	
DEHYDROGENATION				
PROPANE	68.0	1,800		56,000
ISOBUTANE	71.7	1,100	4,600	22,700
MTBE	1.8		225	1,200
AROMATIC EXTRACTION	0.8		150	
HYDROGEN, /MSCFD	0.5	145	24	
SOUR WATER STRIPPER	.8–1.8		100–200	
AMINE TREATER, /MMSCFD				
PHYSICAL	380.0		60	
CHEMICAL	4–6		8,000–10,000	
SULFUR, PER LT/D	40.0		(750)	5,000
TAIL GAS CLEAN UP				

Table 32–3   Summary of Process Utilities

ANNUAL COSTS	PERCENT
UTILITIES	14.0
CATALYST	1.4
VARIABLE COSTS	3.7
OPERATING LABOR	1.3
TOTAL SALARIES & WAGES	3.9
OPERATING SUPPLIES	0.3
MAINTENANCE	5.2
INSURANCE	0.4
FIXED COSTS	10.8
OPERATING COSTS	14.5
ONE TIME COSTS	
INITIAL CATALYST	5.3
ROYALTY	3.7
PRESTARTUP & STARTUP	7.2
WORKING CAPITAL	18.1

Table 32–4 Some Refinery Costs as Percent of B/L Investment

product prices are shown as decimal fraction of the crude price. These have been found to be satisfactory average values.

With the preceding information at hand together with the yield correlations presented earlier in the book, the user is in a position to calculate investment and operating costs, raw material costs, product revenues to arrive at cash flow for the project. Figure 32–4 is a diagram showing the definition of cash flow used by the author. With this number, the user can calculate the cash flow rate of return (CFRR) on the investment. Figure 32–5 shows the typical pattern for the outflow of cash during the engineering and construction of a project. This provides the investors with an approximation of their need for cash.

Some economic history of refining

Figure 32–6 shows how capital expenditures, including expenditures for pollution abatement, in the refining industry have fluctuated over the past 20 plus years. The ROI in the refining/marketing industry is plotted in Figure 32–7 along with a plot for all other businesses. The operating ROI, with and without pollution abatement costs are depicted in Figure 32–8.

391

PROCESS	INITIAL CATALYST	PAID-UP ROYALTY	ANNUAL CATALYST	OPERATING COST
ATM DISTILLATION				108.6
VAC DISTILLATION				79.4
SOLVENT DEASPHALT		5.0		
VISBREAKER				
DELAYED COKER				24.0
FLUID COKER		10.0	4.0	33.0
FLUID CAT CRACKER	0.6	10.0	1.0	24.0
HEAVY OIL CRACKER	0.75		10.0	
HYDROCRACKER				
1,800 SCFB	10.0	4.3	4.0	54.0
2,000 SCFB			5.0	
3,300 SCFB	14.0	15.0	6.0	
HYDROTREATER				
KEROSENE/JET	1.6		2.6	64.0
DIESEL	1.1		2.6	
GAS OIL			2.4	
NAPHTHA HYDROTREATER	2.8	11.0	0.2	
CATALYTIC REFORMER				
SEMIREGENERATIVE	14.6	15.0	2.3	
CONT CAT REGEN	8.5	15.0	1.4	67.6
ISOMERIZER				
BUTANE	4.0		12.3	
NAPHTHA-ONCE THRU	10.0		3.0	
NAPHTHA-RECYCLE	18.0		2.8	
ALKYLATION				
HYDROFLUORIC	0.9	5.0	0.64	64.0
SULFURIC	0.3			
CAT POLYMERIZATION				
DEHYDROGENATION			3.5	
MTBE				
AMINE TREATER				24.3
SULFUR PLANT	1.8	5.0		23.8
SCOT UNIT	1.6	6.0		
HYDROGEN PLANT	3.0	0.5		44.0
AROMATICS EXTRACTION	6.0	2.5	0.5	
MEROX				
LPG	0.7	6.0		
JET FUEL	1.5	10.0		
CAT NAPHTHA	5.0	30.0		
COMPLETE REFINERY	6.0	4.0	0.7	48.0

Table 32–5   Miscellaneous Process Costs as Percent of B/L Investment

UTILITIES	Dollars per unit	
ELECTRIC POWER	0.06 per kWh	
FUEL	3.00 per MMBTU	
STEAM		
HIGH PRESSURE	4.50 per M lb	
MED. PRESSURE	4.00 per M lb	
LOW PRESSURE	3.00 per M lb	
BOILER FEED WATER	0.40 per M lb	
COOLING WATER	0.10 per gal	

FEEDSTOCKS AND PRODUCTS	($ per gal)	($ per bbl)
BUTANES	0.34	14.28
METHANOL	0.36	15.12
MTBE	0.90	37.80
HYDROGEN		
REGULAR GASOLINE		
PREMIUM GASOLINE		
COKE	$100/TON	
SULFUR	$70/TON	

PRODUCTS—RATIO TO CRUDE PRICE	
GASOLINE	1.30
KEROSENE/JET	1.24
DIESEL	1.21
RESIDUAL FUEL OIL	0.78

FINANCIAL	
PROJECT LIFE	20 YRS
RATE OF INTEREST	8%/YR
TYPE OF COMPOUNDING	PERIODIC
UTILITIES/OFFSITES	50% OF B/L
TYPE OF FINANCING	100% EQUITY

Table 32–6   Basis for Economic Evaluation

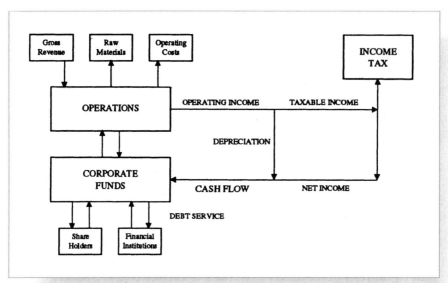

Fig. 32–4 Cash Flow Definition Diagram

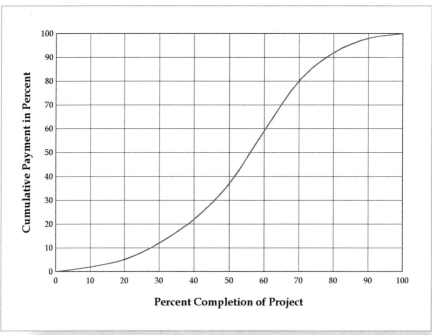

Fig,. 32–5 Schedule of Cash Outflow

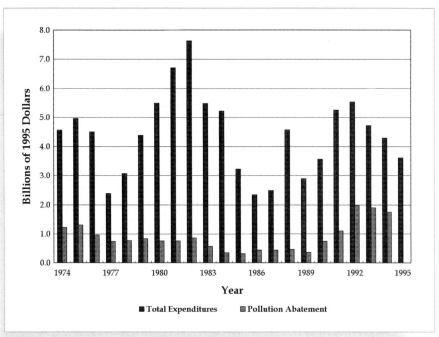

Fig. 32–6 Capital Expenditures (U.S. Refining Industry)

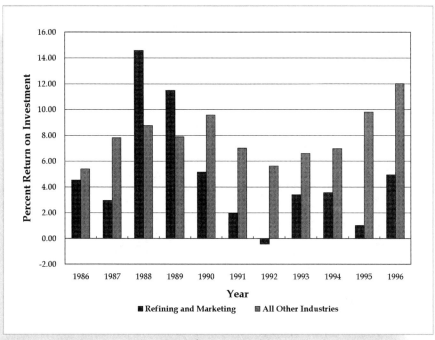

Fig. 32–7 Return on Investment (U.S. Refining and Marketing and Other Business)

395

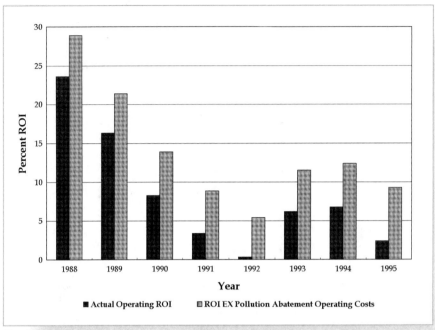

Fig. 32–8   Operating Return on Investment with and without Pollution Abatement Costs

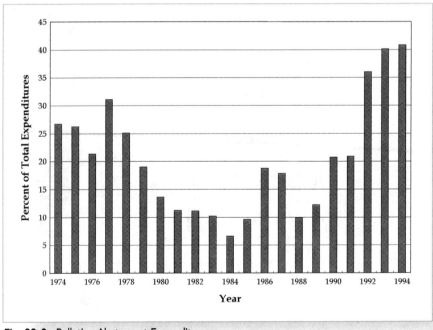

Fig. 32–9   Pollution Abatement Expenditures

396

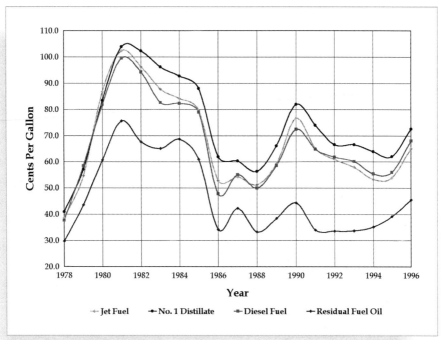

Fig. 32–10   Product Prices Excluding Taxes

Figure 32–9 shows pollution abatement expenditures in the refining industry as percent of total capital expenditures. Refinery product prices in cents per gallon and excluding taxes are pictured in Figure 32–10.[1]

One measure of the fiscal health of a refinery is the margin it realizes. A refinery's gross margin is the difference in dollars per barrel between its product revenue (sum of the barrels of each product times the price of each product) minus the cost of raw materials (primarily crude, but also purchased butane, MTBE, etc.). The net or cash margin is equal to the gross margin minus the operating costs (excluding income taxes, depreciation, and financial charges). Figure 32–11 shows the variation in margins over a 20 year period.[2] Margin values[3] are calculated by Wright Killen and Company for the U.S. Gulf Coast for a high conversion refinery, a medium conversion refinery and the regional average. These data appear in the statistics section of the third issue each month of the *Oil & Gas Journal.*

Another statistic of interest is the crack spread. This is a rough estimate of the gross margin and an indication of the current value of a given crude. A common calculation is the 3–2–1 crack spread. This statistic assumes that

397

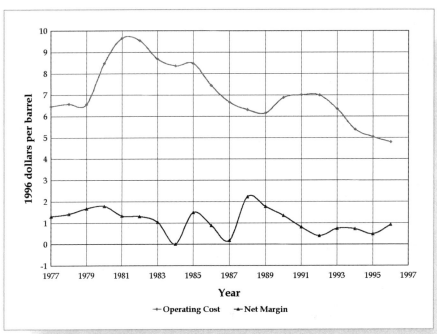

Fig. 32–11   U.S. Refinery Margins

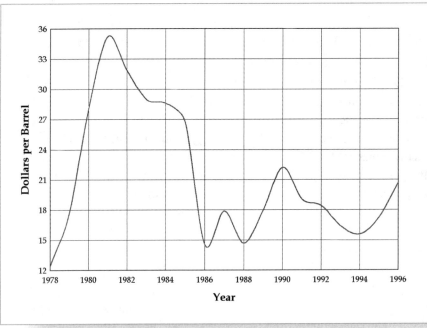

Fig. 32–12   Refiner Acquisition Cost of Crude

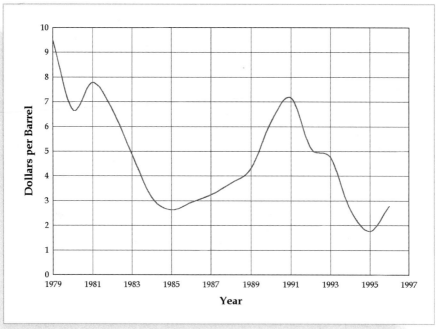

Fig. 32–13 Crude Price Differential—$/bbl (Bonny Light – Arab Heavy)

three barrels of a given crude can produce two barrels of gasoline and one barrel of distillate. This is equal to 2/3 the cost of a barrel of gasoline plus 1/3 the cost of a barrel of distillate minus the cost of a barrel of the given crude. These values appear weekly in the statistics section of the *Oil & Gas Journal.*[4] The dramatic variation in the refiner acquisition cost of crude is illustrated in Figure 32–12.[5]

The variation in the difference in price between a Bonny Light crude and Arab Heavy crude appears in Figure 32–13.[1] Some refineries, to take advantage of the high differential, were designed to process heavy crude, only to find that they were not competitive when the differential shrank.

The following additional figures of interest include: Average Retail Prices for Automotive Fuels (Fig. 32–14),[6] Product Prices Relative to Crude (Fig. 32–15),[7] and Refining Industry Capital Intensity (Fig. 32–16).

Examples

Three examples have been selected to illustrate some of the possibilities presented in this book:

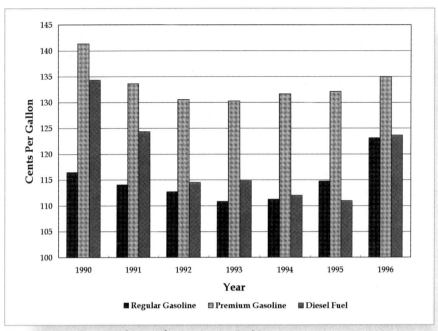

Fig. 32–14   Average Retail Prices (for Automotive Fuels)

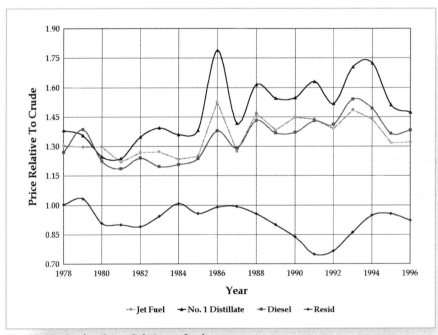

Fig. 32–15   Product Prices Relative to Crude

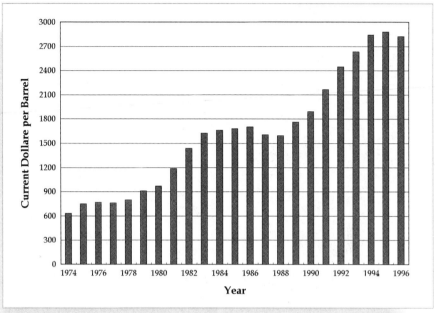

Fig. 32–16 Refining Industry Capital Intensity

- Gasoline reformulation
- Economic viability of a new grass–roots refinery
- Fluid coking vs. delayed coking

Gasoline reformulation. The first example looks at some of the steps that a refiner might take to meet the specifications mandated by the CAA Amendments. The basis chosen was the average of the gasoline compositions reported by U.S. refiners to the NPRA.[8] This is shown in Table 32–7 (base case). Table 32–8 (case 1) shows the results of adding MTBE to the 2 weight percent minimum required and the decrease in butane content to the specified RVP. Aromatic content remains above the maximum specified.

The next step (case 2) was to decrease the reformer severity from 93.5 to 91 average octane. This decreased aromatic content slightly and increased total gasoline volume as shown in Table 32–9.

A parallel trial (case 3) consisted of removing the C_{10+} portion from the FCC gasoline with an attendant reduction in aromatic content as can be seen in Table 32–10.

401

COMPONENT	LV%	RVP	(R + M)/2	% AROM	% OLEF
BUTANE	4.08	60.0	91.9	0.0	2.7
ISOMERATE	3.64	15.6	84.1	0.9	0.4
LT STRAIGHT RUN	6.35	13.3	74.5	4.4	0.8
LT COKER NAPHTHA	0.67	13.0	78.2	4.5	35.2
LT HYDROCRACKATE	2.41	12.5	80.9	2.7	0.2
MTBE	0.80	9.2	106.4	0.1	0.4
POLY GASOLINE	0.42	9.0	88.3	0.3	96.4
OTHER	2.14	8.4	85.9	32.6	20.1
C4= ALKYLATE	12.12	7.9	92.2	0.4	0.5
FBR FCC *	37.66	7.1	86.5	29.0	29.6
RAFFINATE	1.13	6.5	66.7	6.7	2.5
FBR REFORMATE *	26.94	4.6	93.5	66.2	0.7
PYROLYSIS GASO.	0.29	4.8	90.0	68.3	9.4
AROMATICS	1.36	1.2	101.6	90.1	1.5
TOTALS **	100.00	9.5	88.4	31.4	12.7
SURVEY VALUES		9.5	88.8	31.8	12.7

Table 32-7   Reformulated Gasoline Study

COMPONENT	LV%	RVP	(R + M)/2	% AROM	% OLEF
BUTANE	1.87	58.5	91.9	0.0	2.7
ISOMERATE	3.33	15.6	84.1	0.9	0.4
LT STRAIGHT RUN	5.82	13.3	74.6	4.4	0.8
LT COKER NAPHTHA	0.61	13.0	78.3	4.5	35.2
LT HYDROCRACKATE	2.21	12.5	80.9	2.7	0.2
MTBE	11.02	9.2	106.4	0.1	0.4
POLY GASOLINE	0.38	9.0	88.3	0.3	96.4
OTHER	1.96	8.4	85.9	32.6	20.1
C4= ALKYLATE	11.10	7.9	92.2	0.4	0.5
FBR FCC	34.49	7.1	86.5	30.5	28.5
RAFFINATE	1.03	6.5	66.7	6.7	2.5
FBR REFORMATE	24.67	5.3	93.5	62.6	0.7
PYROLYSIS GASO.	0.27	4.8	90.0	68.3	9.4
AROMATICS	1.24	1.2	101.6	90.1	1.5
TOTALS	100.00	8.7	90.2	28.4	11.3
TARGET MAX		8.7		25.0	
TARGET MIN			87.3		

CHANGES FROM BASE:

INCREASE MTBE TO 2 WT % OXYGEN MINIMUM
DECREASE BUTANE TO SPECIFIED RVP
AROMATIC AND OLEFIN CONTENT DECREASED BY DILUTION
VOLUME AS PERCENT OF BASE CASE: 110.3

Table 32-8   Reformulated Gasoline Study—Case [1]

COMPONENT	LV%	RVP	(R+M)/2	% AROM	% OLEF
BUTANE	1.96	58.5	91.9	0.0	2.7
ISOMERATE	3.25	15.6	84.1	0.9	0.4
LT STRAIGHT RUN	5.68	13.3	74.6	4.4	0.8
LT COKER NAPHTHA	0.60	13.0	78.3	4.5	35.2
LT HYDROCRACKATE	2.15	12.5	80.9	2.7	0.2
MTBE	11.02	9.2	106.4	0.1	0.4
POLY GASOLINE	0.37	9.0	88.3	0.3	96.4
OTHER	1.91	8.4	85.9	32.6	20.1
C4= ALKYLATE	10.84	7.9	92.2	0.4	0.5
FBR FCC	33.67	7.1	86.5	30.5	28.5
RAFFINATE	1.01	6.5	66.7	6.7	2.5
FBR REFORMATE	26.07	5.3	91.0	55.0	0.7
PYROLYSIS GASO.	0.26	4.8	90.0	68.3	9.4
AROMATICS	1.21	1.2	101.6	90.1	1.5
TOTALS	100.00	8.7	90.3	27.0	11.0
TARGET MAX		8.7		25.0	
TARGET MIN			87.3		

CHANGES FROM CASE 1:

DECREASE REFORMER SEVERITY RESULTING IN AROMATIC DECREASE AND IN-
CREASED VOLUME OF GASOLINE
VOLUME AS PERCENT OF BASE CASE: 113.

Table 32–9   Reformulated Gasoline Study—Case [2]

COMPONENT	LV%	RVP	(R + M)/2	% AROM	% OLEF
BUTANE	1.95	58.5	91.9	0.0	2.7
ISOMERATE	3.61	15.6	84.1	0.9	0.4
LT STRAIGHT RUN	6.30	13.3	74.6	4.4	0.8
LT COKER NAPHTHA	0.66	13.0	78.3	4.5	35.2
LT HYDROCRACKATE	2.39	12.5	80.9	2.7	0.2
MTBE	11.00	9.2	106.4	0.1	0.4
POLY GASOLINE	0.41	9.0	88.3	0.3	96.4
OTHER	2.12	8.4	85.9	32.6	20.1
C4= ALKYLATE	12.02	7.9	92.2	0.4	0.5
FBR FCC	30.09	9.9	84.5	22.6	25.3
RAFFINATE	1.12	6.5	66.7	6.7	2.5
FBR REFORMATE	26.70	5.3	93.5	62.6	0.7
PYROLYSIS GASO.	0.29	4.8	90.0	68.3	9.4
AROMATICS	1.34	1.2	101.6	90.1	1.5
TOTALS	100.00	8.7	90.3	26.1	9.2
TARGET MAX		8.7		25.0	
TARGET MIN			87.3		

CHANGES FROM CASE 1:

REMOVE C10+ FROM FCC GASOLINE RESULTING IN REDUCED AROMATICS AND
OLEFINS.
VOLUME OF GASOLINE AS PERCENT OF BASE CASE: 101.9

Table 32–10   Reformulated Gasoline Study—Case [3]

COMPONENT	LV%	RVP	(R+M)/2	% AROM	% OLEF
BUTANE	1.95	58.5	91.9	0.0	2.7
ISOMERATE	3.61	15.6	84.1	0.9	0.4
LT STRAIGHT RUN	6.30	13.3	74.6	4.4	0.8
LT COKER NAPHTHA	0.66	13.0	78.3	4.5	35.2
LT HYDROCRACKATE	2.39	12.5	80.9	2.7	0.2
MTBE	11.00	9.2	106.4	0.1	0.4
POLY GASOLINE	0.41	9.0	88.3	0.3	96.4
OTHER	2.12	8.4	85.9	32.6	20.1
C4= ALKYLATE	12.02	7.9	92.2	0.4	0.5
FBR FCC	30.09	9.9	84.5	22.6	25.3
RAFFINATE	1.12	6.5	66.7	6.7	2.5
FBR REFORMATE	26.70	5.3	93.5	62.6	0.7
PYROLYSIS GASO.	0.29	4.8	90.0	68.3	9.4
AROMATICS	1.34	1.2	101.6	90.1	1.5
TOTALS	100.00	8.7	90.3	26.1	9.2
TARGET MAX		8.7		25.0	
TARGET MIN			87.3		

CHANGES FROM CASE 1:

REMOVE C10+ FROM FCC GASOLINE RESULTING IN REDUCED AROMATICS AND OLEFINS.
VOLUME OF GASOLINE AS PERCENT OF BASE CASE: 101.9

Table 32–11 Reformulated Gasoline Study—Case [4]

The two steps, reduced reformer severity and lower FCC gasoline end point, were combined (case 4) with the results seen in Table 32–11. Here we see RVP, (R+M)/2 and aromatics in compliance plus an increase in gasoline pool volume of 3.4% over the base case. This was accomplished without any significant capital expenditure on the part of the refiner. However, it did require the purchase of a very significant quantity of MTBE at a price higher than that of gasoline and the removal of relatively inexpensive butane.

The ratio of premium to regular gasoline obtainable from the case 4 blend was not determined. This would, of course, be of interest to the refiner, and perhaps to the reader.

The following tabulation summarizes the results of this gasoline reformulation study:

Table	Case	Rel Vol	RVP	(R+M)/2	Arom	Olef
7	Base	100.0	9.5	88.4	31.4	12.7
8	1	110.3	8.7	90.2	28.4	11.3

Table	Case	Rel Vol	RVP	(R+M)/2	Arom	Olef
9	2	113.	8.7	90.3	27	11
10	3	101.9	8.7	90.3	26.1	9.2
11	4	103.4	8.6	90.4	24.7	8.9
CAA Target max.			8.7		25.	
min.				87.3		

Many refiners are building or will build MTBE manufacturing facilities in the U.S., if permitted. Some will construct methanol plants. Methanol plants are more likely to be built in foreign locations where there is a plentiful supply of natural gas, but perhaps more importantly, where there is much less difficulty in obtaining permission to do so. There seems to be no lack of groups that seemingly automatically and blindly oppose any proposed new facilities here in the U.S. that involve chemical manufacture or petroleum refining (not to mention nuclear power plants). As we have seen already, the capital cost of an MTBE synthesis unit is a small part (about 17%) of the total outlay if isomerization (24%) and dehydrogenation (59%) are required as well.[9]

EVALU8 is a computer program designed by the author to provide a fast and easy way to EVALU8 refinery processes. The program produces properties and yields of products, utilities required, BL costs, overall project investment cost, gross margin, project cash flow, and DCFRR. It will be used in the next two examples.

Economic viability of new refineries. The Osberg Crude examined in chapter 7 was chosen as the crude charge for this example. The process scheme delineated in Figure 32–17, was selected for this proposed refinery.

Yields and material balances were calculated for each of the processes in the appropriate sequence: crude distillation, delayed coker, cat cracker, cycle oil hydrocracker, catalytic reformer, gas plants (saturate and unsaturate), alkylation, isomerization units, sulfur recovery system plus economic calculations. The results are shown in the following tables: Basis for Economic Evaluation (32–6), Unit Material Balances—Part I (32–12), Overall Material Balance (32–13), Battery Limits Cost (32–14), Owner's Cost (32–15), Total Project Cost (32–16), Utility Summary (32–17), Operating Costs (32–18), Gross Margin (32–19), and Income Statement (32–20).

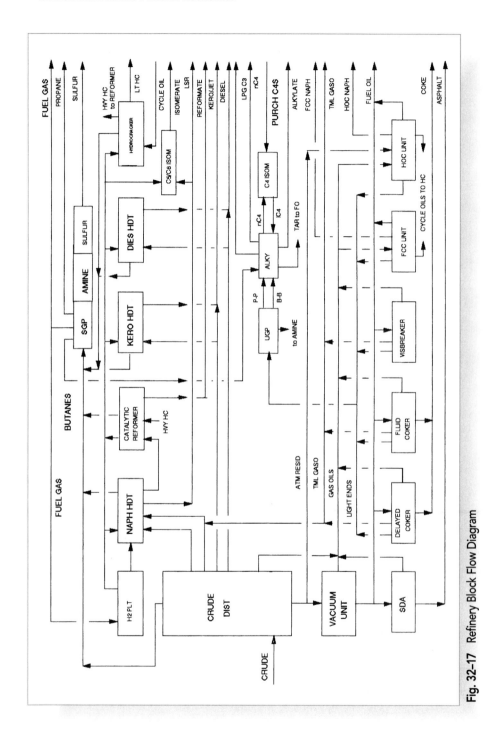

Fig. 32–17 Refinery Block Flow Diagram

	BPCD	LB/HR		BPCD	LB/HR
CRUDE DISTILLATION			**SATURATE GAS PLANT**		
Light straight run	7020	69379	Acid Gas	–	234
Reforming Naphtha	17320	191584	Fuel Gas	–	9739
Kero/Jet	11480	135344	C_3's	–	6342
Diesel	18040	223159	C_4's	–	5969
Atm Gas Oil	10120	131092	Feeds	–	22283
Atm Resid	35960	497770	Products	–	22283
Vac Feed	35960	497770	Loss	–	0
Vac Gas Oil	22070	295590			
Vac Resid	13960	202179	**DELAYED COKER**		
			H_2S	–	1483
Crude Charge	100000	1248328	C_4 & Lighter	936	15755
Products	100010	1248328	Naphtha	3365	36786
Loss	-10	0	Gas Oil	7116	88042
			Coke	–	60113
FLUID CATALYTIC CRACKER			Feed	13960	202179
H_2S	–	1905	Products	11417	202179
Fuel Gas	–	13333	Loss	2543	0
Propane/Propylene	3658	27585			
Butane/Butylene	6585	56159	**UNSATURATE GAS PLANT**		
			Acid Gas	–	3388
FLUID CAT. CRACKER			Fuel Gas	–	27759
FCC Naphtha	22973	254495	C3's	–	28400
Light Cycle Oil	7526	97654	C4's	–	56674
Heavy Cycle Oil	2300	34959	C5's	–	42692
Coke	–	28634	Feeds	–	158913
Cat Cracker Feed	39306	514724	Products	–	158913
			Loss	–	0
Products	43043	514724			
Loss	-3737	0	**CATALYTIC REFORMER**		
			Hydrogen	–	7078
NAPHTHA HYDROTREATER			C_1	–	2363
H_2S	–	12	C_2	–	3916
C_4 & Lighter	21	216	C_3	853	6273
Treated Naphtha	20685	228282	IC_4	278	2267
Naphtha Feed	20685	228370	NC_4	416	3525
Hydrogen	–	139	Reformate	17325	202860
Total Feed	20685	228509	Naphtha Feed	20685	228282
Products	20706	228509	Products	18871	228282
Loss	-21	0	Loss	1814	0

Table 32–12a Unit Material Balances—Delayed Coker Case

407

	BPCD	LB/HR
ALKYLATION UNIT		
LPG Propane	2794	20651
Alky Butane	1605	13658
Alky Pentanes	3979	36306
Propylene Alkylate	4358	45616
Butylene Alkylate	5213	54132
Amylene Alky	2941	30478
Alky Tar	139	1888
SGP Feed	1549	12310
UGP Feed	14936	127766
IC_4 from Isom	1730	14217
Lt Naph	–	–
Purchased IC_4	5907	48435
Total Feeds	24122	202729
Products	21028	202729
Loss	3093	0

	BPCD	LB/HR
BUTANE ISOMERIZATION UNIT		
Fuel Gas	–	72
Propane	9	64
Isobutane	1508	12369
Normal Butane	136	1153
Hydrogen	–	5
Alky Butane	1605	13658
Total Feeds	1605	13664
Products	1653	13659
Loss	-47	5

	BPCD	LB/HR
C_5/C_6 ISOMERIZER		
Fuel Gas	–	142
Propane	25	181
Isobutane	77	631
Normal Butane	143	1216
Isomerate	11039	103669
Hydrogen	–	154
Alky Pentanes	3979	36306
Light Straight Run	7020	69379

	BPCD	LB/HR
C_5/C_6 ISOMERIZER cont'd		
Total Feeds	10999	105839
Products	11284	105839
Loss	-285	0

	BPCD	LB/HR
KERO/JET HYDROTREATER		
H_2S	–	13
C_1	–	477
C_2	–	385
C_3	34	251
IC_4	7	57
NC_4	11	89
Treated Kero/Jet	11612	135003
Kerosene	11480	135344
Hydrogen	–	931
Total Feed	11480	136276
Products	11663	136276
Loss	-183	0

	BPCD	LB/HR
DIESEL HYDROTREATER		
H_2S	–	210
C_1	–	846
C_2	–	681
C_3	60	443
IC_4	12	100
NC_4	19	158
Treated Diesel	18059	22104
Diesel	18040	223159
Hydrogen	–	328
Total Feed	18040	223486
Products	18150	223486
Loss	-110	0

	BPCD	LB/HR
SULFUR RECOVERY		
Sulfur Product	–	3409
H_2 to H_2O	–	213
H_2S Feed	–	3623
Products	–	3623
Loss	–	0

Table 32–12b Unit Material Balances—Delayed Coker Case cont'd

	BPCD	LB/HR
HYDROGEN BALANCE		
H_2 to NHT	–	139
H_2 to KHT	–	931
H_2 to DHT	–	328
H_2 to COHT	–	0
H_2 to HC	–	0
H_2 to C_4 Isom	–	5
H_2 to C_5/C_6 Isom	–	154
Reformer Hydrogen	–	7078
H_2 Consumed	–	1558
Net Product	–	5520

Table 32–12c Unit Material Balances—Delayed Coker Case cont'd

PRODUCTS	BPCD	LV%	LB/HR	WT%
Excess Hydrogen (BFOE)	1271	1.27	5520	0.44
Fuel Gas (BFOE)	2811	2.81	37712	3.02
LPG Propane	2827	2.83	20897	1.67
Excess Butane	–	–	–	–
Gasoline	65655	65.65	712327	57.06
Kero/Jet	11612	11.61	135004	10.81
Diesel	18059	18.06	221047	17.71
No. 1 Fuel Oil	–	–	0	–
No. 2 Fuel Oil	7526	7.53	97654	7.82
No. 6 Fuel Oil	2300	2.30	34959	2.80
Asphalt	139	0.14	1888	0.15
Coke on Catalyst	–	–	28634	2.29
Saleable Coke	–	–	60113	4.82
Sulfur	–	–	3409	0.27
TOTAL	112200	112.20	1359164	108.88
FEEDS				
Crude Oil	100000	100.00	1248328	100.00
Purchased Isobutane	5907	5.91	48435	3.88
Purchased n-Butane	2121	2.12	18050	1.45
Purchased MTBE	4000	4.00	43350	3.47
TOTAL	112028	112.03	1358163	108.80
GAIN	–	172	–	1001
PERCENT GAIN	–	0.15	–	0.07

Table 32–13 Overall Material Balances—Delayed Coker Case

409

UNIT	BPSD	mm$	$/bbl	B/L	Partial
ATM Distillation	105263	39.4	374	1.00	1.00
VAC Distillation	37853	21.7	574	1.53	0.55
Delayed Coker	15511	39.5	2546	6.80	1.00
Fluid Catalyic Cracker	42265	77.8	1840	4.92	1.97
Naphtha Hydrotreater	21774	13.2	606	1.62	0.34
Catalytic Reformer	21774	37.1	1705	4.56	0.94
Kero/Jet Hydrotreater	12084	14.5	1199	3.20	0.37
Diesel Hydrotreater	18989	19.0	1001	2.67	0.48
Saturate Gas Plant	6372	4.4	687	1.84	0.11
Unsaturate Gas Plant	32653	17.3	530	1.42	0.44
Alkylation Unit	13170	29.5	2239	5.98	0.75
C_4 Isomerizer	1690	6.7	3949	10.55	0.17
C_5/C_6 Isomerizer	11578	7.6	660	1.76	0.19
Complexity Index					8.32
TOTAL BATTERY LIMITS COST		338.7			

Table 32–14   Battery Limits Costs—Delayed Coker Case

	% B/L	mm$
Initial Catalysts and Chemicals	5.3	18.0
Paid-up Royalties	3.7	12.5
Prestartup and Startup Expense	10.5	35.6
Working Capital	18.1	61.3
Subtotal		127.4
LAND		50.0
TOTAL OWNER'S COST		177.4

Table 32–15   Owner's Costs—Delayed Coker Case

CHAPTER 32 • ECONOMICS

	mm$
Plant and Equipment	698.6
Interest during construction	0.0
Owner's Cost (excluding land)	127.4
Total Depreciable Cost	825.9
LAND	50.0
TOTAL PROJECT COST	875.9

Table 32–16 Total Project Costs—Delayed Coker Case

PROCESS UNIT	BPCD	ELECTRIC (KWH)	FUEL (mmBtu)	STEAM (k#)	COOLING WATER (kgal)	CATALYST ($)
Atm Distillation	100000	50000	10000	2500	–	–
Vac Distillation	35960	10788	3596	1798	–	–
Solvent Deasphalter	0	0	0	0	–	–
Visbreaker	0	0	0	0	–	–
Fluid Coker	0	0	0	0	0	–
Delayed Coker	13960	50256	1675	-558	14	–
Fluid Cat Cracker	39306	39306	3144	-786	15722	11792
Heavy Oil Cracker	0	0	0	0	–	0
Hydrocracker	0	0	0	–	–	–
Naphtha Hydrotreater	20685	41370	621	310	–	–
Catalytic Reformer	20685	20685	6206	-827	2069	–
Kero/Jet Hydrotreater	11480	19516	103	80	–	–
Diesel Hydrotreater	18040	30668	144	126	–	–
Gas Oil Hydrotreater	0	0	0	0	–	–
Cycle Oil Hydrotreater	0	0	0	0	–	–
Saturate Gas Plant	–	–	–	–	–	–
Unsaturate Gas Plant	–	–	–	–	–	–
Alkylation Unit	12512	38787	39	450	28777	–
Butane Isomerizer	1605	1605	16	58	–	–
C5/C6 Isomerizer	7020	10999	132	396	–	–
Cat Poly Unit	0	0	0	0	0	–
Sulfur Recovery	–	–	–	–	–	–
Hydrogen Plant	–	–	–	–	–	–
Total Process Units	313981	25676	3547	46582	11792	–
Utility Cost ($/Unit)	0.06	3.00	4.00	0.10	1	–
Utility Cost ($/day)	18839	77029	14189	4658	11792	–
Annual Cost (mm$)	6.88	28.12	5.18	1.70	4	–
TOTAL ANNUAL COST (mm$)	46.17					

Table 32–17 Utility Summary—Delayed Coker Case

411

	% B/L	mm$/yr
VARIABLE COSTS		
Utilities	–	46.2
Catalysts & Chemicals	4.7	15.9
Total Variable Costs		62.1
FIXED COSTS		
Payroll	3.9	13.2
Operating Supplies	0.3	1.0
Maintenance	5.2	17.6
Insurance	0.4	1.4
Property Taxes	1	3.4
Total Fixed Costs	10.8	36.6
TOTAL OPERATING COSTS		98.7

Table 32–18   Operating Costs—Delayed Coker Case

PRODUCT SALES	*Units/Day*	*$/Unit*	*$/Day*
Fuel Gas (BFOE)	4082	15.84	64,666
LPG Propane	2827	14.40	40,710
LPG Butane	0	16.20	0
Gasoline	65655	25.20	1654498
Kero/Jet	11612	25.02	290,521
Diesel	18059	23.40	422,586
No. 1 Fuel Oil	0	23.40	0
No. 2 Fuel Oil	7526	18.00	135,477
Residual Fuel Oil	2300	15.84	36,432
Asphalt	139	15.84	2,195
Coke (short tons)	721	100.00	72,135
Sulfur (long tons)	36.5	70.00	2,557
Total Revenues			2,721,777
PURCHASED RAW MATERIALS			
Purchased Isobutane	5907	20.00	118,135
Purchased Normal Butane	2121	18.00	38,187
MTBE	4000	38.00	152,000
Crude	100000	18.00	1800000
Total Raw Materials	2108321		
GROSS MARGIN			613,456

Table 32–19   Gross Margin—Delayed Coker Case

TOTAL PROJECT COST 875.90
DEPRECIABLE COST 825.90

	$/DAY	mm$/yr
TOTAL REVENUES	2,721,777	993.4
Total Raw Materials	2,108,321	769.5
Operating Costs		98.7
TOTAL PRODUCTION COST		868.2
Operating Income		125.2
Less Depreciation (10% straight line)		82.6
Taxable Income		42.6
Less Income Tax (at 40%)		17.1
Net Income		25.6
Add Back Depreciation		82.6
Cash Flow		108.2
Project Cost/Cash Flow		8.10

(This is equal to payout period in years.
Also equals *Present Value Factor.*)

The Discounted Flow Rate of Return (DCFRR) is: 10.9%

The production of "Pro Forma" financial statements is probably not justified at this point in the study.

Table 32–20 Income Statement—Delayed Coker Case

The DCF calculation is based on periodic (annual) cash flow and periodic (annual) interest compounding.

Some workers use continuous cash flow and continuous interest compounding. Others use combinations of the two. Such sophistication is hardly justified in preliminary screening studies such as this.

The present value factor (PVF) for uniform annual cash flow is the same as the present value of an annuity that is readily calculated from the following equation:

$$PV = \frac{(1+i)^n - 1}{i(1+i)^n}$$

413

where

i = interest rate

n = number of years

The relationship between interest rate and present value factor for a 20 year project life is shown in Figure 32–18.

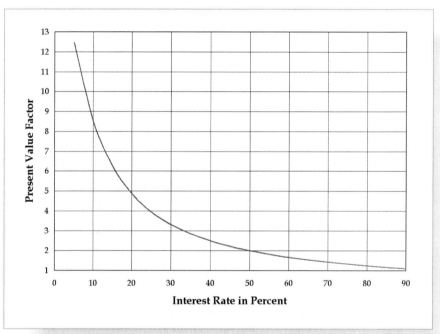

Fig. 32–18   Present Value Factor vs. Interest Rate (Based on a 20-year Project Life)

Fluid coking. While the delayed coker case was being developed, a fluid coker case was developed in parallel. Only the following tables are included for this case: Fluid Coker (32–21), Fluid Catalytic Cracker (32–22), Gasoline Pool Properties (32–23), Battery Limits Costs (32–24), Total Project Cost (32–25), Gross Margin (32–26), and Income Statement (32–26).

	VAC RESID	H_2S	C_3 & Lighter	C_4's	RN (C5-400)	GAS OIL (400+)	COKE	TOTALS
SCFB	–	–	–	–	–	–	–	–
BPCD Avail.	13960	–	–	–	–	–	–	–
Feed Fract.	1	–	–	–	–	–	–	–
BPCD	13960	–	774	397	2913	8229	–	12312
API	10.6	–	–	–	56.2	21.5	–	–
#/BBL	348.2	–	–	–	263.6	323.6	–	–
#/HR	202554	197	17244	3461	31998	110949	41963	205811
Adj. #/HR	–	194	16971	3406	31491	109193	41299	202554
Wt%	100.0	0.10	8.51	1.71	15.55	53.91	20.72	100.5
Adj. Wt %	–	0.10	8.38	1.68	15.55	53.91	20.39	100.0
LV%	100	–	5.55	2.84	20.86	58.94	–	88.2
Wt% S	0.9	–	–	–	0.18	0.98	1.47	–
Wt% CCR	17.6	–	–	–	–	2.25	–	–
Wt Adj. Factor	1.0161	–	–	–	–	–	–	–
P	–	–	–	–	15	–	–	–
O	–	–	–	–	58	–	–	–
N	–	–	–	–	3	–	–	–
A	–	–	–	–	24	–	–	–
RON	–	–	–	–	73.9	–	–	–
MON	–	–	–	–	65.2	–	–	–
RVP	–	–	–	–	–	–	–	–

LIGHT ENDS COMPOSITION	WT % STRM	WT % FD	#/HR	#/BBL	BPD	LV % FD	LV % STRM
H_2S	–	–	194	276.2	17	0.12	
Hydrogren	1.6	0.14	276	–	–	–	–
Methane	29.5	2.51	5087	–	–	–	–
Ethylene	13.3	1.13	2293	–	–	–	–
Ethane	21.9	1.86	3776	–	–	–	–
Propylene	18.9	1.61	3259	182.3	429	3.07	–
Propane	14.8	1.26	2552	177.4	345	2.47	–
Total C_3 & Lighter	100.0	8.51	17244	–	774	5.55	–
Butylenes	–	1.21	2448	212.0	277	1.99	69.9
Isobutane	–	0.08	166	196.8	20	0.14	5.1
Normal Butane	–	0.42	847	204.2	100	0.71	25.1
Total C_4's		1.71	3461		397	2.84	
TOTAL (excluding H_2S)		10.22	20704		1171	8.39	

OPERATING REQUIREMENTS	kWh/d	MBtu/d	k#/d	kgpd CW
	181480	-2792	419	–

BATTERY LIMITS COST	NF INDEX	BPSD	mm$	
	1241.7	15511	38.5	

Table 32–21   Fluid Coker Unit—Fluid Coker Case

	AGO	VGO	TML GO	DAO	Feed	H_2S	FG	C_3s
BPCD Avail.	10120	22070	8229	0				
Feed Fract.	1	1	1	10^{-7}				
BPCD	10120	22070	8229	1.02×10^{-8}	40419			3762
API	27.5	22.2	23.9	16.81	21.9			
#/BBL	311.3	322.0	318.5		318.2			
#/HR	131092	295590	109193	0	535875	1983	13881	28366
Wt%					100	0.37	2.59	5.29
Lv%					100			9.31
Wt% S	0.94	0.48	0.98	0.75	0.94			
Wt% CCR	0.02	0.12	2.25	7.66	0.53			

Light Ends Composition	BBL	#/B	#/HR	Wt%	Lv%
$C_3=$	2744	182.3	20842	3.89	6.79
C_3	1018	177.4	7524	1.40	2.52
Total C_3	3762		28366	5.29	9.31
$C_4=$	3260	212.0	28799	5.37	8.07
IC_4	2990	196.8	24516	4.58	7.40
NC_4	521	204.2	4433	0.83	1.29
Total C_4	6771		57749	10.78	16.75

OPERATING REQUIREMENTS	kWh/d	MBtu/d	k#/d	kgpd CW	#/D cat
	40419	3233	-808	16167	12126

BATTERY LIMITS COST	NF INDEX	BPSD	mm$
	1241.7	43461	79.1

Table 32–22 FCC Unit—Fluid Coker Case

C_4S	Light Naphtha	Heavy Naphtha	TOTAL NAPHTHA	CYCLE OIL TOTAL	Light	Slurry	Coke	TOTALS
6771	4725	18899	23624	10105	7740	2365		44261
	90.458	47.457	54.7	14.9	18.5	4.2		
204.7	223.0	276.6	265.9	338.2	330.1	364.8		
57749	43901	217798	261699	142386	106438	35948	29811	535875
10.78			48.8	26.6	19.9	6.7	5.56	100.0
16.75			58.45	25.0	19.1	5.9		109.5
			0.077		0.986	1.669	1.266	

P	–	–	–	
O	40	46.23	45.0	
N	–	–	–	
A	0	17.133	13.7	
RON	96.0	93.2	93.4	
MON	82.0	80.5	80.6	
(R+M)/2	89.0	86.9	87.0	Cetane No. 21.39

	BPCD	#/BBL	#/HR
O_5	1889.9	229	18032.792
P_5	2834.849	219	25868.001
Light Naphtha	4724.749	223	43900.794
Heavy Naptha	18898.996	276.584	217798.459

	BPCD	RVP	AROM	OLEF	% BZ	AV OCT	10%
n-Butane	2800	58.5	0.0	2.7	—	91.9	29
C_5/C_6 Isomerate	11162	14.4	0.4	0.0	—	81.3	100
Light Hydrocrackate	0.004	14.3	2.7	0.2	—	77.5	99
Light Straight Run	0	13.3	2.8	0.8	—	70.6	104
Methanol	0	60.0	—	—	—	102.0	—
Ethanol	0	18.0	—	—	—	100.0	—
Isopropanol	0	16.0	—	—	—	108.0	—
TBA	0	9.5	—	—	—	101.0	—
MTBE	4000	9.2	0.1	0.4	—	109.0	129
ETBE	0	4.0	—	—	—	110.0	—
DIPE	0	4.0	—	—	—	104.0	154
TAME	0	1.0	—	—	—	104.5	—
Reformate	17091	3.7	64.8	0.7	4.2	89.7	169
FCC Naphtha	0	7.1	13.7	45.0	0.8	87.0	128
LIGHT FCC	0	15.6	0.0	40.0	—	0.0	—
HEAVY FCC	18899	4.0	17.1	46.2	—	86.9	—
HOC Naphtha	0.03	7.0	14.7	42.8	—	86.7	128
C_3= Alkylate	4770	7.9	0.4	0.5	—	91.0	128
C_4= Alkylate	5471	7.9	0.4	0.5	—	94.5	158
C_5= Alkylate	3024	7.9	—	—	—	91.0	—
Cat Poly Gasoline	0	2.0	0.5	95.0	—	89.0	149
Thermal Gasoline	0	13.0	10.0	37.0	—	69.6	110
Blend	67217	9.0	21.4	13.4	1.1	89.3	90.4

Table 32–23   Gasoline Pool Properties—Fluid Coker Case

50%	90%	LV%	RON	MON	WT% O_2	S (ppmw)	#/BBL	#/HR	API
34	40	4.17	94.2	89.6	–	15.6	204.2	23823	110.9
115	146	16.61	82.3	80.2	–	2.3	225.5	104816	88.0
122	172	0.00	78.5	76.5	–	0.0	237.1	0	77.3
125	164	0.00	71.1	70.0	–	216.6	237.5	0	76.9
148	–	0.00	112.0	92.0	49.9	–	278.5	0	46.2
174	–	0.00	110.0	90.0	34.7	–	277.8	0	46.7
180	–	0.00	0.0	0.0	26.0	–	275.9	0	47.8
181	–	0.00	109.0	93.0	21.6	–	276.8	0	47.3
131	138	5.95	118.0	100.0	18.2	–	260.1	43350	58.8
160	–	0.00	118.0	102.0	15.7	–	260.5	0	58.5
155	156	0.00	0.0	0.0	15.7	–	253.6	0	63.7
187	–	0.00	111.0	98.0	15.7	–	269.3	0	52.3
256	334	25.43	95.0	84.4	–	46.7	282.3	199395	43.8
220	366	0.00	93.4	80.6	–	788.8	265.9	–	54.7
–	–	–	0.0	0.0	–	–	–	–	–
–	–	28.12	93.2	80.5	–	–	276.6	217798	47.5
220	366	0.00	92.9	80.5	–	–	–	0	67.8
216	289	7.10	92.0	90.0	–	25.5	243.7	49934	71.6
216	289	8.14	96.0	93.0	–	25.5	243.7	56806	71.6
–	–	4.50	92.0	90.0	–	25.5	248.8	31341	67.5
236	346	–	95.0	83.0	–	125.0	252.0	0	64.8
139	184	0.00	73.9	65.2	–	2194.0	259.5	0	59.3
126.3	163.1	100.0	93.5	85.1	1.08	18.49	259.7	727263	59.1

419

UNIT	BPSD	mm$	$/bbl	COMPLEXITY Battery Limits	Partial
Atm. Distallation	105263	39.4	374	1.00	1.00
Vacuum Distallation	37853	21.7	574	1.53	0.55
SDA Unit	0	0.0	–	–	–
Visbreaker	0	0.0	–	–	–
Fluid Coker	15511	38.5	2482	6.63	0.98
Delayed Coker	0	0.0	–	–	–
Fluid Catalytic Cracker	43461	79.1	1819	4.86	2.01
Heavy Oil Cracker	0	0.0	–	–	–
Hydrocracker	0	0.0	–	–	–
Naphtha Hydrotreater	21298	13.0	612	1.63	0.33
Catalytic Refromer	21298	36.6	1720	4.60	0.93
Kerosene/Jet Hydrotreater	12084	14.5	1199	3.20	0.37
Diesel Hydrotreater	18989	19.0	1001	2.67	0.48
Gas Oil Hydrotreater	0	0.0	–	–	–
Cycle Oil Hydrotreater	0	0.0	–	–	–
Saturate Gas Plant	5919	4.2	702	–	–
Unsaturate Gas Plant	35238	18.3	518	1.38	0.46
Alkylation Unit	13963	30.5	2187	5.85	0.78
C_4 Isomerizer	1621	6.5	4015	10.73	0.17
C_5/C_6 Isomerizer	11713	7.7	657	1.76	0.20
Catalytic Polymerization Unit	0	0.0	–	–	–
Sulfur Recovery	–	10.7	–	–	–
Complexity Index					8.25
TOTAL BATTERY LIMITS COST		339.7			

Table 32–24   Battery Limits Cost—Fluid Coker Case

	mm$
Plant and Equipment	702.4
Interest during construction	0.0
Owner's Cost (excluding land)	127.7
Total Depreciable Cost	830.2
Land	50.0
TOTAL PROJECT COST	880.2

Table 32–25   Total Project Cost—Fluid Coker Case

PRODUCT SALES	Units/Day	$/Unit	$/Day
Fuel Gas (BFOE)	4335	15.84	68,660
LPG Propane	2881	14.40	41,481
LPG Butane	0	16.20	0
Gasoline	67217	25.20	1693865
Kerosene/Jet	11612	25.02	290,521
Diesel	18059	23.40	422,586
No. 1 Fuel Oil	0	23.40	0
No. 2 Fuel Oil	7740	18.00	139,312
Residual Fuel Oil		2365	15.84
37,464			
Asphalt	147	15.84	2,330
Coke (short tons)	0	100.00	0
Sulfur (long tons)	24.3	70.00	1,702
Total Revenues			2697921
PURCHASED RAW MATERIALS			
Purchased Isobutane	6590	20.00	131796
Purchased Normal Butane	2525	18.00	45455
MTBE	4000	38.00	152000
Crude	100000	18.00	1800000
Total Raw Materials			2129251
GROSS MARGIN			568670

Table 32–26   Gross Margin—Fluid Coker Case

421

| TOTAL PROJECT COST | | 880.20 |
| DEPRECIABLE COST | | 830.20 |

	$/DAY	mm$/yr
TOTAL REVENUES	2,697,921	984.7
Total Raw Materials	2,129,251	777.2
Operating Costs		96.8
TOTAL PRODUCTION COST		874.0
Operating Income		110.7
Less Depreciation (10% straight line)		63.7
Taxable Income		47.0
Less Income Tax (at 40%)		18.8
Net Income		28.2
Add Back Depreciation		63.7
Cash Flow		91.9
Project Cost/Cash Flow		7.48

(This is equal to payout period in years.
Also equals *Present Value Factor.*)

The Discounted Flow Rate of Return (DCFRR) is: 11.76%

The production of "Pro Forma" financial statements is probably not justified at this point in the study.

Table 32–27   Income Statement—Fluid Coker Case

The results of these two cases may be summarized as follows:

	Delayed Coker	Fluid Coker
Battery limits costs ($mm)	338.7	339.7
Total project cost ($mm)	875.9	880.2
Gross margin ($m/day)	613.5	568.7
Present value factor (yr)	8.10	7.48
DCFRR (%)	10.9	11.76

The calculated DCFRR may be disappointing, but should not be a surprise. Rather than building new refineries, the industry is shutting down marginal existing refineries in the U.S. As a result, the utilization rate (charge rate as percent of capacity) of U.S. refineries keeps ratcheting up into the mid 90% range. Since this is into the maximum range for sound planning, unless new refineries are built in the U.S., importation of products will increase as product demand increase.

Though neither of the example cases is attractive as an investment, the delayed coker does show a slight edge over the fluid coker. It has a lower total project cost and a higher cash flow.

The phrase *feasibility study* is somewhat ambiguous in meaning. It may refer to a simple comparison, based on non–confidential information, of salient features of two or more technologies for performing a given task. Or, it may envision a thorough examination of licensor information, obtained under non–disclosure agreements, of these same processes, culminating in a selection of a recommended technology. An even more extensive study might include input regarding market potential and possible sites plus visits to plants practicing the various technologies. These possibilities are illustrated in Figure 32–19.[10] An example outline of a report for such a study can be found in Table 32–28.

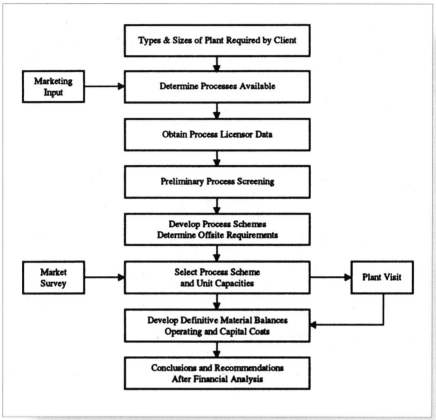

Fig. 32–19 Typical Process Selection Procedure

1.0 INTRODUCTION

2.0 SUMMARY

3.0 EVALUATION METHODOLOGY/CRITERIA

4.0 DISCUSSION OF EACH TECHNOLOGY

    4.1 PROCESS DESCRIPTION
    4.2 PROCESS FLOW DIAGRAMS
    4.3 TYPICAL PLOT PLAN
    4.4 OVERALL AND DETAILED MATERIAL BALANCES
    4.5 CONSUMABLES

Table 32–28a Example Outline of Technology Evaluation Report

424

4.6   OTHER OPERATING COST ELEMENTS
     4.6.1   OPERATING LABOR
     4.6.2   MAINTENANCE COST
     4.6.3   WORKING CAPITAL
4.7   COMMERCIAL EXPERIENCE
     4.7.1   LIST OF LICENSED UNITS **
     4.7.2   PLANT VISITS
4.8   ENVIRONMENTAL CONSIDERATIONS
4.9   COST DATA FROM LICENSOR
     4.9.1   BUDGET ESTIMATE OF CAPITAL COST
     4.9.2   COST OF INITIAL CATALYST FILL
     4.9.3   ANNUAL CATALYST COST
     4.9.4   ROYALTY BASES
     4.9.5   BASIC ENGINEERING DESIGN
4.10 LICENSOR'S GUARANTEES
4.11 TRAINING AVAILABLE

5.0   COMPARISON OF TECHNOLOGIES

5.1   PRODUCT YIELDS
5.2   PRODUCT QUALITY
5.3   OPERATING COSTS
5.4   CAPITAL COSTS
5.5   COMMERCIAL EXPERIENCE
5.6   LICENSOR SERVICES
5.7   INTANGIBLES
5.8   RATE OF RETURN CALCULATIONS
5.9   RISK EVALUATION

6.0   CONCLUSIONS AND RECOMMENDATIONS

7.0   ATTACHMENTS

7.1   REQUESTS FOR LICENSORS PROPOSALS
7.2   RESPONSES FROM LICENSORS

---

*   Cost, life, sources, replacement, precautions, dispoal, etc.
** Size, location, startup date, operating history, etc.

Table 32–28b Example Outline of Technology Evaluation Report

Notes

1. Johnson, D., *Oil & Gas Journal,* March 18, 1996, ff. 74

2. Rasmussen, J.A., "The Impact of Economic Compliance Costs on U.S. Refining Profitability," Energy Information Administration, October, 1997

3. EIA/Petroleum 1996: Issues and Trends, "U.S. Refining Cash Margin Trends: Factors Effecting the Margin—Component of Price," September 15, 1997

4. Dudley, J.A., Killen, P.J., and Ory, R.E., *Oil & Gas Journal,* November 19, 1984, pp. 110–115

5. EIA Crude Oil Price Summary, Table 9.1

6. American Automobile Manufacturers Association, "Motor Vehicle Facts and Figures," Annual Issues

7. EIA, Tables 9.6 and 12

8. Anon., "NPRA Survey of U.S. Refining Industry Capacity to Produce Reformulated Gasolines—Part A," January 1991

9. Bakas, S.T., Gregor, J.H., and Cottrell, P.R., "Integration of Technologies for the Conversion of Butanes into MTBE," 1991 NPRA Annual Meeting, San Antonio

10. Maples, R.E., and Hyland, M.J., "What is Involved in Major Venture Financing," *Chemical Engineering Progress,* January, 1980, pp. 24–28

APPENDIX

This Appendix contains tables fom the 1989 book of ASTM Standards for each of the following designations:

Table	Method	Detailed Requirements for:
A–1	D 1835	Liquefied Petroleum Gases
A–2	D 4814	Gasoline
A–3	D 3699–96a	Kerosine
A–4	D 1655	Aviation Turbine Fuels
A–5	D 2880	Gas Turbine Fuel Oils
A–6	D 975	Diesel Fuel Oils
A–7	D 2069	Marine Distillate Fuels
A–8	D 396	Fuel Oils

Note: These tables are reprinted with permission
of ASTM, holder of the copyright.

	Product Designation				
	Commercial Propane	Commercial Butane	Commercial PB Mixtures	Special-Duty Propane[A]	ASTM Test Methods (see Section 2)
Vapor pressure at 100°F (37.8°C), max, psig	208	70	[B]	208	D 1267 or
kPa	1430	485		1430	D 2598[C]
Volatile residue:					
evaporated temperature, 95 %, max, °F	−37	36	36	−37	D 1837
°C	−38.3	2.2	2.2	−38.3	
or					
butane and heavier, max, vol %	2.5	...	...	2.5	D 2163
pentane and heavier, max, vol %	...	2.0	2.0	...	D 2163
Propylene content, max, vol %	...	...	...	5.0	D 2163
Residual matter:					
residue on evaporation 100 mL, max, mL	0.05	0.05	0.05	0.05	D 2158
oil stain observation	pass[D] [E]	pass[D] [E]	pass[D] [E]	pass[D]	D 2158
Relative density at 60/60°F (15.6/15.6°C)					D 1657 or D 2598
Corrosion, copper, strip, max	No. 1	No. 1	No. 1	No. 1	D 1838[G]
Sulfur, ppmw	185	140	140	123	D 2784
Hydrogen sulfide	pass	pass	pass	pass	D 2420
Moisture content	pass	pass	pass	pass	D 2713
Free water content	...	none[F]	none[F]	...	

[A] Equivalent to Propane HD-5 of GPA Standard 2140.

[B] The permissible vapor pressures of products classified as PB mixtures must not exceed 208 psig (1430 kPa) and additionally must not exceed that calculated from the following relationship between the observed vapor pressure and the observed relative density:

Vapor pressure, max = 1167 − 1880 (relative density 60/60°F) or 1167 − 1880 (density at 15°C)

A specific mixture shall be designated by the vapor pressure at 100°F in pounds per square inch gage. To comply with the designation, the vapor pressure of the mixture shall be within +0 to −10 psi of the vapor pressure specified.

[C] In case of dispute about the vapor pressure of a product, the value actually determined by Test Method D 1267 shall prevail over the value calculated by Practice D 2598.

[D] An acceptable product shall not yield a persistent oil ring when 0.3 mL of solvent residue mixture is added to a filter paper, in 0.1-mL increments and examined in daylight after 2 min as described in Test Method D 2158.

[E] Although not a specific requirement, the relative density must be determined for other purposes and should be reported. Additionally, the relative density of PB mixture is needed to establish the permissible maximum vapor pressure (see Footnote B).

[F] The presence or absence of water shall be determined by visual inspection of the samples on which the relative density is determined.

[G] This method may not accurately determine the presence of reactive materials (for example, H₂S, S°) in liquefied petroleum gas if the product contains corrosion inhibitors or other chemicals which diminish the reaction with the copper strip.

Table A-1 ASTM D 1835 Liquefied Petroleum Gases (© ASTM; reprinted with permission)

Vapor Pressure/ Distillation Class	Vapor Pressure,A max, kPa(psi)	Distillation Temperatures, °C(°F), at % Evaporated, maxB					Distillation Residue, vol %, max
		10 vol %, max	50 vol % min	50 vol % max	90 vol %, max	End Point, max	
AA	54(7.8)	70(158.)	77(170.)	121(250.)	190(374.)	225(437.)	2.
A	62(9.0)	70(158.)	77(170.)	121(250.)	190(374.)	225(437.)	2.
B	69(10.0)	65(149.)	77(170.)	118(245.)	190(374.)	225(437.)	2.
C	79(11.5)	60(140.)	77(170.)	116(240.)	185(365.)	225(437.)	2.
D	93(13.5)	55(131.)	66(150.)	113(235.)	185(365.)	225(437.)	2.
E	103(15.0)	50(122.)	66(150.)	110(230.)	185(365.)	225(437.)	2.

A Consult EPA for approved test methods for compliance with EPA vapor pressure regulations.
B At 101.3 kPa pressure (760 mm Hg).

Table A-2 ASTM D 4814 Gasoline, part 1: Vapor Pressure and Distillation Class Requirements (© ASTM; reprinted with permission)

Vapor Lock Protection Class	Vapor/Liquid Ratio (V/L)A,B	
	Test Temperature, °C(°F)	V/L, max
1	60(140.)	20
2	56(133.)	20
3	51(124.)	20
4	47(116.)	20
5	41(105.)	20
6	35(95.)	20

A At 101.3 kPa pressure (760 mm Hg).
B The mercury confining fluid procedure of Test Method D 2533 shall be used for gasoline-oxygenate blends. Either glycerin or mercury confining fluid may be used for gasoline. Test Method D 5188 may be used for all fuels. The procedure for estimating temperature-V/L (see Appendix X2) may only be used for gasoline.

Table A-2 ASTM D 4814 Gasoline, part 2: Vapor Lock Protection Class Requirements
(© ASTM; reprinted with permission)

Lead Content, max, g/L (g/U.S. gal)[B]		Copper Strip Corrosion, max	Solvent-washed Gum Content, mg/100 mL, max	Sulfur, max, mass %		Oxidation Stability, Minimum, minutes	Water Tolerance
Unleaded	Leaded			Unleaded	Leaded		
0.013(0.05)	1.1(4.2)	No. 1	5	0.10	0.15	240.	c

[A] See Appendix X1 for information on Antiknock Index.
[B] See Appendix X3 for U.S. EPA maximum limits for lead and phosphorus contents in unleaded gasoline (X3.2.1) and maximum average lead limits for leaded gasoline (X3.2.2).
[C] Water tolerance limits in terms of maximum temperature for phase separation are given in Table 5 (consult Annex A1).

Table A-2 ASTM D 4814 Gasoline, part 3: Detailed Requirements for all Volatility Classes (© ASTM; reprinted with permission)

NOTE 1—This schedule, subject to agreement between purchaser and seller, denotes the volatility properties of the fuel at the time and place of bulk delivery to the fuel dispensing facilities for the end user. For Sept. 16 through April 30 (the time period not covered by EPA Phase II vapor pressure requirements), volatility properties for the previous month or the current month are acceptable for the end user from the 1st through the 15th day of the month. From the 16th day through the end of the month, volatility properties of the fuel delivered to the end user shall meet the requirements of the specified class(es). To ensure compliance with EPA Phase II vapor pressure requirements, vapor pressure for finished gasoline tankage at refineries, importers, pipelines, and terminals during May and for the entire distribution system, including retail stations, from June 1 to Sept. 15 shall meet only the current month's class. Shipments should anticipate this schedule.

NOTE 2—Where alternative classes are listed, either class or intermediate classes are acceptable; the option shall be exercised by the seller.

NOTE 3—See Appendix X2 of Research Report: D02-1347 (available from ASTM Headquarters) for detailed description of areas. Contact EPA for the latest information on areas requiring reformulated fuel.

State	Jan.	Feb.	Mar.	Apr.	May[B]	June	July	Aug.	Sept. 1-15	Sept. 16-30	Oct.	Nov.	Dec.
Alabama	D-4	D-4	D-4/C-3	C-3/A-3	A-3 (C-3)	A-3[D]	A-3[D]	A-2[E]	A-2[E]	A-2/C-3	C-3	C-3/D-4	D-4
Alaska	E-6	E-6	E-6	E-6	E-6/D-4	D-4	D-4	D-4	D-4	D-4/E-6	E-6	E-6	E-6
Arizona[K]													
N 34° Latitude and E 111° Longitude	D-4	D-4	D-4/C-3	C-3/A-2	A-2 (B-2)	A-1	A-1	A-1	A-2	A-2/B-2	B-2/C-3	C-3/D-4	D-4
Remainder of State	D-4	D-4/C-3	C-3/B-2	B-2/A-2	A-2 (B-2)	A-1[F]	A-1[F]	A-1[F]	A-1[E]	A-1	A-1/B-2	B-2/C-3	C-3/D-4
Arkansas	D-4	D-4	D-4/C-3	C-3/A-3	A-3 (C-3)	A-3	A-2	A-2	A-2	A-2/C-3	C-3/D-4	D-4	D-4/E-5
California[K]													
North Coast	E-5/D-4	D-4	D-4	D-4/A-3	A-3 (C-3)	A-3[D]	A-2[E]	A-2[E]	A-2[E]	A-2/B-2	B-2/C-3	C-3/D-4	D-4/E-5
South Coast	D-4	D-4	D-4/C-3	C-3/A-3	A-3 (C-3)	A-2[E,I]	A-2[E,I]	A-2[E,I]	A-2[E,I]	A-2/B-2	B-2/C-3	C-3/D-4	D-4
Southeast	D-4	D-4/C-3	C-3/B-2	B-2/A-2	A-2 (B-2)	A-1[F]	A-1[F]	A-1[F]	A-1[F]	A-1	A-1/B-2	B-2/C-3	C-3/D-4
Interior	E-5/D-4	D-4	D-4	C-3/A-3	A-3 (C-3)	A-2[E,I]	A-2[E,I]	A-2[E,I]	A-2[E,I]	A-2/B-2	B-2/C-3	C-3/D-4	D-4/E-5
Colorado	E-5	E-5/D-4	D-4/C-3	C-3/A-3	A-3 (C-3)	A-2[E]	A-2[E]	A-2[E]	A-2[E]	A-2/B-2	B-2/C-3	C-3/D-4	D-4/E-5
Connecticut	E-5	E-5	E-5/D-4	D-4/A-4	A-4 (D-4)	A-3[G]	A-3[G]	A-3[G]	A-3[G]	A-3/D-4	D-4	D-4/E-5	E-5
Delaware	E-5	E-5	E-5/D-4	D-4/A-4	A-4 (D-4)	A-3[G]	A-3[H]	A-3[H]	A-3[G]	A-3/C-3	C-3/D-4	D-4/E-5	E-5
District of Columbia	E-5	E-5/D-4	D-4	D-4/A-3	A-3 (C-3)	A-3[G]	A-3[H]	A-3[H]	A-3[H]	A-3/C-3	C-3/D-4	D-4/E-5	E-5
Florida	D-4	D-4	D-4/C-3	C-3/A-3	A-3 (C-3)	A-3[D]	A-3[D]	A-3[D]	A-3[D]	A-3/C-3	C-3	C-3/D-4	D-4
Georgia[K]	D-4	D-4	D-4/C-3	C-3/A-3	A-3 (C-3)	A-3[D]	A-3[D]	A-2[E]	A-2[E]	A-2/C-3	C-3	C-3/D-4	D-4
Hawaii	C-3	C-3	C-3	C-3	C-3	C-3	C-3	C-3	C-3	C-3	C-3	C-3	C-3
Idaho:[K]													
N 46° Latitude	E-5	E-5	E-5/D-4	D-4/A-4	A-4 (D-4)	A-3	A-2	A-2	A-2	A-2/C-3	C-3/D-4	D-4/E-5	E-5
S 46° Latitude	E-5	E-5/D-4	D-4	D-4/A-3	A-3 (C-3)	A-2	A-2	A-2	A-2	A-2/B-2	B-2/C-3	C-3/D-4	D-4/E-5
Illinois:[K]													
N 40° Latitude	E-5	E-5	E-5/D-4	D-4/A-4	A-4 (D-4)	A-3[G]	A-3[G]	A-3[G]	A-3[G]	A-3/C-3	C-3/D-4	D-4/E-5	E-5
S 40° Latitude	E-5	E-5	E-5/D-4	D-4/A-3	A-3 (C-3)	A-3	A-3	A-3	A-3	A-3/C-3	C-3/D-4	D-4	D-4/E-5
Indiana[K]	E-5	E-5	E-5/D-4	D-4/A-4	A-4 (D-4)	A-3[G]	A-3[G]	A-3[G]	A-3[G]	A-3/C-3	C-3/D-4	D-4/E-5	E-5
Iowa	E-5	E-5	E-5/D-4	D-4/A-3	A-3 (C-3)	A-3	A-3	A-3	A-3	A-3/C-3	C-3/D-4	D-4/E-5	E-5

Table A-2 ASTM D 1835, part 4: Schedule of Seasonal and Geographical Volatility Classes (© ASTM; reprinted with permission)

State	Jan.	Feb.	Mar.	Apr.	May[B]	June	July	Aug.	Sept. 1–15	Sept. 16–30	Oct.	Nov.	Dec.
Kansas	E-5	E-5/D-4	D-4/C-3	C-3/A-3	A-3 (C-3)	A-2^E	A-2^E	A-2^E	A-2^E	A-2/B-2	B-2/C-3	C-3/D-4	D-4/E-5
Kentucky	E-5	E-5/D-4	D-4	D-4/A-3	A-3 (C-3)	A-3^G	A-3^G	A-3^G	A-3^G	A-3/C-3	C-3/D-4	D-4/E-5	E-5
Louisiana	D-4	D-4	D-4/C-3	C-3/A-3	A-3 (C-3)	A-3^D	A-3^D	A-2^E	A-2^E	A-2/C-3	C-3	C-3/D-4	D-4
Maine	E-5	E-5	E-5/D-4	D-4/A-4	A-4 (D-4)	A-3^G	A-3^G	A-3^G	A-3^G	A-3/D-4	D-4	D-4/E-5	E-5
Maryland	E-5	E-5	E-5/D-4	D-4/A-4	A-4 (D-4)	A-3^{G,H}	A-3^{G,H}	A-3^{G,H}	A-3^{G,H}	A-3/C-3	C-3/D-4	D-4/E-5	E-5
Massachusetts	E-5	E-5	E-5/D-4	D-4/A-4	A-4 (D-4)	A-3^G	A-3^G	A-3^G	A-3^G	A-3/D-4	D-4	D-4/E-5	E-5
Michigan	E-5	E-5	E-5/D-4	D-4/A-4	A-4 (D-4)	A-3	A-3	A-3	A-3	A-3/D-4	D-4	D-4/E-5	E-5
Minnesota	E-5	E-5	E-5/D-4	D-4/A-4	A-4 (D-4)	A-3	A-3	A-3	A-3	A-3/C-3	C-3/D-4	D-4/E-5	E-5
Mississippi	D-4	D-4	D-4/C-3	C-3/A-3	A-3 (C-3)	A-3	A-3	A-2	A-3	A-3/C-3	C-3	C-3/D-4	D-4
Missouri^K	E-5	E-5/D-4	D-4	D-4/A-3	A-3 (C-3)	A-3^D	A-2^E	A-2^E	A-2^E	A-2/C-3	C-3/D-4	D-4	D-4/E-5
Montana	E-5	E-5	E-5/D-4	D-4/A-3	A-3 (C-3)	A-2	A-2	A-2	A-2	A-2/C-3	C-3/D-4	D-4/E-5	E-5
Nebraska	E-5	E-5	E-5/D-4	D-4/A-3	A-3 (C-3)	A-2	A-2	A-2	A-2	A-2/B-2	B-2/C-3	C-3/D-4	D-4/E-5
Nevada:													
N 38° Latitude	E-5	E-5/D-4	D-4	D-4/A-3	A-3 (C-3)	A-2^E	A-2^E	A-2^E	A-2^E	A-2/B-2	B-2/C-3	C-3/D-4	D-4/E-5
S 38° Latitude	D-4	D-4/C-3	C-3/B-2	B-2/A-2	A-2 (B-2)	A-1	A-1	A-1	A-1	A-1	A-1/B-2	B-2/C-3	C-3/D-4
New Hampshire	E-5	E-5	E-5/D-4	D-4/A-4	A-4 (D-4)	A-3^G	A-3^G	A-3^G	A-3^G	A-3/D-4	D-4	D-4/E-5	E-5
New Jersey	E-5	E-5	E-5/D-4	D-4/A-4	A-4 (D-4)	A-3^G	A-3^G	A-3^G	A-3^G	A-3/D-4	D-4	D-4/E-5	E-5
New Mexico:													
N 34° Latitude	E-5/D-4	D-4	D-4/C-3	C-3/A-2	A-2 (B-2)	A-1	A-1	A-2	A-2	A-2/B-2	B-2/C-3	C-3/D-4	D-4
S 34° Latitude	D-4	D-4/C-3	C-3/B-2	B-2/A-2	A-2 (B-2)	A-1	A-1	A-1	A-1	A-1/B-2	A-1/B-2	B-2/C-3	C-3/D-4
New York	E-5	E-5	E-5/D-4	D-4/A-4	A-4 (D-4)	A-3^G	A-3^G	A-3^G	A-3^G	A-3/D-4	D-4	D-4/E-5	E-5
North Carolina	E-5/D-4	D-4	D-4	D-4/A-3	A-3 (C-3)	A-3^D	A-3^D	A-2^E	A-2^E	A-2/C-3	C-3/D-4	D-4	D-4/E-5
North Dakota	E-5	E-5	E-5/D-4	D-4/A-4	A-4 (D-4)	A-3	A-2	A-2	A-2	A-2/C-3	C-3/D-4	D-4/E-5	E-5
Ohio	E-5	E-5	E-5/D-4	D-4/A-4	A-4 (D-4)	A-3	A-3	A-3	A-3	A-3/C-3	C-3/D-4	D-4/E-5	E-5
Oklahoma	E-5/D-4	D-4	D-4/C-3	C-3/A-3	A-3 (C-3)	A-2	A-2	A-2	A-2	A-2/B-2	B-2/C-3	C-3/D-4	D-4/E-5
Oregon:													
E 122° Longitude	E-5	E-5/D-4	D-4	D-4/A-4	A-4 (D-4)	A-3	A-2	A-2	A-2	A-2/C-3	C-3/D-4	D-4	D-4/E-5
W 122° Longitude	E-5	E-5/D-4	D-4	D-4/A-4	A-4 (D-4)	A-3^D	A-3^D	A-3^D	A-3^D	A-3/C-3	C-3/D-4	D-4/E-5	E-5
Pennsylvania	E-5	E-5	E-5/D-4	D-4/A-4	A-4 (D-4)	A-3^G	A-3^G	A-3^G	A-3^G	A-3/D-4	D-4	D-4/E-5	E-5
Rhode Island	E-5	E-5	E-5/D-4	D-4/A-4	A-4 (D-4)	A-3^G	A-3^G	A-3^G	A-3^G	A-3/D-4	D-4	D-4/E-5	E-5
South Carolina	D-4	D-4	D-4	D-4/A-3	A-3 (C-3)	A-3	A-3	A-2	A-2	A-2/C-3	C-3/D-4	D-4	D-4/E-5
South Dakota	E-5	E-5	E-5/D-4	D-4/A-3	A-3 (C-3)	A-2	A-2	A-2	A-2	A-2/C-3	B-2/C-3	C-3/D-4	D-4/E-5
Tennessee	E-5/D-4	D-4	D-4/C-3	C-3/A-3	A-3 (C-3)	A-2^E	A-2^E	A-2^E	A-2^E	A-2/C-3	C-3/D-4	D-4	D-4/E-5
Texas:^{K}													
E 99° Longitude	D-4	D-4	D-4/C-3	C-3/A-3	A-3 (C-3)	A-3^{D,H}	A-2^{E,J}	A-2^{E,J}	A-2^{E,J}	A-2/B-2	B-2/C-3	C-3/D-4	D-4
W 99° Longitude	D-4	D-4/C-3	C-3/B-2	B-2/A-2	A-2 (B-2)	A-1^F	A-1^F	A-1^F	A-1^F	A-1/B-2	B-2/C-3	C-3/D-4	E-5
Utah	E-5	E-5/D-4	D-4	D-4/A-3	A-3 (C-3)	A-2^E	A-2^E	A-2^E	A-2^E	A-2/B-2	B-2/C-3	C-3/D-4	D-4/E-5

Table A-2 ASTM D 1835, part 4: Schedule of Seasonal and Geographical Volatility Classes, cont'd (© ASTM; reprinted with permission)

State	Jan.	Feb.	Mar.	Apr.	May[B]	June	July	Aug.	Sept. 1–15	Sept. 16–30	Oct.	Nov.	Dec.
Vermont	E-5	E-5	E-5/D-4	D-4/A-4	A-4 (D-4)	A-3	A-3	A-3	A-3	A-3/A-4	D-4	D-4/E-5	E-5
Virginia	E-5	E-5/D-4	D-4	D-4/A-3	A-3 (C-3)	A-3[D,H]	A-3[D,H]	A-3[D,H]	A-3[D,H]	A-3/C-3	C-3/D-4	D-4/E-5	E-5
Washington:													
E 122° Longitude	E-5	E-5	E-5/D-4	D-4/A-4	A-4 (D-4)	A-3	A-2	A-2	A-2	A-2/C-3	C-3/D-4	D-4/E-5	E-5
W 122° Longitude	E-5	E-5	E-5/D-4	D-4/A-4	A-4 (D-4)	A-3	A-3	A-3	A-3	A-3/C-3	C-3/D-4	D-4/E-5	E-5
West Virginia	E-5	E-5	E-5/D-4	D-4/A-4	A-4 (D-4)	A-3	A-3	A-3	A-3	A-3/C-3	C-3/D-4	D-4/E-5	E-5
Wisconsin	E-5	E-5	E-5/D-4	D-4/A-4	A-4 (D-4)	A-3[G]	A-3[G]	A-3[G]	A-3[G]	A-3/C-3	C-3/D-4	D-4/E-5	E-5
Wyoming	E-5	E-5	E-5/D-4	D-4/A-3	A-3 (C-3)	A-2	A-2	A-2	A-2	A-2/B-2	B-2/C-3	C-3/D-4	D-4/E-5

[A] For the period May 1 through September 15, the specified vapor pressure classes comply with 1992 U.S. EPA Phase II volatility regulations and the Federal RFG "Simple Model" requirements. EPA regulations (under the Phase II regulations) allow 1.0 psi higher vapor pressure for gasoline-ethanol blends containing 9 to 10 vol % ethanol for the same period. See Appendix X3 for additional federal volatility regulations.

[B] Values in parentheses are permitted for retail stations and other end users.

[C] Details of State Climatological Division by county as indicated:
California, North Coast—Alameda, Contra Costa, Del Norte, Humbolt, Lake, Marin, Mendocino, Monterey, Napa, San Benito, San Francisco, San Mateo, Santa Clara, Santa Cruz, Solano, Sonoma, Trinity
California, Interior—Lassen, Modoc, Plumas, Sierra, Siskiyou, Alpine, Amador, Butte, Calaveras, Colusa, El Dorado, Fresno, Glenn, Kern (except that portion lying east of Los Angeles County Aqueduct), Kings, Madera, Mariposa, Marced, Placer, Sacramento, San Joaquin, Shasta, Stanislaus, Sutter, Tehama, Tulare, Tuolumne, Yolo, Yuba, Nevada
California, South Coast—Orange, San Diego, San Luis Obispo, Santa Barbara, Ventura, Los Angeles (except that portion north of the San Gabriel Mountain range and east of the Los Angeles County Aqueduct)
California, Southeast—Imperial, Riverside, San Bernardino, Los Angeles (that portion north of the San Gabriel Mountain range and east of the Los Angeles County Aqueduct), Mono, Inyo, Kern (that portion lying east of the Los Angeles County Aqueduct)

[D] AA-3 for the following ozone nonattainment areas:
(See Federal Register 56, 215, 56694, November 6, 1991, for description of the geographic boundary for each area.)

Alabama—Birmingham
California—Monterey Bay
California—San Francisco-Bay Area
District of Columbia—Washington
Florida—Jacksonville
Florida—Miami-Fort Lauderdale-West Palm Beach
Florida—Tampa-St. Petersburg-Clearwater
Georgia—Atlanta

Maryland—Washington Area
Missouri—Kansas City
Missouri—St. Louis
North Carolina—Charlotte-Gastonia
North Carolina—Greensboro-Winston Salem-High Point
North Carolina—Raleigh-Durham
Oregon—Portland-Vancouver AQMA
Oregon—Salem

Table A-2 ASTM D 1835, part 4: Schedule of Seasonal and Geographical Volatility Classes (© ASTM; reprinted with permission)

Louisiana—Baton Rouge
Louisiana—Beauregard Parish
Louisiana—Grant Parish
Louisiana—Lafayette
Louisiana—Lafourche Parish
Louisiana—Lake Charles
Louisiana—New Orleans
Louisiana—St. James Parish
Louisiana—St. Mary Parish
Maryland—Baltimore
Maryland—Kent and Queen Anne's Counties
Maryland—Philadelphia-Wilmington-Trenton Area

Tennessee—Memphis
Tennessee—Nashville
Texas—Beaumont-Port Arthur
Texas—Dallas-Fort Worth
Texas—Houston-Galveston-Brazoria
Texas—Victoria
Virginia-Norfolk-Virginia Beach-Newport News
Virginia—Richmond
Virginia—Smyth County
Virginia—Washington Area

E AA-2 for the following ozone nonattainment areas:
(See Federal Register 56, 215, 56694, November 6, 1991, for description of the geographic boundary for each area.)

Alabama—Birmingham
Arizona—Phoenix
California—Chico
California—Los Angeles-South Coast Air Basin
California—Monterey Bay
California—Sacramento Metro
California—San Diego
California—San Francisco-Bay Area
California—San Joaquin Valley
California—Santa Barbara-Santa Maria-Lompoc
California—Ventura County
California—Yuba City
Colorado—Denver-Boulder
Georgia—Atlanta
Kansas—Kansas City
Louisiana—Baton Rouge
Louisiana—Beauregard Parish
Louisiana—Grant Parish
Louisiana—Lafayette
Louisiana—Lafourche Parish

Louisiana—Lake Charles
Louisiana—New Orleans
Louisiana—St. James Parish
Louisiana—St. Mary Parish
Missouri—Kansas City
Missouri—St. Louis
Nevada—Reno
North Carolina—Charlotte-Gastonia
North Carolina—Greensboro-Winston Salem-High Point
North Carolina—Raleigh-Durham
Tennessee—Memphis
Tennessee—Nashville
Texas—Beaumont-Port Arthur
Texas—Dallas-Forth Worth
Texas—Houston-Galveston-Brazoria
Texas—Victoria
Utah—Salt Lake City

Table A-2 ASTM D 1835, part 4: Schedule of Seasonal and Geographical Volatility Classes, cont'd (© ASTM; reprinted with permission)

435

F AA-1 for the following ozone nonattainment areas:
(See Federal Register 56, 215, 56694, November 6, 1991, for description of the geographic boundary for each area.)

Arizona—Phoenix
California—Imperial County
California—Southeast Desert Modified AQMA
Texas—El Paso

G AA-3 and federal RFG areas requiring 57.2 kPa (8.3 psi) per-gallon maximum vapor pressure. No waiver for gasoline-ethanol blends.

Connecticut—Hartford-New Britain-Middletown-New Haven-Meriden-Waterbury
Connecticut—Remainder[J]
Delaware—New Castle (Wilmington), Kent Counties
Delaware—Sussex County[J]
Illinois—Chicago Area
Indiana—Gary-Lake and Porter Counties
Kentucky—Louisville[J]
Maine—Knox and Lincoln Counties[J]
Maine—Lewiston-Auburn[J]
Maine—Portland[J]
Massachusetts—Boston-Lawrence-Worcester[J]
Massachusetts—Springfield[J]
Maryland—Cecil County
New Hampshire—Manchester[J]
New Hampshire—Portsmouth-Dover-Rochester[J]
New Jersey—Northern
New Jersey—Trenton Area
New Jersey—Atlantic City[J]

New York—New York City-Long Island Areas
New York—Portions of Exxex County[J]
New York—Poughkeepsie[J]
Pennsylvania—Philadelphia Area

Rhode Island[J]
Wisconsin—Milwaukee-Racine

Table A–2   ASTM D 1835, part 4:   Schedule of Seasonal and Geographical Volatility Classes cont'd   (© ASTM; reprinted with permission)

[H] AA-3 and federal RFG areas requiring 51.0 kPa (7.4 psi) per-gallon maximum vapor pressure. No waiver for gasoline-ethanol blends.

District of Columbia—Washington[J]
Maryland—Baltimore
Maryland—Kent and Queen Anne's Counties[J]
Maryland—Washington DC Area[J]

Texas—Dallas-Fort Worth[J]
Texas—Houston-Galveston-Brazoria
Virginia—Norfolk-Virginia Beach-Newport News[J]
Virginia—Richmond[J]
Virginia—Washington DC Area[J]

[I] AA-2 and federal RFG areas requiring 51.0 kPa (7.4 psi) per-gallon maximum vapor pressure. No waiver for gasoline-ethanol blends.

California—Los Angeles-South Coast Air Basin
California—Sacramento Metro (1996)
California—San Diego
California—Ventura County

Texas—Dallas-Fort Worth[J]
Texas—Houston-Galveston-Brazoria

[J] Opt-in federal RFG areas.

[K] The following states have more restrictive vapor pressure limits in some areas. In some cases, these limits have been approved by the U.S. EPA as a part of a State Implementation Plan (SIP). In other cases, states are awaiting official EPA approval for the more restrictive local vapor pressure limits.

Atlanta, GA area—48.2 kPa (7.0 psi) max June 1–Sept. 15 (approval pending)
St. Louis, MO area—48.2 kPa (7.0 psi) max June 1–Sept. 15
Phoenix, AZ area—48.2 kPa (7.0 psi) max May 31–Sept. 30 (approval pending)
Madison, Monroe, and St. Clair counties, IL area—49.6 kPa (7.2 psi) max June 1–Sept. 15
Clark, Floyd Counties, IN area—53.8 kPa (7.8 psi) max May 1–Sept. 15
California—48.26 kPa (7.00 psi) max April 1, May 1, or June 1–Sept. 30 or Oct. 31 depending on air basin
El Paso, TX area—48.2 kPa (7.0 psi) max May 1 (terminal)/June 1 (retail)–Sept. 15

Table A-2   ASTM D 1835, part 4:   Schedule of Seasonal and Geographical Volatility Classes cont'd   (© ASTM; reprinted with permission)

437

Temperature Conversion °F = (°C × 1.8) + 32°

State	Jan.	Feb.	March	April	May	June	July	Aug.	Sept.	Oct.	Nov.	Dec.
Alabama	-4.	-3.	0.	5.	10.	10.	10.	10.	10.	6.	0.	-4.
ᵃAlaska:												
Southern Region	-27.	-26.	-23.	-11.	1.	7.	9.	7.	1.	-9.	-19.	-23.
South Mainland	-41.	-39.	-31.	-14.	-1.	7.	9.	5.	-2.	-18.	-32.	-41.
N of 62° Latitude												
Arizona:												
N of 34° Latitude	-11.	-7.	-7.	-2.	2.	6.	10.	10.	6.	1.	-6.	-9.
S of 34° Latitude	-2.	-1.	2.	7.	10.	10.	10.	10.	10.	9.	2.	-1.
Arkansas	-9.	-6.	-2.	6.	10.	10.	10.	10.	10.	4.	-2.	-6.
ᶜCalifornia												
North Coast	-2.	0.	1.	4.	5.	8.	9.	9.	8.	6.	2.	-2.
South Coast	-2.	-1.	2.	4.	7.	9.	10.	10.	9.	6.	1.	-2.
Southeast	-7.	-3.	-1.	3.	8.	10.	10.	10.	9.	4.	-3.	-6.
Interior	-4.	-3.	-3.	-1.	3.	9.	10.	10.	10.	6.	0.	-2.
Colorado:												
E of 105° Longitude	-14.	-12.	-9.	-3.	4.	10.	10.	10.	7.	1.	-8.	-11.
W of 105° Longitude	-24.	-20.	-12.	-6.	-1.	4.	8.	6.	1.	-6.	-14.	-21.
Connecticut	-14.	-13.	-8.	-1.	5.	10.	10.	10.	7.	1.	-4.	-12.
Delaware	-9.	-8.	-3.	0.	8.	10.	10.	10.	10.	4.	-1.	-8.
District Columbia	-8.	-7.	-3.	3.	9.	10.	10.	10.	10.	5.	0.	-7.
Florida:												
N of 29° Latitude	-1.	1.	4.	9.	10.	10.	10.	10.	10.	9.	3.	-1.
S of 29° Latitude	4.	7.	8.	10.	10.	10.	10.	10.	10.	10.	9.	5.
Georgia	-5.	-2.	1.	6.	10.	10.	10.	10.	10.	6.	0.	-3.
Hawaii	10.	10.	10.	10.	10.	10.	10.	10.	10.	10.	10.	10.
Idaho	-17.	-16.	-11.	-3.	-5.	4.	10.	9.	3.	-2.	-11.	-15.
Illinois:												
N of 40° Latitude	-18.	-16.	-9.	-1.	4.	10.	10.	10.	7.	1.	-7.	-16.
S of 40° Latitude	-15.	-12.	-7.	1.	7.	10.	10.	10.	9.	3.	-6.	-13.
Indiana	-16.	-13.	-7.	-1.	4.	10.	10.	10.	7.	1.	-6.	-14.
Iowa	-23.	-19.	-13.	-3.	4.	10.	10.	10.	6.	0.	-12.	-20.
Kansas	-17.	-12.	-9.	-3.	5.	10.	10.	10.	7.	0.	-8.	-13.
Kentucky	-12.	-9.	-4.	1.	8.	10.	10.	10.	9.	3.	-4.	-11.
Louisiana	-3.	0.	3.	8.	10.	10.	10.	10.	10.	7.	2.	-1.

Table A-2   ASTM D 1835, part 5:   Maximum Temperature for Phase Separation, °C   (© ASTM; reprinted with permission)

Temperature Conversion °F = (°C × 1.8) + 32°

State	Jan.	Feb.	March	April	May	June	July	Aug.	Sept.	Oct.	Nov.	Dec.
Maine	-24.	-22.	-16.	-4.	1.	7.	10.	8.	3.	-2.	-8.	-20.
Maryland	-9.	-8.	-3.	3.	9.	10.	10.	10.	10.	4.	-2.	-8.
Massachusetts	-15.	-14.	-7.	-1.	4.	10.	10.	10.	6.	0.	-4.	-13.
Michigan:												
Lower Michigan	-18.	-17.	-12.	-3.	1.	7.	10.	9.	5.	0.	-6.	-14.
Upper Michigan	-21.	-20.	-15.	-6.	-1.	6.	9.	9.	4.	-1.	-9.	-18.
Minnesota	-31.	-28.	-20.	-7.	0.	6.	10.	8.	1.	-3.	-16.	-28.
Mississippi	-3.	-1.	2.	7.	10.	10.	10.	10.	10.	7.	1.	-3.
Missouri	-14.	-11.	-6.	1.	8.	10.	10.	10.	10.	3.	-5.	-12.
Montana	-28.	-24.	-19.	-6.	1.	5.	9.	8.	1.	-5.	-17.	-23.
Nebraska	-19.	-14.	-11.	-3.	4.	9.	10.	10.	5.	-2.	-10.	-16.
Nevada:												
N of 38° Latitude	-18.	-13.	-8.	-3.	1.	5.	9.	7.	2.	-3.	-11.	-14.
S of 38° Latitude	-9.	-5.	-1.	1.	9.	10.	10.	10.	10.	4.	-3.	-6.
New Hampshire	-18.	-17.	-9.	-2.	3.	9.	10.	9.	3.	-1.	-6.	-16.
New Jersey	-10.	-9.	-4.	2.	7.	10.	10.	10.	10.	4.	-1.	-8.
New Mexico:												
N of 34° Latitude	-14.	-11.	-7.	-2.	1.	7.	10.	10.	7.	1.	-8.	-12.
S of 34° Latitude	-7.	-5.	-1.	6.	10.	10.	10.	10.	10.	7.	-2.	-5.
New York:												
N of 42° Latitude	-21.	-20.	-13.	-3.	2.	9.	10.	10.	4.	-1.	-6.	-18.
S of 42° Latitude	-13.	-13.	-7.	1.	6.	10.	10.	10.	8.	2.	-3.	-12.
North Carolina	-9.	-7.	-3.	1.	7.	10.	10.	10.	8.	1.	-5.	-8.
North Dakota	-29.	-27.	-11.	-6.	1.	8.	10.	10.	3.	-2.	-17.	-24.
Ohio	-14.	-13.	-8.	-2.	6.	10.	10.	10.	7.	1.	-5.	-13.
Oklahoma	-12.	-6.	-5.	1.	7.	10.	10.	10.	10.	4.	-4.	-9.
Oregon:												
E of 122° Longitude	-17.	-12.	-6.	-3.	0.	4.	6.	6.	2.	-3.	-8.	-12.
W of 122° Longitude	-5.	-3.	-1.	2.	5.	8.	10.	10.	7.	2.	-3.	-3.
Pennsylvania:												
N of 41° Latitude	-17.	-19.	-13.	-4.	1.	6.	9.	8.	2.	-1.	-6.	-16.
S of 41° Latitude	-13.	-14.	-9.	-1.	5.	10.	10.	10.	7.	2.	-4.	-12.
Rhode Island	-11.	-11.	-5.	1.	6.	10.	10.	10.	8.	3.	-2.	-10.
South Carolina	-3.	-2.	0.	6.	10.	10.	10.	10.	10.	7.	1.	-3.
South Dakota	-24.	-21.	-16.	-4.	3.	10.	10.	10.	4.	-2.	-12.	-21.
Tennessee	-9.	-7.	-3.	2.	9.	10.	10.	10.	10.	2.	-3.	-8.

Table A-2 ASTM D 1835, part 5: Maximum Temperature for Phase Separation, °C, cont'd (© ASTM; reprinted with permission)

Temperature Conversion °F = (°C × 1.8) + 32°

State	Jan.	Feb.	March	April	May	June	July	Aug.	Sept.	Oct.	Nov.	Dec.
Texas:												
N of 31° Latitude	-11	-8	-4	2	8	10	10	10	10	5	-3	-7
S of 31° Latitude	-1	1	4	10	10	10	10	10	10	10	3	1
Utah	-15	-11	-7	-2	2	8	10	10	7	2	-11	-12
Vermont	-20	-21	-12	-2	2	9	10	10	5	0	-6	-17
Virginia	-8	-7	-3	3	9	10	10	10	10	4	-2	-7
Washington:												
E of 122° Longitude	-13	-6	-3	1	4	7	10	10	7	1	-5	-7
W of 122° Longitude	-6	-2	-2	1	4	7	9	9	6	2	-2	-2
West Virginia	-13	-12	-7	-2	4	9	10	10	5	-2	-7	-12
Wisconsin	-25	-21	-15	-3	3	8	10	10	5	-1	-11	-21
Wyoming	-23	-17	-14	-6	0	5	10	10	3	-2	-13	-16

A A maximum phase separation temperature of 10°C (50°F) is specified, even if the 6-h 10th percentile minimum temperature for the area and month can be higher.

B The designated areas of Alaska are divided as follows:

Southern Region—The Aleutians, Kodiak Island, the coastal strip East of Longitude 141°, and the Alaskan Peninsula South of Latitude 52°.

South Mainland—The portion of Alaska South of Latitude 62°, except the Southern Region.

North of Latitude 62°—The specification test temperature must be agreed between the vendor and purchaser having regard to equipment design, expected weather conditions, and other relevant factors.

C The designated areas of California are divided by county as follows:

North Coast—Alameda, Contra Costa, Del Norte, Humbolt, Lake, Marin, Mendocino, Monterey, Napa, San Benito, San Francisco, San Mateo, Santa Clara, Santa Cruz, Solano, Sonoma, Trinity.

Interior—Lassen, Modoc, Plumas, Sierra, Siskiyou, Alpine, Amador, Butte, Calaveras, Colusa, El Dorado, Fresno, Glenn, Kern (excepting that portion lying east of the Los Angles County Aqueduct). Kings, Madera, Mariposa, Merced, Placer, Sacramento, San Joaquin, Shasta, Stanislaus, Sutter, Tehama, Tulare, Tuolumne, Yolo, Yuba, Nevada.

South Coast—Orange, San Diego, San Luis Obispo, Santa Barbara, Ventura, Los Angeles (except that portion lying north of the San Gabriel Mountain range and east of the Los Angeles County Aqueduct).

Southeast—Imperial, Riverside, San Bernadino, Los Angeles (that portion lying north of the San Gabriel Mountain range and east of the Los Angeles County Aqueduct), Mono, Inyo, Kern (that portion lying east of the Los Angeles County Aqueduct).

D The designated areas of Michigan are divided as follows:

Lower Michigan—That portion of the state lying East of Lake Michigan.

Upper Michigan—That portion of the state lying North of Wisconsin and of Lake Michigan.

Table A-2 ASTM D 1835, part 5: Maximum Temperature for Phase Separation, °C, cont'd (© ASTM; reprinted with permission)

Property	ASTM Test[A] Method	Limit[B]
Flash Point °C, min	D 56	38
Distillation temperature °C	D 86	
10 % volume recovered, max		205
Final boiling point, max		300
Kinematic viscosity at 40°C, mm²/s	D 445	
min		1.0
max		1.9
Sulfur, % mass	D 1266	
No. 1-K, max		0.04
No. 2-K, max		0.30
Mercaptan sulfur, % mass, max[C]	D 3227	0.003
Copper strip corrosion rating max., 3 h at 100°C	D 130	No. 3
Freezing point, °C, max	D 2386	−30
Burning quality, min	D 187	pass (see 4.2)
Saybolt color, min	D 156	+16

[A] The test methods indicated are the approved referee methods. Other acceptable methods are indicated in Section 2.

[B] To meet special operating conditions, modifications of individual limiting requirements, except sulfur, can be agreed upon among purchaser, seller and manufacturer.

[C] The mercaptan sulfur determination can be waived if the fuel is considered sweet by Test Method D 4952.

Table A-3 ASTM D 3699–96a Detailed Requirements for Kerosene
(© ASTM; reprinted with permission)

441

Property		Jet A or Jet A-1	Jet B	ASTM Test Method[B]
COMPOSITION				
Acidity, total mg KOH/g	max	0.10	...	D 3242
Aromatics, vol %	max	25	25	D 1319
Sulfur, mercaptan,[C] weight %	max	0.003	0.003	D 3227
Sulfur, total weight %	max	0.30	0.3	D 1266, D 1552, D 2622, D 4294 or D 5453
VOLATILITY				
Distillation temperature, °C:				
10 % recovered, temperature	max	205	...	D 86
20 % recovered, temperature	max	...	145	
50 % recovered, temperature	max	report	190	
90 % recovered, temperature	max	report	245	
Final boiling point, temperature	max	300	...	
Distillation residue, %	max	1.5	1.5	
Distillation loss, %	max	1.5	1.5	
Flash point, °C	min	38	...	D 56 or D 3828[H]
Density at 15°C, kg/m³		775 to 840	751 to 802	D 1298 or D 4052
Vapor pressure, 38°C, kPa	max	...	21	D 323 or D 5191[K]
FLUIDITY				
Freezing point, °C	max	−40 Jet A[D] / −47 Jet A-1[D]	−50[D]	D 2386, D 5901, or D 5972[L]
Viscosity −20°C, mm²/s[J]	max	8.0		D 445
COMBUSTION				
Net heat of combustion, MJ/kg	min	42.8[E]	42.8[E]	D 4529, D 3338, or D 4809
One of the following requirements shall be met:				
(1) Luminometer number, or	min	45	45	D 1740
(2) Smoke point, mm, or	min	25	25	D 1322
(3) Smoke point, mm, and	min	18	18	D 1322
Naphthalenes, vol, %	max	3.0	3.0	D 1840
CORROSION				
Copper strip, 2 h at 100°C	max	No. 1	No. 1	D 130

Table A–4 ASTM D 1655 Detailed Requirements for Aviation Turbine Fuels (© ASTM; reprinted with permission)

Property		Jet A or Jet A-1	Jet B	ASTM Test Method[B]
STABILITY				
Thermal:				
Filter pressure drop, mm Hg	max	25[I] Code 3	25[I] Code 3	D 3241[F]
Tube deposit less than		No Peacock or Abnormal Color Deposits		
CONTAMINANTS				
Existent gum, mg/100 mL	max	7	7	D 381
Water reaction:				
Interface rating	max	1b	1b	D 1094
ADDITIVES		See 5.2	See 5.2	
Electrical conductivity, pS/m[G]		...	...	D 2624

A For compliance of test results against the requirements of Table 1, see 6.2.

B The test methods indicated in this table are referred to in Section 10.

C The mercaptan sulfur determination may be waived if the fuel is considered sweet by the doctor test described in Test Method D 4952.

D Other freezing points may be agreed upon between supplier and purchaser.

E For all grades use either Eq 1 or Table 1 in Test Method D 4529 or Eq 2 in Test Method D 3338. Test Method D 4809 may be used as an alternative. In case of dispute, Test Method D 4809 shall be used.

F Thermal stability test (JFTOT) shall be conducted for 2.5 h at a control temperature of 260°C, but if the requirements of Table 1 are not met, the test may be conducted at 245°C. Results at both temperatures shall be reported in this case. Tube deposits shall always be reported by the Visual Method: a rating by the Tube Deposit Rating (TDR) optical density method is desirable but not mandatory.

G If electrical conductivity additive is used, the conductivity shall not exceed 450 pS/m at the point of use of the fuel. When electrical conductivity additive is specified by the purchaser, the conductivity shall be 50 to 450 pS/m under the conditions at point of delivery.

$$1 \text{ pS/m} = 1 \times 10^{-12} \ \Omega^{-1} \text{ m}^{-1}$$

H Results obtained by Test Methods D 3828 may be up to 2°C lower than those obtained by Test Method D 56 which is the preferred method. In case of dispute, Test Method D 56 will apply.

I Preferred SI units are 3.3 kPa, max.

J 1 mm²/s = 1 cSt.

K Cyclohexane and toluene, as cited in 7.2 and 7.7 of Test Method D 5191, shall be used as calibrating reagents. Test Method D 5191 shall be the referee method.

L Test Method D 5972 may produce a higher (warmer) result than that from Test Method D 2386 on wide-cut fuels such as Jet B and JP-4. In case of dispute, Test Method D 2386 shall be the referee method.

Table A–4 ASTM D 1655 Detailed Requirements for Aviation Turbine Fuels, cont'd   (© ASTM; reprinted with permission)

Property	ASTM Test Method[c]	Grade[D]				
		No. 0-GT	No. 1-GT	No. 2-GT	No. 3-GT	No. 4-GT
FLASH POINT °C (°F) min	D 93	[E]	38 (100)	38 (100)	55 (130)	66 (150)
WATER AND SEDIMENT % vol max	D 2709	0.05	0.05	0.05	...	...
	D 1796	...	...	...	1.0	1.0
DISTILLATION TEMPERATURE °C (°F) 90 % VOL. RECOVERED min	D 86	...	...	282	...	...
max		...	288	338	...	...
KINEMATIC VISCOSITY mm²/s[F] AT 40°C (104°F) min	D 445	[E]	1.3	1.9	5.5	5.5
max			2.4	4.1	...	...
AT 100°C (212°F) max			...	...	50.0	50.0
RAMSBOTTOM CARBON RESIDUE on 10 % DISTILLATION RESIDUE % mass, max	D 524	0.15	0.15	0.35	...	...
ASH % MASS, max	D 482	0.01	0.01	0.01	0.03	...
DENSITY at 15°C kg/m³ max	D 1298	...	850	876	...	...
POUR POINT[F] °C (°F) max	D 97	...	-18	-6	...	...

[A] To meet special operating conditions, modifications of individual limiting requirements may be agreed upon between purchaser, seller, and manufacturer.

[B] Gas turbines with waste heat recovery equipment may require fuel sulfur limits to prevent cold end corrosion. Environmental limits may also apply to fuel sulfur in selected areas in the United States and in other countries.

[C] The test methods indicated are the approved referee methods. Other acceptable methods are indicated in 6.1.

[D] No. 0-GT includes naphtha, Jet B fuel and other volatile hydrocarbon liquids. No. 1-GT corresponds in general to specification D 396 Grade No. 1 fuel and D 975 Grade 1-D diesel fuel in physical properties. No. 2-GT corresponds in general to specification D 396 Grade No. 2 fuel and D 975 Grade 2-D diesel fuel in physical properties. No. 3-GT and No. 4-GT viscosity range brackets specification D 396 Grades No. 4, No. 5 (light), No. 5 (heavy), and No. 6, and D 975 Grade No. 4-D diesel fuel in physical properties.

[E] When the flash point is below 38°C (100°F) or when kinematic viscosity is below 1.3 mm²/s at 40°C (104°F) or when both conditions exist, the turbine manufacturer should be consulted with respect to safe handling and fuel system design.

[F] For cold weather operation, the pour point should be specified 6°C below the ambient temperature at which the turbine is to be operated except where fuel heating facilities are provided. When a pour point less than −18°C is specified for Grade No. 2-GT, the minimum viscosity shall be 1.7 mm²/s and the minimum 90 % recovered temperature shall be waived.

Table A–5 ASTM D 2880 Detailed Requirements for Gas Turbine Fuels at Time and Place of Custody Transfer to User (© ASTM; reprinted with permission)

Property	ASTM Test Method[B]	Grade Low Sulfur No. 1-D[C]	Grade Low Sulfur No. 2-D[C]	Grade No. 1-D[C]	Grade No. 2-D[C]	Grade No. 4-D
Flash Point, °C, min.	D 93	38	52	38	52	55
Water and Sediment, % vol, max	D 2709 D 1796	0.05	0.05	0.05	0.05	0.50
Distillation Temperature, °C 90 % vol Recovered	D 86					
min		...	282[D]	...	282[D]	...
max		288	338	288	338	...
Kinematic Viscosity, mm^2/S at 40°C	D 445					
min		1.3	1.9	1.3	1.9	5.5
max		2.4	4.1	2.4	4.1	24.0
Ash % mass, max	D 482	0.01	0.01	0.01	0.01	0.10
Sulfur, % mass, max[E]	D 2622[F] D 129	0.05	0.05	0.50	0.50	2.00
Copper strip corrosion rating max 3 h at 50°C	D 130	No. 3	No. 3	No. 3	No. 3	...
Cetane number, min[G]	D 613	40[H]	40[H]	40[H]	40[H]	30[H]
One of the following properties must be met:						
(1) Cetane index, min.	D 976[F]	40	40	...	...	...
(2) Aromaticity, % vol, max	D 1319[F]	35	35	...	...	...
Cloud point, °C, max	D 2500	[I]	[I]	[I]	[I]	[I]
Ramsbottom carbon residue on 10 % distillation residue, % mass, max	D 524	0.15	0.35	0.15	0.35	...

A To meet special operating conditions, modifications of individual limiting requirements may be agreed upon between purchaser, seller and manufacturer.

B The test methods indicated are the approved referee methods. Other acceptable methods are indicated in 4.1.

C Grades Low Sulfur No. 1-D, Low Sulfur No. 2-D No. 1-D, and No. 2-D may be required to contain red dye as specified by the IRS (CFR 26 Part 48) and by EPA (CFR 40 Part 80) regulations.

D When a cloud point less than −12°C is specified, the minimum flash point shall be 38°C, the minimum viscosity at 40°C shall be 1.7 mm^2/s and the minimum 90 % recovered temperature shall be waived.

E Other sulfur limits can apply in selected areas in the United States and in other countries.

F These test methods are specified in CFR 40 Part 80.

G Where cetane number by Test Method D 613 is not available, Test Method D 4737 can be used as an approximation.

H Low ambient temperatures as well as engine operation at high altitudes may require the use of fuels with higher cetane ratings.

I It is unrealistic to specify low temperature properties that will ensure satisfactory operation at all ambient conditions. However, satisfactory operation should be achieved in most cases if the cloud point (or wax appearance point) is specified at 6°C above the tenth percentile minimum ambient air temperature for the area in which ambient temperatures for U.S. locations are shown in Appendix X2. This guidance is general. Some equipment designs or operation may allow higher or require lower cloud point fuels. Appropriate low temperature operability properties should be agreed upon between the fuel supplier and purchaser for the intended use and expected ambient temperatures.

Table A–6 ASTM D 975 Detailed Requirements for Diesel Fuel Oils (© ASTM; reprinted with permission)

Property	DMX	DMA	DMB	DMC
Density at 15°C, kg/m³ᴬ max	ᴮ	890.0	900.0	920.0
Kinematic viscosity at 40°C, cStᶜ, min	1.40	1.50	...	...
max	5.50	6.00	11.0	14.0
Flash point, °C, min	43	60	60	60
Pour pointᴰ (upper), °C				
Winter quality, max	...	−6	0	0
Summer quality, max	...	0	6	6
Cloud point, °C	−16ᴱ	...	...	...
Carbon residue on 10 % btms, Ramsbottom, % mass, max	0.20	0.20	...	...
Carbon residue, Ramsbottom, % mass, max	...	...	0.25	2.50
Ash, % mass, max	0.01	0.01	0.01	0.05
Sediment by extraction, % mass, max	...	...	0.07	...
Water, % vol, max	...	...	0.30	0.30
Cetane number, min	45	40	35	...
Visual inspection	ᶠ	ᶠ	...	...
Sulfur, % mass	1.0	1.5	2.0	2.0
Vanadium, mg/kg, max	...	...	...	100

ᴬ Density in kg/L at 15°C should be multiplied by 1000 for comparison with these values.
ᴮ At some locations (ports) there will be a maximum limits.
ᶜ One cSt = mm²/s.
ᴰ Purchasers should ensure that this pour point is suitable for equipment on board the vessel.
ᴱ This fuel is suitable for use at ambient temperatures down to −15°C without heating the fuel.
ᶠ This fuel shall be visually clear and bright.

Table A–7 ASTM D 2069, part 1: Detailed Requirements for Marine Distillate Fuels (© ASTM; reprinted with permission)

Property	RMA-10	RMB-10	RMC-10	RMD-15	RME-25	RMF-25	RMG-35	RMH-35	RMK-35	RML-35	RMH-45	RMK-45	RML-45	RMH-55	RML-55
Density at 15°C, kg/m³[A], max	975.0	991.0	991.0	991.0	991.0	991.0	991.0	991.0	...	...	991.0	...	...	991.0	...
Kinematic viscosity at 100°C, cSt[B], max	10.0	10.0	10.0	15.0	25.0	25.0	35.0	35.0	35.0	35.0	45.0	45.0	45.0	55.0	55.0
Flash point, °C, min	60	60	60	60	60	60	60	60	60	60	60	60	60	60	60
Pour point[C] (upper), °C Winter quality, max	0	24	24	30	30	30	30	30	30	30	30	30	30	30	30
Summer quality, max	6	24	24	30	30	30	30	30	30	30	30	30	30	30	30
Carbon residue, Conradson, % mass, max	10	10	14	14	15	20	18	22	22	...	22	22	...	22	...
Ash, % mass, max	0.10	0.10	0.10	0.10	0.10	0.15	0.15	0.20	0.20	0.20	0.20	0.20	0.20	0.20	0.20
Water, % vol, max	0.50	0.50	0.50	0.80	1.0	1.0	1.0	1.0	1.0	1.0	1.0	1.0	1.0	1.0	1.0
Sulfur, % mass, max	3.5	3.5	3.5	4.0	5.0	5.0	5.0	5.0	5.0	5.0	5.0	5.0	5.0	5.0	5.0
Vanadium, mg/kg, max	150	150	300	350	200	500	300	600	600	600	600	600	600	600	600

[A] Density in kg/L at 15°C should be multiplied by 1000 for comparison with these values.
[B] One cSt = mm²/s.
[C] Purchasers should ensure that this pour point is suitable for the equipment on board the vessel.

Table A–7   ASTM D 2069, part 2: Detailed Requirements for Marine Residual Fuels   (© ASTM; reprinted with permission)

447

Property	ASTM Test Method[B]	No. 1	No. 2	Grade No. 4 (Light)	No. 4	No. 5 (Light)	No. 5 (Heavy)	No. 6
Flash Point °C, min	D 93	38	38	38	55	55	55	60
Water and sediment, % vol, max	D 2709 / D 95 + D 473	0.05	0.05	(0.50)[C]	(0.50)[C]	(1.00)[C]	(1.00)[C]	(2.00)[C]
Distillation temperature °C	D 86							
10 % vol recovered, max		215	...	...	...	...	...	...
90 % vol recovered, min		...	282	...	...	...	...	...
max		288	338	...	...	...	...	...
Kinematic viscosity at 40°C, mm²/s	D 445							
min		1.3	1.9	1.9	...	...	...	...
max		2.1	3.4	5.5	...	...	...	...
Kinematic viscosity at 100°C, mm²/s	D 445							
min		...	...	...	>5.5	5.0	9.0	15.0
max		...	...	...	24.0[D]	8.9[D]	14.9[D]	50.0[D]
Ramsbottom carbon residue on 10 % distillation residue % mass, max	D 524	0.15	0.35	0.05	...	...	...	...
Ash, % mass, max	D 482	...	...	0.05	0.10	0.15	0.15	...
Sulfur, % mass max[E]	D 129	0.50	0.50	...	...	...	...	...
Copper strip corrosion rating, max, 3 h at 50°C	D 130	No. 3	No. 3	...	...	...	...	...
Density at 15°C, kg/m³	D 1298							
min		850	876	...	...	...	...	...
max		...	...	>876[F]	...	...	...	...
Pour Point °C, max[G]	D 97	−18	−6	−6	−6	...	...	H

[A] It is the intent of these classifications that failure to meet any requirement of a given grade does not automatically place an oil in the next lower grade unless in fact it meets all requirements of the lower grade. However, to meet special operating conditions modifications of individual limiting requirements may be agreed upon among the purchaser, seller and manufacturer.

[B] The test methods indicated are the approved referee methods. Other acceptable methods are indicated in Section 2 and 5.1.

[C] The amount of water by distillation by Test Method D 95 plus the sediment by extraction by Test Method D 473 shall not exceed the value shown in the table. For Grade No. 6 fuel oil, the amount of sediment by extraction shall not exceed 0.50 mass %, and a deduction in quantity shall be made for all water and sediment in excess of 1.0 mass %.

[D] Where low sulfur fuel oil is required, fuel oil falling in the viscosity range of a lower numbered grade down to and including No. 4 can be supplied by agreement between the purchaser and supplier. The viscosity range of the initial shipment shall be identified and advance notice shall be required when changing from one viscosity range to another. This notice shall be in sufficient time to permit the user to make the necessary adjustments.

[E] Other sulfur limits may apply in selected areas in the United States and in other countries.

[F] This limit assures a minimum heating value and also prevents misrepresentation and misapplication of this product as Grade No. 2.

[G] Lower or higher pour points can be specified whenever required by conditions of storage or use. When a pour point less than −18°C is specified, the minimum viscosity at 40°C for grade No. 2 shall be 1.7 mm²/s and the minimum 90 % recovered temperature shall be waived.

[H] Where low sulfur fuel oil is required, Grade No. 6 fuel oil will be classified as Low Pour (+15°C max) or High Pour (no max). Low Pour fuel oil should be used unless tanks and lines are heated.

† This table is currently on ballot for revision.

Table A–8   ASTM D 396 Detailed Requirements for Fuel Oils   (© ASTM; reprinted with permission)

GLOSSARY

Acid gas removal. Removal of acid gases, primarily SO_2 and CO_2.

Additive Properties. A property is considered additive when the property of a blend is the average of that same property of each component in the blend averaged on a weight basis or a volume basis.

AIT. Autoignition temperature—the temperature at which a vapor will spontaneously ignite in the absence of spark or flame.

Alkylate. The result of addition of an alkyl group to another molecule—such as addition of isobutylene to an olefin.

Alkylation. Process for combining isobutane with an olefin.

Amylenes. Olefins containing five carbon atoms.

Aniline Point. The minimum equilibrium solution temperature for equal volumes of aniline and the hydrocarbon.

Antiknock Agent. A material added to gasoline to suppress knock tendency.

API. American Petroleum Institute.

Aquaconversion. A new hydrovisbreaking process that uses a catalyst system to achieve higher conversion as well as lower asphaltene and carbon residue content in the residue than is possible in conventional visbreaking.

Asphaltene. High molecular weight colloidal materials that are insoluble in light petroleum naptha but soluble in benzene or chloroform.

ASTM. American Society for Testing and Materials.

BFOE. Barrels of Fuel Oil Equivalent—a means of calculating a liquid volume equivalent for gases. A heating value of 6 to 6.3 million Btus is used by various companies.

Blending Number. An index assumed for the particular property of a component in a blend so that the property of the blend can be calculated.

BOD. Biochemical oxygen demand—measure of the oxygen required for the biodegradation of water contaminants.

BOV. Blending octane value—the apparent octane value of a component in a gasoline blend.

BPCD. Barrels per calendar day—measure of flow rate on a calendar day basis.

BPSD. Barrels per stream day—measure of flow rate on a stream day basis.

BTX. Benzene, toluene, and xylenes—an aromatics extraction plant products.

Butylenes. An olefin containing four carbon atoms.

B/L. Battery limits—used to define the area occupied by a process unit.

CAFE. Corporate average fuel economy—sales weighted average fuel economy (in mpg) for new cars.

Carbon Residue. Residue remaining after evaporation and pyrolysis of an oil—an indication of the relative coke-forming propensity of the oil.

Carom. A process for the extraction of aromatics using ethylene glycols and licensed by UOP.

Catofin. A catalytic process for dehydrogenation of isobutane licensed by Houdry.

CCR. Continuous catalyst regeneration process—a reforming process licensed by UOP. *also:*

Conradson Carbon Residue—a specific means for determining carbon residue.

CDU. Crude distillation unit—the first separation process in the refinery.

Cetane Index. An estimate of the cetane number of an oil calculated from its API gravity and mid-boiling point.

Cetane Number. A measure of the ignition quality of a diesel fuel determined by engine test.

CF. Cash flow—the monies realized from an operation after deduction of costs and taxes.

CFRR. Cash flow rate of return—rate of return based on the calculated present value factor and the specified project life.

Claus Process. A process by which H_2S is converted to elemental sulfur.

Clean Air Act. Under this law, EPA sets limits on how much of a pollutant can be in the air anywhere in the United States. The law allows individual states to have stronger pollution controls, but states are not allowed to have weaker pollution controls than those set for the whole country.

Cloud Point. The temperature at which a haze is first observed in a layer.

CNG. Compressed natural gas.

CO. Carbon monoxide.

COD. Chemical oxygen demand.

Coefficient of Determination. A measure of how well a chosen equation is satisfied by regression analysis—denoted as R^2.

Coke. Solid material remaining after carbonization of petroleum residue.

Complexity. *see* Complexity Index.

Complexity Index. A measure of the complexity of a process or a refinery based on relative cost of the process(es).

Condensation. Chemical combination of small molecules to produce larger molecules as in polymerization or alkylation.

Correlation. The relationship between independent and dependent variables resulting from regression analysis.

CR. Catalytic Reforming—a process for upgrading a naphtha with respect to octane rating.

Crack Spread. A rough estimate of the gross margin to be realized by a refinery when processing a certain crude.

Cracking. A process in which relatively heavy hydrocarbons are broken up into lighter products.

Curve Estimate. An estimate obtained from a plot of cost vs. capacity.

Cut-back Oil. A diluent oil added to a viscous fuel oil to reduce its viscosity.

Cycle Oil. A heavy gas oil product of cracking—often recycled.

DAO. Deasphalted oil—produced by solvent deasphalting.

DC. Delayed coking process.

Dealkylation. Removal of an alkyl group from a molecule—usually an alkylbenzene.

Dehydrocyclization. A process occurring in catalytic reforming in which hydrogen is removed from a paraffin and a naphthene is then formed.

Dehydrogenation. A process in which hydrogen is removed from a molecule (usually propane or isobutane).

Demethyllation. A dealkylation process where the alkyl group removed is a methyl group.

Desalter. A process unit for the removal of salt and brine from crude oil.

DI. Diesel index—an early estimate of diesel ignition quality calculated from aniline point and API gravity.

Dimer. A polymer resulting from the joining of two identical molecules.

Dimersol. A process for forming a dimer from an olefin—licensed by IFP.

Distillate. A general term for a distilled product—usually in the kerosene or diesel boiling range.

Distillation. A process for the separation of materials by boiling temperature.

DMB. Dimethyl butane.

DMF. Dimethyl formamide—an aromatic extraction solvent.

DMSO. Dimethyl sulfoxide—an aromatic extraction solvent.

DON. Distribution octane number—an octane number designed to simulate the mal-distribution of fuel that can occur in an engine.

Driveability Index. A number calculated from a gasoline's distillation characteristics which provides an indication of the driveability performance of the gasoline.

Ebbulated catalyst bed process. A process in which the catalyst in slurry form is circulated through the reactor.

ETAE. Ethyl tertiary amyl ether—an oxygenate gasoline additive.

ETBE. Ethyl tertiary butyl ether—an oxygenate gasoline additive.

Ethanol. Ethyl alcohol—usually obtained by fermentation.

Ether. An organic compound having the general molecular form: R-Q-R', where R and R' represent alkyl groups.

EVALU8. Software developed by the author to permit the economic evaluation of a process or an entire process scheme.

FBP. Final boiling point—final temperature on a distillation.

FC. Fluid coking—a coking process in which coke particles are circulated through the reactor in a fluidized manner until the particles attain a certain size.

FCC. Fluid catalytic cracking—a process in which the catalyst is circulated through the reactor and the regenerator in a fluidized manner.

Feasibility Study. A somewhat ambiguous reference to studies varying widely in complexity but designed to determine the feasibility of a proposed project.

FEON. Front end octane number—is derived by determining the octane of a portion of the lower boiling part of a gasoline and comparing that with the octane of the whole gasoline.

Flash Point. The temperature at which sufficient vapor is generated to support combustion when a flame is applied.

FM. N-formylmorpholine—an aromatics extraction solvent.

FOEB. *see* BFOE.

Gas Oil. A petroleum distillate with viscosity intermediate between that of kerosene and lubricating oil.

Gravity. The relative density of a material—usually referred to water.

HC. Hydrocracker—a catalytic cracking process occurring in a hydrogen environment.

HDS. Hydrodesulfurization—a catalytic process for the removal of sulfur and utilizing hydrogen.

HOC. Heavy oil cracking—an FCC process for handling heavier oils with more CCR content.

454

Hot Butamer. A process licensed by UOP for the isomerization of normal butane to isobutane.

Hot Penex. A process licensed by UOP for the isomerization of normal pentane to isopentane.

Hydrocrackate. The gasoline produced by hydrocracking.

Hydrotreating. A catalytic process for treating stocks with hydrogen.

see HDS.

IBP. Initial Boiling Point—the temperature at which the first drop is received in a distillation.

IFP. Institute Francaise du Petrole (French Petroleum Institute).

Isomerization. The skeletal rearrangement of a molecule as in the conversion of normal butane into isobutane.

K Factor. The Watson characterization factor—a calculated value widely used as a parameter for correlating properties of petroleum products.

Lang Exponent. An exponent applied to the radio of the capabilities of two units of a given process to approximate the ratio of their costs.

Lead Alkyl. Additive used to boost the octane of a gasoline—such as tetraethyl lead and tetramethyl lead.

Light Ends. The gases produced in various processes—usually ethane and lighter or propane and lighter.

LNG. Liquefied natural gas.

LPG. Liquefied Petroleum Gas.

Margins. Gross margin = revenue minus cost of raw materials.

Net margin = gross margin minus operating costs.

MCP. Methyl cyclopentane—a benzene precursor in catalytic reforming.

Merox. A process licensed by UOP for the "sweetening" of a hydrocarbon stream by converting mercaptan to disulfide.

Methanation. A process for purification of hydrogen by conversion of CO and CO_2 to CH_4.

Methanol. Methyl alcohol—also known as wood alcohol.

MON. Motor octane number—one of the methods for rating gasoline.

MP. Methyl pentane—an isopentane.

MTBE. Methyl tertiary butyl ether—an oxygenate additive.

Naphtha. A generic term applied to petroleum liquids boiling approximately between 122°F and 400°F.

NFRCI. Nelson Farrar Refinery Construction Index—an index used to adjust the cost of a process from one time to another.

NOX. Nitrogen oxide.

Oleflex. A catalytic dehydrogenation process licensed by UOP for the production of light olefins from the corresponding paraffins.

Oxygenate. A compound added to gasoline to increase its oxygen content.

Platformate. Trade name for reformate produced by UOP catalytic reforming process.

Polymer. Compound resulting from condensation of two or more olefin molecules.

Polymerization. The process of condensing two or more olefin molecules together.

PONA. Paraffin-olefin-naphthene-aromatic—an acronym to denote the composition of a naphtha with respect to type of hydrocarbon—a value used in correlating reforming results.

Pour Point. The lowest temperature at which an oil is observed to flow under prescribed test conditions.

Propylene. An olefin containing three carbon atoms.

PVF. Present value factor—for a uniform cash flow situation, this is the same as the present value of an annuity which is a function of interest rate and term of the investment.

Rankine. An absolute temperature scale equal to degrees Fahrenheit plus 459.69 .

Reformate. The naphtha resulting from a reforming process.

Reforming. A process for upgrading the octane of a naphtha—primarily by converting naphthenes to aromatics.

Refutas. A method for estimating the viscosities of blends.

Regression Analysis. The determination of the equation of a given type to relate the independent and dependent variables in a set of data.

Resin. The portion of a residual oil which is soluble in light naphtha.

RFG. Reformulated gasoline—gasoline required to meet requirements of the amendment to the Clean Air Act.

RON. Research octane number.

RSH. General molecular formula for mercaptans.

RSSR. General molecular formula for a disulfide.

RVP. Reid vapor pressure—vapor pressure as determined by ASTM test.

SCFB. Standard cubic feet per barrel.

Standard Cubic Feet per Barrel. An expression of the ratio of hydrogen to feed in a process.

SDA. Solvent deasphalting—a process for removing asphalt from a heavy oil generally to obtain more feed for the FCC unit.

SEE. Standard error of the estimate—a statistic produced by a regression program indicating the possible variance in the quantity calculated.

Sensitivity. RON minus MON.

Smoke Point. The maximum height of flame that can be achieved in the smoke point lamp without smoking—for kerosenes and jet fuels.

SOX. Sulfur oxides.

Splash Blending. Where a component is added at a terminal rather than at the refinery.

Star. A catalytic dehydrogenation process licensed by Phillips.

Sulfolane. An aromatic extraction solvent.

Susceptibility. The appreciation in octane due to addition of an additive.

Sweetening. A process in which mercaptans are converted to disulfides.

SWS. Sour water stripper—a process for the removal of acidic compounds from waste water.

Tail Gas. The gas leaving a Claus plant.

TAME. Tertiary amyl ether—an oxygenate additive.

TBP. True boiling point—the temperature at which a certain material evaporates in a TBP distillation.

TDS. Total dissolved solids—a measure of solids dissolved in a water sample.

TEL. Tetra ethyl lead—a lead alkyl gasoline additive.

Tetra. A proprietary process for aromatics extraction.

TML. Tetra methyl lead—a lead alkyl gasoline additive.

TOC. Total organic carbon—a water purity measure similar to COD.

Trimer. A polymer resulting from the condensation of three identical olefin molecules.

Udex. A proprietary process for extraction of aromatics.

UOP. Universal Oil Products—a licensor of refinery processes.

VABP. Volumetric average boiling point—an average of certain distillation temperatures used in correlating properties.

Vaporlock. Occurs when too much vapor is generated so that fuel lines and fuel pump are so full of vapor that flow of liquid fuel is inhibited.

VB. Visbreaking—a mild thermal cracking operation to reduce the viscosity of residual oil and produce gas oil.

Viscosity. Resistance of a fluid to internal flow.

VOC. Volatile organic compound—one of the factors of concern in air quality.

V/L. Vapor/liquid ratio—a measure of the volatility of a naphtha. An indication of vapor lock tendency.

Watson Characterization Factor. The K factor used in correlating properties of petroleum products.

INDEX

A

J

Jet fuel, 30, 68–73, 95–96, 211, 213, 227, 300, 307, 335, 344, 375, 379

K

K factors, 27–28, 134, 360

Kerosene, 30, 90, 236–238, 245–246, 335, 375, 427, 441

L

Lang cost-capacity exponents, 5, 385, 387

Lead additives, 78

Lead alkyl, 51

Lead content, 83, 85, 253

Light cycle oil, 172, 185–186, 199, 204, 206

Light distillate processing, 251–282
  naphtha desulfurization, 253–262
  catalytic reforming, 263–281

Light ends, 90, 158, 272

Light gas oil, 229–230

Light hydrocarbon processing, 283–312
  isomerization, 285–292
  alkylation, 293–298
  catalytic polymerization, 299–305
  catalytic dehydrogenation, 307–312

Linear program optimization, 5

Liquefied natural gas, 87

Liquefied petroleum gas, 30, 46–47, 300–301, 303, 339, 343, 427, 429

Liquid-liquid extraction, 335

Lubricants, 43, 61

Luminometer number, 375, 379

M

Marine distillate fuel, 427, 446–447

Marine residual fuel oil, 73

Material balances, 405, 407–409

Mercaptans, 343–344

Merox treating process, 343–344

Metal deactivators, 61

Methanation, 340

Methane, 144, 158, 221, 308

Methanol, 47, 315–317, 321

Methanol fuel, 47, 315–317

Methylcyclopentane, 263

Molecular weight, 34

Motor octane number, 49, 128–129, 161, 184, 199, 203, 224–225, 271, 275–276, 300, 303, 362, 364–365

Moving catalyst bed, 211

MTBE, 83, 285, 294, 296, 299–300, 307, 315–317, 321, 365, 397, 401, 404–405

N

Naphtha, 90, 95–96, 123–124, 170, 235, 253–262, 266, 271, 319, 339, 341, 344

Petroleum, 19–21, 45
  supply and demand, 19–21
Petroleum coke, 30
Phenols, 341
Phosphoric acid, 301, 304, 318
Physical solvent, 347
Pipelines/terminals, 43
Platinum-bearing catalyst, 83, 287
Pollutant/contaminant, 77–88
Pollution abatement, 391, 396
Polymerization, 299–305, 307
Polymers, 300, 303
Polypropylene, 307
PONA analysis, 93, 96, 144, 226, 229, 257, 260, 273
Pour point, 66–67, 123, 360, 371, 375, 377, 379
Power recovery, 170
Present value factor, 413–414, 422
Process configurations, 37–40
Process cost, 391–392
Process descriptions, 102, 123–124, 139–140, 153–154, 170–171, 197, 211–213, 236, 254–255, 266–267, 286–287, 289–291, 294–296, 301–302, 308–309
Process economics, 383–426
  refinery economic factors, 385–425
Process prices, 397, 399–400
Process selection, 424
Process utility requirements, 390
Processing trends, 40, 43
Product properties, 9, 30

Product slate, 40–42
Product yield, 9
Project cost, 405, 411, 414, 421
Propane, 90, 102, 144, 156, 158, 172, 181, 199, 201, 221, 294, 307–308, 311
Propylene, 144, 158, 169, 172, 181, 199, 201, 293–294, 299–301, 307–308, 310

R

Raffinate, 319
Ramsbottom carbon residue, 206
Refinery cost, 401, 405–414
Refinery economic factors, 385–425
  economic history (refining), 391, 393–401
  examples, 399, 401–425
  gasoline reformulation, 401–405
  new refinery, 405–414
  fluid coking vs. delayed coking, 414–425
Refinery margin, 397–398, 405, 412, 414, 421–422
Refinery model, 5
Refinery process capacities, 34–35
Refinery process data, 3–4
refinery process schemes/capacities, 37–43
  product slate, 40–52
  processing trends, 40, 43
  lubricants, 43
  pipelines/terminals, 43
Refinery processing overview, 33–44
  organizational changes, 36–37
  process schemes/capacities, 37–43

Refinery products, 19–31
  petroleum supply/demand, 19–21
  product properties, 30

Reformate, 263–264, 266–271, 273–275, 278

Reforming, 263–281, 339

Reformulated gasoline, 46–47, 82, 319, 401–405

Refractory stocks, 211

Refutas method, 379

Regression analysis, 10, 13–14

Regulations (environmental), 78, 81–82

Reid vapor pressure, 50, 53–54, 81–82, 128, 296, 303, 316, 360, 364–367, 404

Renewable sources, 45

Research octane number, 49, 52, 80, 128–129, 144, 159–160, 224–225, 263, 271, 299–300, 303, 362, 364–365

Residual fuel oil, 29, 102, 123–125, 128, 131–132, 139, 148, 153, 197

residual oil processing, 99–165
  solvent deasphalting, 101–122
  visbreaking, 123–137
  delayed coking, 139–152
  fluid coking, 153–165

Return on investment, 36–37, 391, 395–396

Riser cracker, 169–170

S

Sacrificial catalyst, 254

Salt content, 22

Saturate gas plant, 92

Selectivity, 287

Sensitivity, 50–52, 271, 300, 362

Sewerage, 356

Silica-alumina catalyst, 172, 187

Simulation, 5

Smoke point, 227, 300, 360, 371, 375, 379

Soaker process, 123

Solar energy, 45–46
  photovoltaic generator, 46

Solid adsorbents, 347

Solid waste, 87

Solution loss, 238

Solvent, 102–113, 335, 347

Solvent deasphalting, 101–122
  process description, 102
  yield data correlation, 102–114
  comparison (correlations), 113, 115–118
  operating requirements, 118
  capital cost, 119

Sour water stripper/stripping, 341–342, 355

Specific gravity, 359–360

Splash blending, 43

Standard error of estimate, 112

Star technology, 308–309

Steam coil, 197

Steam generator, 154

Virgin gas oil, 55–56, 213, 222, 238,
241–242, 245

Visbreaking, 101, 123–137
process description, 123–124
yield data correlation, 124–133
comparison (correlations), 128,
134
operating requirements, 134
capital cost, 134
aquaconversion, 135

Viscosity, 67, 123, 360, 371, 379
blending index, 379

Viscosity blending index, 379

Volatile combustible matter, 139

Volatile organic compounds, 61, 82

Volatility, 51, 53–54, 66, 82, 367–371

Volumetric average boiling point, 187,
368

W

Waste heat recovery, 170

Waste treatment/disposal, 355–35

Waste water, 341–342, 355–356

Water content, 253

Water disposal, 355–356

Water treating, 341–342, 355–356

Wind energy, 45

X

Xylene, 335

Y

Yield data correlation, 4–5, 102–114,
124–133, 140–147, 155–162,
172–187, 197–205, 213–229,
236–245, 255–260, 267–277,
287–288, 296, 303, 385, 391

Z

Zeolite, 172